CREAK

CREAK

Theories and Practices of Pulse Phonation

edited by
Francesco Venturi

JENNY STANFORD
PUBLISHING

Published by

Jenny Stanford Publishing Pte. Ltd.
101 Thomson Road
#06-01, United Square
Singapore 307591

Email: editorial@jennystanford.com
Web: www.jennystanford.com

British Library Cataloguing-in-Publication Data
A catalogue record for this book is available from the British Library.

Creak: Theories and practices of pulse phonation

ISBN 978-981-5129-29-8 (Hardcover)
ISBN 978-1-003-61606-1 (eBook)

Creak

This groundbreaking publication brings together leading scholars and practitioners to explore what pulse phonation is, what it can do, and how it develops into a cultural practice. The volume employs a wide array of approaches, from sociolinguistics and voice anatomy to acoustic ecology and performance studies, including:

- Case studies of creaky voices across cultures and media
- Physiology and acoustics of the pulse register
- Creak singing: possibilities, perspectives, pedagogies
- Pulse phonation, embodiment, and gender
- The "phenomenon of extreme vocal fry"
- Vocology, somatics, and the disease condition
- The use of pulse phonation from live arts to film studies
- Composition, improvisation, and creation with creak
- The pulse register in the animal world

This is the first thorough investigation of pulse phonation to jointly take into consideration its sociocultural, bioacoustic, and creative dimensions. Merging theoretical frameworks with embodied practice, this multidisciplinary inquiry explores the processes of producing and perceiving pulse phonation, its use and significance in contemporary discourse, and its place in a broader reflection on voice and sound production. The volume concludes with a multifaceted series of testimonies from users and listeners of creaky voices.

Contents

PART II
Narratives

PART III
Experiences

List of figures

List of contributors

Nassima Abdelli-Beruh is a speech scientist. Formerly an assistant professor at New York University, she is Associate Professor of Communication Science and Disorders at Long Island University. Her research interests include gender differences in vocal fry, Parisian French voicing contrast, and the relationship between language and speech.

Tristam V. Adams is a writer and theorist with a PhD in Visual Cultures from Goldsmiths University. Author of "The Psychopath Factory: How Capitalism Organises Empathy," he researches the relationships between voice and horror across media, psychopathy, sociopathy, social technologies, and the cultural symptoms of capitalism.

Poe M. Allphin is a musicology PhD student at the CUNY Graduate Center. He is exploring trans involvement in the U.S. women's music movement and the intersection of disability, gender, and voice. A board member of QORDS, his work has been published in *Transgender Studies Quarterly* and the *Anthology of New Music: Trans & Nonbinary Voices.*

Nicol Bana is a music producer, singer-songwriter, and specialized music educator for neonatal and preschool ages. Certified by Aigam and Audiation Institute, she promotes music education according to Edwin E. Gordon's Music Learning Theory.

Dario Buccino is a composer, multi-instrumentalist, music theorist, and a designer of musical instruments. In the early 1990s, he devised the "HN System," a unique approach integrating composition, notation, and performance techniques. His instruments are designed to fully engage the performer's body in sound creation.

Whitney Chappell is an Associate Professor of Hispanic Linguistics at the University of Texas, San Antonio, a former Lutcher Brown Distinguished Professor and a Fulbright Scholar.

Recognized as UTSA College of Liberal and Fine Arts' 2021 Researcher of the Year, her research focuses on sociophonetic variation in the Spanish-speaking world.

Andrew Cheng is an Assistant Professor in the Department of Linguistics at the University of Hawai`i at Mānoa. Previously a postdoctoral researcher at Simon Fraser University, his research applies sociophonetic approaches to bilingual child language development.

Francesca M. Dovetto, a Professor of Historical and General Linguistics at the University of Naples Federico II, coordinates the Research Center Text Rhetoric and Pragmatics of Communication. She authored the Corpus CIPPS (corpus of schizophrenic spontaneous speech) and published extensively in the fields of spoken and written language.

Mark Ølholm Eaton is an Associate Professor in British and Commonwealth History, Society and Culture at Aarhus University. His primary research areas include the study of "political voices" in crisis contexts, as well as the role of metaphor, history, literary quotation, and other features of language in political communication.

Franz Goller is a Professor Emeritus at the University of Utah. His work revolves around the behavioral physiology of birds' sound production and song learning. Recent projects cover sound production mechanisms, motor coordination in singing, vocal and visual display coordination, vocal development, song syntax, and song production energetics.

Robert James Hellyer is a doctoral student and Teaching Associate at Queen Mary University of London in the School of Languages, Linguistics and Film. His work investigates emerging frontiers of queer visibility, particularly in adolescence, their intersection with linguistic performativity, and the discursive navigation of queer identities.

Míša (Michaela) Hejná is an Associate Professor in English Language at Aarhus University. Her primary research areas are sociolinguistics, phonetics and phonology, and interdisciplinary studies. In particular, she is interested in the human voice and phonatory phenomena and the relationship between the social and the biological.

Carlos T. Ishi is the leader of the Department of Sound Environment Intelligence at the University of Tokyo. He is a member of the ATR Intelligent Robotics and Communication Labs and the RIKEN Guardian Robot Project. His research includes speech prosody, voice quality, multi-modal information processing, and human–robot interaction.

Leah Kardos is a senior lecturer in music at Kingston University London, where she cofounded the Visconti Studio with music producer Tony Visconti. She is the author of "Blackstar Theory: The Last Works of David Bowie." A composer and pianist, her music combines live instrumental performance with technology and found sounds.

Ilona Krawczyk, a Lecturer in Acting at Norwich University of the Arts, is a performer and vocalist. Her PhD centered on performer wellbeing in voicework. She explores overlaps between physical, musical theater, experimental music, and sound art, drawing from post-Grotowskian practice. She is the founder of DreamVoice and the Insoundout Collective.

Joan La Barbara is a singer, composer, sound artist, and educator renowned for pushing the boundaries of vocal expression. Recognized with the John Cage Award and the Guggenheim Fellowship in Music Composition, she created works for voices, ensembles, and interactive technology. She collaborates widely across cinema and live arts.

Madeleine Masters is a Research Assistant at the University of Oxford and works as an English and Science Tutor. She previously served as a Linguistics Research Assistant at the University of Warwick, focusing on Cockney cultures and archiving the Cockney variety of English.

Edoardo Mozzanega is a performer and director whose work intersects theater, dance, film making, somatic practices, and collaborative processes. He cofounded the experimental theater company Kokoschka Revival, the curatorial project Neutopica, and has performed with influential companies and choreographers.

Elise McClay is the manager of the Language Learning and Development Laboratory at Simon Fraser University. She holds an MA in Linguistics at the University of British Columbia, focusing on indigenous language revitalization. Her research includes English imperative intonation and focus expression in Ktunaxa question–answer pairs.

Alexsandro Meireles is an Associate Professor at the Federal University of Espírito Santo. With a joint Unicamp and University of South California PhD, his research involves phonetics and phonology, specifically speech rate, rhythm, and acoustic phonetics. He coordinates the phonetics labs of the Center for Studies on Phonetics and Phonology.

Katherine Meizel, an Associate Professor of Ethnomusicology at Bowling Green State University, specializes in voice, popular music, media, religion, identity, and disability. She authored "Multivocality: Singing on the Borders of Identity," coedited the "Oxford Handbook of Voice Studies," and coproduced the album "Raise Your Voice."

John Nix is Professor of Voice and Voice Pedagogy at the University of Texas, San Antonio. His research spans voice pedagogy, literature, and acoustics. He edited works including "From Studio to Stage: Repertoire for the Voice," the "Oxford Handbook of Music Education," the "Oxford Handbook of Singing," and the "Oxford Handbook of Voice Pedagogy."

Pedro J. S. Vieira de Oliveira is a sound artist and researcher working within decolonial thinking. He holds a PhD from the Universität der Künste Berlin and is a former teaching and research associate at the Heinrich-Heine Universität Düsseldorf. His work investigates the articulations of sonic violence and the policing of bodies in public spaces.

Roberto Panzanelli, PhD, is a musician, saxophonist, and voice scholar. Founder of the UmbriainCanto school, he integrates vocal teaching with Gestalt Therapy and Dance Movement Therapy. The creator of the "vocal fry method," his publications focus on voice physiology and personal growth related to singing.

Joe Pearce is a PhD researcher at the University of Glasgow, exploring the production and perception of voice quality in contemporary Scotland from a trans-focused perspective. An educator and activist, he conducts workshops on language in the LGBTQ+ community.

Margaret Pikes is a founding member of the Roy Hart Theatre who trained with Roy Hart and participated in all of the Roy Hart Theatre's early experimental performances. She has been teaching the Wolfsohn-Hart approach to vocal expression internationally for more than 50 years and regularly leads workshops in the UK, France, and Germany.

Lisa Quoresimo, an Assistant Professor of Theater at Southern Utah University, is a performer, composer, playwright, director. Cofounder of Catalyst: A Theatre Think Tank, she previously served as the artistic director of the Kairos Theatre in New York City. Her research focuses on the intersection of voice, gender, and theater.

Bojana Radovanović is a research assistant at the SASA Institute of Musicology in Belgrade. Her interests include contemporary music and art, voice, metal music, art and new media, and transdisciplinary research. She has authored three books and serves as the editor-in-chief of the *INSAM Journal of Contemporary Music, Art and Technology*.

Diana Sidtis is Professor Emerita of Communicative Sciences and Disorders at New York University and Research Scientist at the Nathan Kline Institute for Psychiatric Research. She is the coauthor of "Foundations of Voice Studies" and published extensively on voice, aphasia, motor speech, prosody, and formulaic language.

Gino Sitson is a vocalist and voice trainer who holds a PhD in Musicology. With eight albums to his name, he has collaborated with renowned artists and composed music for films and commercials. In his roles as a UNICEF Goodwill Ambassador and a cultural arts ambassador, he is engaged in music cognition research.

Christopher Strelluf is an Assistant Professor of Linguistics at the University of Warwick, focusing on English varieties and sound changes affected by socio-linguistic factors. Editor of the "Routledge Handbook of Sociophonetics," he engages in projects on historical sociophonetics and sentiment analysis, as well as English teaching in African settings.

Francesco Venturi is the director of the Center for Interdisciplinary Research on Voice (CRIV) in Bologna, Italy. A musician-researcher and PhD candidate at Kingston University London, he specializes in voice theory and education. He collaborates with cultural and educational institutions, has received major commissions, scored award-winning productions, and performed solo and in ensembles across Europe.

Henny Yeung directs the Language Learning and Development Lab at Simon Fraser University, where he explores the impact of language learning on sensory perception, articulatory-motor abilities, and cognition. Previously, he held a research position at the Laboratoire Psychologie de la Perception at the Université Paris Descartes.

Acknowledgments

The deepest gratitude goes to the thirty-two authors who enthusiastically accepted the invitation to contribute to this investigation into the pulsed voice, enriching this book with their knowledge and experience. Appreciation extends to all the people who have, along the way, experimented with me using creak. Their experiences have informed the research leading to this book. A debt of gratitude is owed to Ronald Scherer, Marco Aime, Leah Kardos, and Oded-Ben Tal for their insightful advice. Nicol Bana, Camilla Crispino, and Alvaro Nuñez Carbullanca deserve special mention for infusing this research with the gifts of their omnivorous minds and omniscient hearts. Particular gratitude goes to Raffaele Rezzonico for the inspiring and electrifying conversations. Our discussions have been immeasurably beneficial to this research. I also wish to thank James Turrell and Vittoria De Franchis for the inspiration in the search for a cover picture. The work leading to this volume is shaped by the voices of many people, some never met in person, and others encountered along the way, regardless of the immediacy of their contribution. Among them, special thanks go to Harry Hollien, who passed away while this book was in production. Although he was not directly involved in this research, his pioneering work underpins it. This book is dedicated to his memory.

Francesco Venturi

Foreword

My days and nights with creaky voice

Diana Sidtis

In a modest adaptation of the personal observation by George Miller (1956), who claimed to be pursued by an integer, I have been continuously visited by a certain voice quality—one which, until recently, had garnered little scientific attention. These encounters were with creaky voice, and they have popped up across the entire linguistic horizon. Its status in functional descriptions of speech was not well understood until recently, when, over a long period, a steady sprinkling of studies has begun to yield revealing insights. Following a brief effort at defining this elusive phenomenon, this essay presents a personal history of anecdotal and professional chance meetings with creaky voice—beginning with my early doctoral studies at the University of California, Los Angeles (UCLA)—noted with curiosity in casual conversation—emerging vividly while researching for a text on the normal voice (Kreiman and Sidtis, 2011); discernible nights in film and television; from the street and in the classroom; realized in student projects, and finally, posing a striking clinical presence in the Brain and Behavior Research Laboratory at the Nathan Kline Institute.

Definitions and characteristics

Acoustic properties found in different kinds of nonmodal phonation—i.e., vocal production not consistently using elemental laryngeal vibratory processes (Keating, Garellek, and Kreiman 2015; Gerratt and Kreiman 2001)—allow for a distinct category called "creaky voice" (Gordon and Ladefoged 2001; Ladefoged 1971; Clark and Yallop 1995), as rendered in this simple continuum (Fig. F.1):

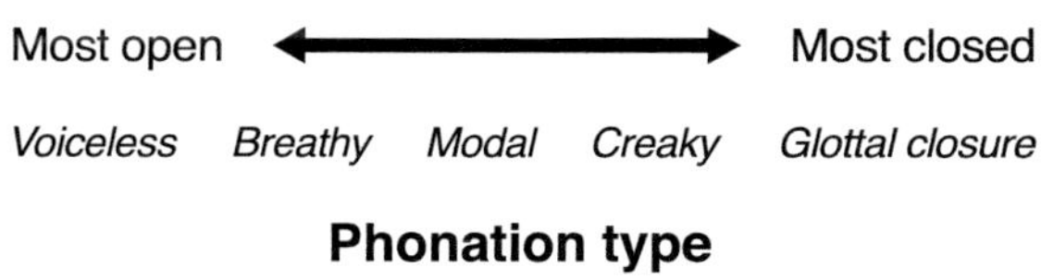

Figure F.1 Graphic continuum of five phonation types placed along a parameter of degrees of openness of the vocal folds. Adapted from Gordon and Ladefoged 2001 and Ladefoged 1971. Figure by J. Sidtis.

Terms for this quality are prolific (see Abdelli-Beruh, this collection), and convincing distinctions between creaky voice and vocal fry, using sophisticated acoustic analyses, have been advanced (Meireles, this collection). Nix and Chappell (this collection) propose a similar distinction for singing. These and other perspectives serve to further differentiate various other versions of this voice quality (Ishi, this collection). In this personal overview, the term "creaky voice" is used generically and interchangeably with "vocal fry" to refer to these many subtly differing but essentially similar manifestations.

How do we describe creaky voice (or its moniker, vocal fry)? Now that "vocal fry" is a household word and given the explosion in the past decade of sociolinguistic commentary on the use of this vocal mode, one can safely say, "you know it when you hear it." Consistent acoustic cues, although somewhat difficult to describe (Clark and Yallop 1995: 21), do serve to highlight the sound (Davidson 2019; Gerratt and Kreiman 2001; Borden, Harris, and Raphael 2003). More precisely, the creaky mode features strikingly lowered pitch, ranging between 20 and 50 pulses per second—note that 50 Hz is the lowest limit of the normally phonating male voice—produced by slowed airflow through irregularly or quasi-regularly vibrating vocal folds, which are shortened and thickened (Davidson 2020); the folds are held together loosely "like two pieces of calves liver, and air bubbles up through them" (Johnson 1997, p. 127). The phonatory effect might be called "raspy," "gravelly," or more tellingly, "creaky." You cannot miss it. The practice occurs more in spontaneous than read speech (Irons and Alexander 2016), resonating with the proposed "task effect," described for Parkinson's speech (Kempler and Van Lancker 2002; Van Lancker Sidtis, Hanson, Jackson, *et al.* 2005; Van Lancker Sidtis, Rogers, Godier, *et al.* 2010) and other neurological conditions, and more recently verified in normal speech (Lee and Kreiman 2022). In another inquiry, creaky voice was found to be more prevalent in the native language (Benoist-Lucy and Pillot-Loiseau 2013)—I cannot imagine producing this vocal mode in my second language, German.

First close encounter

The first chance meeting with creaky voice occurred while recording tonal stimuli from native speakers of the Thai language for a dichotic listening experiment (Van Lancker and Fromkin 1973). The five tones were easily distinguished, as was expected and known, by height and direction of pitch—with a contribution of duration contrasts. But, in addition, one tone, rising falling, as elicited from native Thai speaker informants, who traveled to the UCLA Phonetics Laboratory to help with a research project, consistently featured a clear, terminal portion of glottal fry, or creaky voice. Perceptually, this vocal material, appearing reliably, seemed to be intrinsic to the raising falling tone; this observation evoked questions in my mind about its phonemic status. I wondered: What was the contribution of creaky voice to a description of minimally distinctive elements of the tones? It was gratifying to see comments by Thai expert A. Abramson (1986), who acknowledged that vocal cues other than pitch may play a role in the production and perception of tonal contrasts (1986: 105).

Since then, pertinent linguistic analyses have emerged. Voice quality details of the creaky variety have recently been described as factors in the perception of Mandarin (Huang 2020), Cantonese

(Yu and Lam 2014), and of both Thai and Mandarin (Rhee and Kuang 2020). Pham (2003) reported that creaky elements overrode pitch information in a falling tone—similar to the original Thai observation—suggesting that creakiness and breathiness may serve as primary cues for signaling tone in Vietnamese. Similarly, in an earlier study, it was proposed that creaky voice "is the only feature that differentiates the Northern Vietnamese falling tone and drop tone" (Phuong, 1981, p. 23). Thus, in some of the tone languages that have been studied, the "phonetic" material of nonmodal phonation, as it was previously viewed, can be viewed, at least in trade-off relationships, as functioning phonologically.

An earlier, popular book on phonetics (Heffner 1950), noting that voice quality details are important only when contrastive, did not include creaky phonation in the array of representative phonetic elements. This view prevailed for many years in phonetic and linguistic literature. Over the years (2001–2011) of performing literature research for our book *Foundations of Voice Studies* (Kreiman and Sidtis 2011), my coauthor Jody and I found and reported numerous instances of the creaky voice modality serving to yield minimal contrasts on phonemes. Thus, beyond revelations of the phonological status on tones, the creaky sound on vowels and consonants is now known, on its own, to signal distinctive contrasts in languages such as Hupa, Hausa, Jalapa Mazatec (Kirk, Ladefoged, and Ladefoged 1993), Zapotec (Ladefoged and Maddieson 1996), and Montana Salish (Davidson 2020; Kreiman, Keating, and Berry 2006). The International Phonetic Alphabet chart includes a diacritic, a tilde (Clark and Yallop 1995: 239), for laryngealized sounds (Ladefoged 1975; Gordon and Ladefoged 2001), which is placed under the transcription letter, as shown for words for nine and day in the Northwest Indian language, Kwakw''ala:

n̰an̰əm̰a "nine"
n̰ala "day"

These facts denote that the mode of creaky voice is perceptible and linguistically significant, including when vocalized on the smallest of phonological elements (Gordon and Ladefoged 2001: 387).

Structuring talk and conversational pragmatics

The curious, phonetically tuned ear picks up the effulgent presence of creaky voice across spoken material. The voice often trails softly at the end with a gravelly sound. This phonation type performs a structuring task in ongoing talk embodying a crucial process that allows for orderliness, for phrasing, in utterances. Creaky voice serves as a phrase boundary marker, as has been described for English (Drugman, Kane, and Gobl 2014; Drugman, Kane, Raitio, *et al.* 2014; Garellek and Keating 2015). With a modicum of linguistic awareness, one hears how the voice quality shift signals phrases and, even more clearly, ends of utterances. In a similar fashion, parenthetical segments in fluent speech are called out using creaky voice (Lee 2015). The role of creaky voice in signaling prosodic boundaries has also been described for the Spanish language (González, Weissglass, and Bates 2022).

In conversation, turn-taking can be signaled using creaky voice sounds, as has been documented for Finnish (Ogden 2001). In a study of nonlexical or semi-lexical sounds, which serve to govern turn-taking in the conversation, creakiness appears often, for example, on *ummm, yeah, okay, um, hm,* and *aa,* either spanning the entire item or present only toward the end (Ward 2006: 11). The alert listener, consciously or unconsciously, presumably in a well-executed conversation, appreciates the flow of talk and responds appropriately in part through identifying these signals (Grivicic and Nilep 2004).

Semantics and meaning

The influence on grammatical and conversational structure is crystal clear, but one struggles to imagine a case where creaky voice quality influences semantic content at the lexical level. Meanings of words such as *chair, purple,* or *ambulate* do not change their semantic content when pronounced first with modal and then on a creaky voice. As versatile as the laryngeal phenomenon is, it cannot do everything; it has no presence in referencing and connoting in the classical semantic system. But meaning on a broader scope—speaker's meaning—welcomes manipulation through voice quality varieties. A prime example lies in sarcasm.

One of our student projects involved recording utterances such as *great, that was a good lecture, my apartment is clean,* and *so fun* in two modalities: literal and sarcastic. Multiple auditory-acoustic features, including pitch contour and height separately and in combination, served to distinguish the two meaning types. We noted subjectively that creakiness appeared more often in the sarcastic exemplars, as was also reported in an acoustic and electroglottographic study of Mandarin (Li, Gu, Liu, *et al.* 2020). A listeners' study of Korean voice quality signaling sarcasm found a consistent (but not statistically significant) presence of creaky voice, one of the rating parameters, differentiating sarcastic from literal utterances spoken by native speakers of Korean (Yang 2021).

The denizen of attitudes and emotions

Our talk is swamped in attitude: the signaling of obvious, refined, intentional, and unintentional stance and belief systems relating to discourse content. Creaky voice is more manifest in attitudinal than emotional conveyance, and predictably few studies of its role in classical emotional signaling can be found.

Two likely candidates from the emotions repertory for the devolvement of voice quality into the creaky realm are sadness and fear. Acoustic analyses report that key indicators of sadness are slower rate and reduced pitch variability in sadness (Breitenstein, Van Lancker, and Daum 2001; see Fig. F.2). But, as in other domains of speech, the parameter of creaky voice has not been examined.

Although not documented in research reports of sad utterances, elicited or natural that I know of, memory serves to recall instances when affective sadness in a talker included stretches of creaky vocalization, especially when occurring as precursor to the laryngeal stricture associated with crying. A careful look at fearful utterances may yield similar insights.

Attitude signaling utilizes creaky voice to considerable benefit. The talker can sound sincere, doubtful, guarded, light-hearted, empathetic, hesitant, nonaggressive; this list can hardly be completed, and the responsible auditory-acoustic cues, also legion and complexly associated and intertwined with utterance content, are barely a gleam in the speech scientist's eye (Kreiman

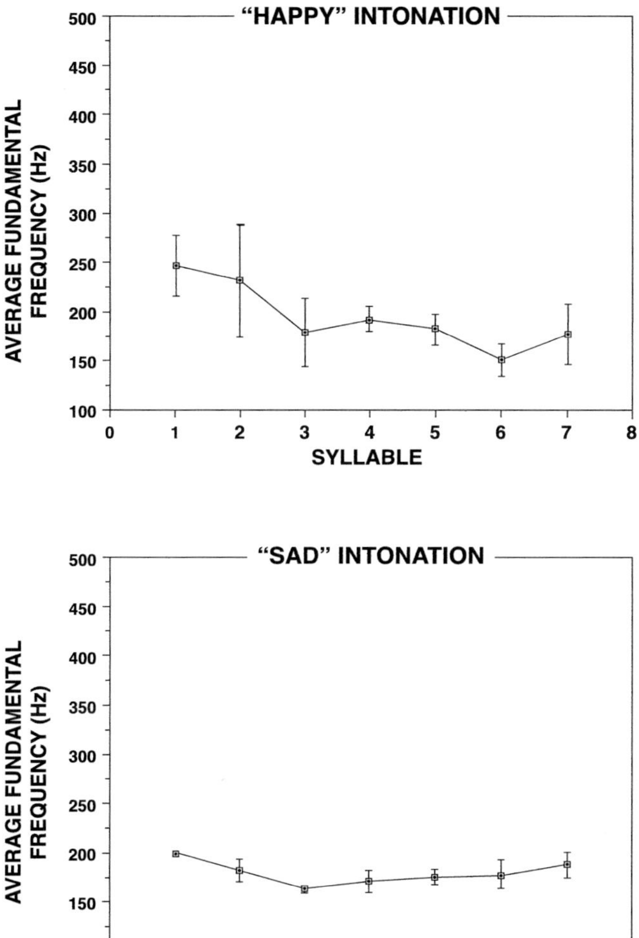

Figure F.2 Mean syllable fundamental frequencies on the vertical axis of the utterance "Johnny is walking his dog" from four productions of each emotion type, *happy* and *sad*. Note lower pitch variability on *sad*. Figure by J. Sidtis.

and Sidtis 2011: Chapter 9). This is a continuous process in talk—one-to-one mapping between cue and content cannot be expected (Van Lancker Sidtis 2018). These comments are consistent with those of Gobl and Chasaide (2003), who observe creaky voice quality effectively communicates the milder mental states, such as bored, relaxed, intimate, and contented. Studies in Swedish found creaky voice to impart uncertainty and indecisiveness (Horne 2009) and others reported its tendency to occur with low activation and expectation as well as high power and valence (Cullen, Kane, Drugman, *et al.* 2013), an impressive range of operational acumen in a single voice quality. These ubiquitous signals of attitudinal stance, so important in communication and enabled by creaky voice, sail past the listener barely engaging conscious awareness. Their deft use in the media, including in film and advertising, deserves careful study.

A compelling personal deportment—shading into actual emotion—radiates forth from utterances consisting entirely of strong creaky vocalizing: This is "victim voice" (Sidtis, J., personal communication), proclaiming self-pity, tragic circumstances, helplessness, and hopelessness. When I hear this unmistakable sound on one of my own utterances, it dawns on me that I should better quickly regroup. I know of no studies on victim voice, but the phenomenon is unmistakable in oneself and others and in the media and it is strangely easy to imitate.

Sociolinguistic explosion

Here is where the supremacy of creakiness emerges. Creaky voice is pervasive in some speakers, often used deliberately (Clark and Yallop 1995), presumably to signal group identity among other intentions. While teaching at New York University, I noted that the classroom was awash in creaky vocal quality, usually terminally (as documented by Abdelli-Beruh, Drugman, and Red Owl 2015), but sometimes pervasively throughout utterances in the talk of the female students (there were no males in the classes I taught). In chatting with one of the women, I mentioned to her that nearly her entire spoken repertory was on creaky voice. She smiled cheerfully in acknowledgement, appearing to be glad to hear this. The phenomenon has spread like proverbial wildfire, and for those with awareness of their speech, the practice seemed fully acceptable, even welcome. Public awareness has also grown, as seen in numerous media articles and jokes generated on the internet, referencing vocal fry (in this essay the terms "creaky voice" and "vocal fry" are used interchangeably) and the "California accent" (the term "fried voice" is used; Blum 2016); the vocal fry phenomenon is often coupled with another heated sociolinguistic trend, "uptalk," whereby most or all utterances terminate in rising intonation.[1] A joyful celebration of uptalk is rendered in the footnote; a like creative effort might be considered for creaky voice.

Observations on incidence and prevalence—who does it and how much—as well as evaluations—who notices it, who likes or dislikes it—vary greatly and span extremes of valence. Talker awareness of the phenomenon has been documented in speakers in a Latino gang (Mendoza-Denton 2011) and by Borrie and Delfino (2017), who observed imitation of the vocal phenomenon in interlocutors. Some say the prevalence is lower in males (Abdelli-Beruh, Wolk, and Slavin 2014; Wolk, Abdelli-Beruh, and Slavin 2012). Others refute the common myth of a high degree of its presence in general (Dallaston and Docherty 2020). Evaluation of the practice is even more divided. My response ranges between amusement and annoyance. Vocal fry in everyday speech has been designated as cool (Greer 2015), irritating (Postal 1921), lazy and annoying (Blum 2016), and in an eminently balanced view, hesitant and nonaggressive, but also urban-oriented, educated, and upwardly mobile (Yuasa 2010). Clinically, Blum (2016) reviews at what point the speech pathologist might consider the potential harm to the voice caused by vocal fry. In earlier days of speech-language pathology, vocal emissions sounding creaky were deemed potentially injurious to the laryngeal mechanisms (reviewed by Abdelli-Beruh, this collection). More recently, professional perspectives have turned toward viewing the usage of creaky voice as a style, or "register," meaning use of the vocal vehicle for a specific sociological purpose.

Clinical territory

Clinically, when laryngeal muscles and vocal folds are compromised, as in Parkinson's disease, the voice becomes low in amplitude (hypophonic) and breathy (Rusz, Cmejla, Ruzickova, *et al.* 2011). Other neurological disorders, such as ataxia associated with cerebellar dysfunction, yield a harsh, louder voice. Huntington's disease yields voice breaks and irregular pitch fluctuations (Rusz, Klempíř, Baborová, *et al.* 2013). Various neurological and psychiatric disorders, when voice is affected, cause the afflicted to speak more softly, loudly, harshly, or with irregular, strained, or breathy quality. Despite this colorful range of vocal pathologies, speech pathological depictions of the disordered voice in neurological disease do not include a finding of creaky voice quality, and creaky voice does not have any purchase in the arsenal of articulatory features that are used to describe pathological speech (e.g., Darley, Aronson, and Brown 1975; Velleman 2016; Duffy 2013). In an interesting turn of events, such usage is encountered in speech-language pathology as a therapeutic tool for fine-tuning the speaking voice and renovating singing technique (Nix and Chappell, this collection).

Only one time in our speech research at the Brain and Behavior Laboratory, where neurological speech disorders were routinely studied, did we encounter creaky voice following neurological insult (Van Lancker Sidtis, Cameron, *et al.* 2012). Diagnosed with Parkinson's disease, the individual had undergone deep brain stimulation (DBS: the placement of electrodes in the lower portions of the brain) to relieve the physical tremor that occurs in this condition. Adjustment of the DBS electrodes by the neurologist resulted almost immediately in a voice disorder, interpreted as an interference with coordination between the three primary systems of motor speech: respiratory, laryngeal, and articulatory mechanisms. It is unknown how and why this problem occurred, and it has not emerged at any other times that we know of.

To speak, airflow must be exactingly controlled as it passes through the larynx between the vocal folds, which open and close with great speed to accommodate voiced and voiceless sounds making up words and phrases. It was inferred from the speech style of this individual that coordination between these three systems of motor speech had been compromised: His utterances were produced entirely on creaky voice. His strategy was to fill the lungs, hold the breath, and to proceed to articulate the intended utterance with minimal, baseline, low-frequency, irregular vibration of the vocal folds; allowing tiny packets of air to pass through. At the end of each utterance, a large volume of air, which had been mostly held throughout the speaking event, rushed out from his lungs. In other words, all spoken utterances were on held breath and creaky voice, with the associated lack of periodic (resonant) acoustic material and leading to very poor intelligibility. We worked with this client on establishing successful airflow, first, through elongated vocal sigh supported by "belly" breathing, and then on predominantly voiced words such as "hello." These practices proved successful in the clinical sessions, but the treatment was discontinued without successful carryover to intentional speech.

To sum up: Creaky voice has raised its conspicuous head and shown its fresh face, in my experience, first in an apparent phonetic context, which proved to comprise important phonological cues for tonal and phonemic contrasts; broadly as part of pragmatic discourse, notably in attitudinal and extreme emotion conveyance, and most compellingly and ubiquitously, in correspondence with sociolinguistic variables. Interestingly, although creaky voice reigns supreme in conveying speakers' attitudes, this vocal quality has not been found to play a role in communicating semantic content. Formal studies of emotional intent are lacking, but examples of extreme sadness, fear, and "victim voice" are observed. Some of these lacunae may be filled when creaky voice serves more often as a selected variable in research designs. Creaky vocalization does not appear in the known pantheon of features used to describe the vocal pathologies accompanying motor speech disorders, but it has reportedly been employed as a therapeutic or pedagogical tool in speaking and singing. Only uniquely, in very striking form and in an unusual context, did we encounter this phenomenon in neurolinguistic research.

In the past, the vocal phenomenon of creaky voice was deemed to be outside the realm of linguistically important themes and it follows that there had been little scientific examination of this interesting and increasingly recognizable phenomenon. This perspective has changed. More will surely be revealed. It is gratifying to see creaky voice come into its own, receiving proper attention in this edited volume.

Endnote

1. Uptalk inspired a verse, sung by the NYU B&B Singers, to the tune of J. Durante's "I'll never forget the day I read a book" (words by D Sidtis): "We'll never forget the day we upped our talk? / It wasn't intentional; it just seemed conventional. / Where a sentence should decline / We intoned it with a whine. / The day...we upped our talk. / This is not a fad we can easily shake. / We make no apology; we're talking like our friends. / It's just sociology; and new linguistic trends. / The day we upped our talk, we can't remember when; / But one of these days, we're gonna do it again?"

References

Abdelli-Beruh, N. B., Wolk, L., and Slavin, D. (2014). Prevalence of vocal fry in young adult male American English speakers. *Journal of Voice*, **28**: 185–190.

Abdelli-Beruh, N. B., Drugman, T., and Red Owl, R. H. (2015). Occurrence frequencies of acoustic patterns of vocal fry in American English speakers. *Journal of Voice*, **30**: 759.e11–759.e20.

Abramson, A. (1986). The Thai tonal space. In I. Mattingly (Ed.), *Status Report on Speech Research*. New Haven, CT: Haskins Laboratories: 105–114.

Benoist-Lucy, A. and Pillot-Loiseau, C. (2013). The influence of language and speech task upon creaky voice use among six young American women learning French. *Interspeech*: 2395–2399. Retrieved from https://www.isca-speech.org/archive/interspeech_2013/i13_2395.html.

Blum, H. (2016). Totally fried: What do you need to know about vocal fry? *The ASHA Leader*, **21**(2): 50–56.

Borden, G. J., Harris, K. S., and Raphael, L. J. (2003). *Speech Science Primer*. Baltimore, MD: Lippencott Williams & Wilkins.

Borrie, S. A. and Delfino, C. R. (2017). Conversational entrainment of vocal fry in young adult female American English speakers. *Journal of Voice*, **31**: 513.e25–513.e32.

Breitenstein, C., Van Lancker, D., and Daum, I. (2001). The contribution of speech rate and pitch variation to the perception of vocal emotions in a German and an American sample. *Cognition & Emotion*, **15**(1): 57–79.

Cullen, A., Kane, J., Drugman, T., and Harte, N. (2013). Creaky voice and the classification of affect. *Proceedings of WASSS*, Grenoble, France.

Clark, J. and Yallop, C. (1995). *An Introduction to Phonetics and Phonology*. Oxford, UK: Blackwell.

Dallaston, K. and Docherty, G. (2020). The quantitative prevalence of creaky voice (vocal fry) in varieties of English: A systematic review of the literature. *PLoS ONE*, **15**(3): e0229960.

Darley, F. I., Aronson, A. E., and Brown, J. R. (1975). *Motor Speech Disorders*. Philadelphia: W. B. Saunders.

Davidson, L. (2019). Perceptual coherence of creaky voice qualities. *Proceedings of the 19th International Congress of Phonetic Sciences*. Canberra, Australia: Australasian Speech Science and Technology Association Inc.: 147–151.

Davidson, L. (2020). The versatility of creaky phonation: Segmental, prosodic, and sociolinguistic uses in the world's languages. *WIREs Cognitive Science*, **12**(3): e1547.

Duffy, J. R. (2013). *Motor Speech Disorders*. New York: Elsevier.

Drugman, T., Kane, J., and Gobl, C. (2014). Data-driven detection and analysis of the patterns of creaky voice. *Computer Speech and Language*, **28**(5): 1233–1253.

Drugman, T., Kane, J., Raitio, T., and Gobl, C. (2014). Prediction of creaky voice from contextual factors. *2013 IEEE International Conference on Acoustics, Speech and Signal Processing*, Vancouver, BC, Canada, 2013: 967–7971, doi: 10.1109/ICASSP.2013.6639216.

Garellek, M. and Keating, P. (2015). Phrase-final creak: Articulation, acoustics, and distribution. Paper presented at the *89th meeting of the Linguistic Society of America, Portland*, Oregon, 8–11 January.

Gerratt, B. R. and Kreiman, J. (2001). Toward a taxonomy of nonmodal phonation. *Journal of Phonetics*, **29**: 365–381.

Gobl, C. and Chasaide, A. N. (2003). The role of voice quality in communicating emotion, mood and attitude. *Speech Communication*, **40**(1–2): 189–212.

González, C., Weissglass, C., and Bates, D. (2022). Creaky voice and prosodic boundaries in Spanish: An acoustic study. *Studies in Hispanic and Lusophone Linguistics*, **15**(1): 33–65.

Gordon, M. and Ladefoged, P. (2001). Phonation types: A cross-linguistic overview. *Journal of Phonetics*, **29**: 383–406.

Greer, S. (2015). The perception of coolness: Voice quality and its social uses and interpretations, Master's thesis, Graduate Studies.

Grivicic, T. and Nilep, C. (2004). When phonation matters: The use and function of yeah and creaky voice. *Colorado Research in Linguistics*, **17**(1): 1–11.

Heffner, R.-M. S. (1950). *General Phonetics*. Madison, WI: University of Wisconsin.

Horne, M. (2009). Creaky fillers and speaker attitude: Data from Swedish. Where prosody meets pragmatics. In *Studies in Pragmatics*. Bingley, UK: Emerald: 277–288.

Huang, Y. (2020). Different attributes of creaky voice distinctly affect Mandarin tonal perception. *The Journal of the Acoustical Society of America*, **147**(3): 1441–1458.

Irons, S. T. and Alexander, J. E. (2016). Vocal fry in realistic speech: Acoustic characteristics and perceptions of vocal fry in spontaneously produced and read speech. *The Journal of the Acoustical Society of America*, **140**(4): 3397.

Johnson, K. (1997). *Acoustic & Auditory Phonetics*. Cambridge, MA: Blackwell.

Keating, P., Garellek, M., and Kreiman, J. (2015). Acoustic properties of different kinds of creaky voice. In *The Scottish Consortium for ICPhS 2015. Proceedings of the 18th International Congress of Phonetic Science*. Glasgow, UK: The University of Glasgow: 1–5.

Kirk, P. L., Ladefoged, J., and Ladefoged, P. (1993). Quantifying acoustic properties of modal, breathy and creaky vowels in Jalapa Mazatec. In *American Indian Linguistics and Ethnography in Honor of Laurence C. Thompson*: 435–450.

Kempler, D. and Van Lancker, D. (2002). The effect of speech task on intelligibility in dysarthria: Case study of Parkinson's disease. *Brain and Language*, **80**: 449–464.

Kreiman, J. and Sidtis, D. (2011). Foundations of Voice Studies: Interdisciplinary Approaches to Voice Production and Perception. Boston: Wiley-Blackwell.

Kreiman, J. E., Keating, P. A., and Berry, D. A. (2006). The human voice. *Acoustics Today*, **2**(1): 51–52.

Ladefoged, P. and Maddieson, I. (1996). *The Sounds of the World's Languages*. London: Blackwell.

Ladefoged, P. (1971). *Preliminaries to Linguistic Phonetics*. Chicago: University of Chicago.

Ladefoged, P. (1975). *A Course in Phonetics*. New York: Harcourt Brace Jovanovich.

Lee, S. (2015). Creaky voice as a phonational device marking parenthetical segments in talk. *Journal of Sociolinguistics*, **19**(3): 275–302.

Lee, Y. and Kreiman, J. (2022). Acoustic voice variation in spontaneous speech. *The Journal of the Acoustical Society of America*, **151**(5): 3462–3472.

Li, S., Gu, W., Liu, L., and Tang, P. (2020). The role of voice quality in Mandarin sarcastic speech: An acoustic and electroglottographic study. *Journal of Speech, Language, and Hearing Research*, **63**(8): 2578–2588.

Mendoza-Denton, N. (2011). The semiotic hitchhiker's guide to creaky voice: Circulation and gendered hardcore in a Chicana/o gang persona. *Journal of Linguistic Anthropology*, **21**: 261–280.

Miller, G. A. (1956). The magical number seven, plus or minus two: Some limits on our capacity for processing information. *The Psychological Review*, **63**: 81–97.

Ogden, R. (2001). Turn transition, creak and glottal stop in Finnish talk-in-interaction. *Journal of the International Phonetic Association*, **31**(1): 139–152.

Pham, A. H. (2003). Vietnamese tone: A new analysis. In L. Horn (Ed.), *Outstanding dissertations in linguistics*. London: Routledge.

Phuong, V. T. (1981). The acoustic and perceptual nature of tone in Vietnamese. Doctoral dissertation. Australian National University.

Postal, K. (2021). The Art and Science of Expert Witness Testimony. London: Routledge.

Rhee, N. and Kuang, J. (2020). The different enhancement roles of covarying cues in Thai and Mandarin Tones. In *INTERSPEECH*: 2407–2411.

Rusz, J., Klempíř J., Baborová, E., Tykalová, T., Majerová, V., Čmejla, R., Růžička, E., and Roth, J. (2013). Correction: Objective acoustic quantification of phonatory dysfunction in Huntington's Disease. *PLoS ONE*, **8**(9).

Rusz, J., Cmejla, R., Ruzickova, H., and Ruzicka, E. (2011). Quantitative acoustic measurements for characterization of speech and voice disorders in early untreated Parkinson's disease. *The Journal of the Acoustical Society of America*, **129**(1): 350–367.

Van Lancker Sidtis, D. (2018). Ancient of days: The vocal pattern as primordial big bang of communication. In P. Belin and S. Frühholz (Eds.), *The Oxford Handbook of Voice Perception*. Oxford: Oxford University Press: 15–36.

Van Lancker Sidtis, D., Hanson, W., Jackson, C., Lanto, A., Kempler, D., and Metter, E. J. (2005). Fundamental frequency (F0) measures comparing speech tasks in aphasia and Parkinson's disease. *Journal of Medical Speech-Language Pathology*, **12**(4): 207–212.

Van Lancker Sidtis, D., Cameron, K., and Sidtis, J. J. (2012). Dramatic effects of speech task on motor and linguistic planning in severely dysfluent parkinsonian speech. *Journal of Clinical Linguistics and Phonetics*, **26**: 695–711.

Van Lancker Sidtis, D., Rogers, T., Godier, V., Tagliati, M., and Sidtis, J. J. (2010). Voice and fluency changes as a function of speech task and deep brain stimulation. *Journal of Speech Language and Hearing Research*, **53**(5): 1167–1177.

Van Lancker, D. and Fromkin, V. A. (1973). Hemispheric specialization for pitch and 'tone': Evidence from Thai. *Journal of Phonetics*, **1**: 101–109.

Velleman, S. L. (2016). *Speech Sound Disorders*. Philadelphia, PA: Wolters Kluwer.

Ward, N. (2006). Non-lexical conversational sounds in American English. *Pragmatics and Cognition*, **14**: 129–182.

Wolk, L., Abdelli-Beruh, N. B., and Slavin, D. (2012). Habitual use of vocal fry in young adult female speakers. *Journal of Voice*, **26**: e111–e116.

Yang, S.-Y. (2021). Listener's ratings and acoustic analyses of voice qualities associated with English and Korean sarcastic utterances. *Speech Communication*, **129**: 1–6.

Yu, K. M. and Lam, H. W. (2014). The role of creaky voice in Cantonese tonal perception. *The Journal of the Acoustical Society of America*, **136**(3): 1320–1333.

Yuasa, I. P. (2010). Creaky voice: A new feminine voice quality for young urban-oriented upwardly mobile American Women? *American Speech*, **85**: 315–337.

PART I

Frames

Introduction

What is the matter with pulse phonation?

Francesco Venturi

A good place to start considering the problem of pulse register phonation, or simply pulse phonation, is perhaps one of the birthplaces of "vocal fry," the flowering of studies published between the 1960s and 1970s that understood this phenomenon, previously seen as pathological, as normal for any human voice (e.g., Wendahl, Moore and Hollien 1963; Hollien *et al.* 1966; Hollien and Michel 1968; Hollien and Wendahl 1968; Hollien, Damsté and Murry 1969). A new era of voice research was inaugurated; a new branch of linguistics was launched, bearing the somewhat whimsical name of a cooking method. The clinicians who promoted the term eventually replaced it with "pulse register" (Hollien 1974), but the classic onomatopoeia "fry" was to remain, and so was its controversial nature.

The word "problem" subtly presumes a certain petition of principle. Speaking of "the problem of pulse phonation" is to postulate that pulse phonation is a problem. However, such an idea is a cornerstone assumption propelling the exploration of this phenomenon. It draws from a similar premise that pulse phonation evolved "from deviance to the birth of a [vocal] register" (Abdelli-Beruh, this collection). While it was earlier seen as a voice disorder or a symptom of damage, the pulse register is now accepted as a legitimate vocal behavior. This shift hinged on the very question: "Is it a voice problem?" Pulse phonation is innate in humans (e.g., Buder, Warlaumont and Oller 2013; Oller *et al.* 2013; 2019), but scientific focus was scant until that question was posed (see Chapter 1).

If the orientation to pulse phonation as "deviant" was crucial to early studies (e.g. Voelker 1935) and the reorientation as "non-deviant" marked the second wave of research (e.g., Shipp and McGlone 1970), a reorientation as "deviant" is key to recent developments. Another petition of principle initiated a broad strand of research that strived to determine whether and why pulse phonation might be harmful (e.g., Elias *et al.* 2015) or unfavorable (e.g., Ligon *et al.* 2019), especially when it evolves into a vocal trend, or even a *whole way of life* (Malinowski 1923). Once again, a petition of principle triggered what would become "The great vocal fry dustup" (Reynolds 2015), when a hyper-quoted article on *PLOS One* advised young women to avoid pulse phonation to enhance their opportunities in the job market (Anderson *et al.* 2014). Pulse phonation made it twice into the *Time* magazine and surfaced in a flood of media content on how to "get rid of vocal fry" (Van Edwards 2021) or why it "might be your new best friend" (Ramsey 2020).

Over sixty years, between the birth of a register (1960s) and the great dustup (2010s), the amplified perceptibility of pulse phonation involved the establishment of new standards for its perception, changing the way it is used today. The "problem" proved transformative.

How can something ephemeral such as a petition of principle "affect the very meat of the body and prove transformative?" (Vicky Kirby, in McLoughlin 2019: 3). One way to answer is to take the clue from the question, "what is the problem with vocal fry?" and ask, "what happens when we say that vocal fry is a problem?" The epistemological grounding that this change entails is a matter of scholarly observation: It means that we no longer just ask, "what is vocal fry?" but rather, "what criteria do we use to define it?" This entails a fundamental change: The question is not only, "what are the proprieties of pulse phonation?" but rather, "what does pulse phonation do?" and *"how do I do what I do as an observer in observing"* it? (Maturana 2002: 6, original emphasis).

Such epistemological change represents the starting point of this collection, which brings together various theories and practices of pulse phonation in the attempt to situate this "complex, multifaceted phenomenon" (Proctor *et al.* 2024: 13). Pulse phonation cannot be defined solely as an acoustic or physiological phenomenon, although it is constituted by specific properties; rather, it necessitates an analysis that incorporates cultural practices (what we do with the pulse register) and the perspectives of different disciplines and media (how they shape it as conceptual object). Thus, before presenting the entire collection, a closer examination of this problem will provide a framework for this volume and establish two other positions that have led to its development.

Situating pulse phonation research

Pulse phonation involves a multitude of vocal qualities and vocal gestures in an intricate, multifaceted intertwining of acoustic patterns and cultural practices, at a certain time and place. As already suggested in recent work (Proctor *et al.* 2024; Davidson 2020), these pattern-meaning constructions interact with the means of their observation, and even the basic measurement in acoustic analysis "entirely depends on what kind(s) of creak the investigator wants to identify" (Keating, Garellek, and Kreiman 2015: 4). Acknowledging the interplay between tangible underpinnings and definitional practices then becomes vital. It is no coincidence, in fact, that pulse phonation has many names (see Chapter 1: 21; Chapter 11: 305). Among these, "creak" is adopted as the overarching onomatopoeia for this volume, but it is not necessarily preferred in each chapter.

If naming is a way of understanding things and making them relatable and usable, then extreme polyonymy is an indication of the difficulty of arriving at a single understanding (e.g., Gordon and Ladefoged 2001; Laver 1980). To borrow from the work of Edgar Morin (1990), the pulse register implies "complexity," because it names phenomena that cannot be reduced to a single cause or explanation. The study of these phenomena involves at least three levels of analysis (MacDougall-Shackleton 2011): the physical level, the level of cultural practices and resources, and the disciplinary level of how pulse phonation is defined as a conceptual object. While acknowledging their interconnectedness, the three dimensions will be presented in this order.

The pulse register, or "M0" (mechanism zero), names a vocal feature of terrestrial mammals, non-avian reptiles, amphibians, marine mammals, and birds. It produces patterns of pulse phonation, which refer to a series of discrete, acoustic excitations, generated by the larynx (terrestrial mammals, non-avian reptiles, and amphibians), the phonic lips (marine mammals), or the syrinx (birds). The definition of a "register" is based on the spectrum of frequencies it generates and the way the vibratory mechanism operates (Titze 1994). According to acoustic parameters, the pulse register mostly covers the low-end frequencies, which is why the terms "pulse" and "M0" have been used both for human and nonhuman vocalization (e.g., primates: Riede and Zuberbuhler 2003; odontocetes: Madsen, Siebert, and Elemans 2023; Au 1993; birds: Jensen *et al.* 2007; humans: Roubeau, Henrich, and Castellengo 2009), even if each register is sitting on a different range. In humans, the pulse register lies below 80 Hz, while for birds, it is near 300 Hz. For a sperm whale (*Physeter macrocephalus*), it goes down to 0.5 Hz, while for a harbour porpoise (*Phocoena phocoena*), it goes up to 1000 Hz. In all these cases, however, the frequencies generated by the pulse register are lower than the lowest available in any other vibratory mechanism. Though diverse, these vocal sounds are all encompassed in the universe of pulse phonation.

If we take the larynx, the phenomenon can be sketched thus: When the folds are close enough, they are sucked together by the forces caused by a pressure drop and change in flow direction (known as Bernoulli forces), creating a pressurized chamber under the glottis. The subglottal air pressure from the lungs collects under the closed folds. Once this pressure reaches a certain level, it pushes the folds outward, opening the glottis and ejecting a single "puff" (van den Berg 1958). This is the "atom" of pulse phonation. The acoustic analysis shows a rapid shift in signal amplitude from a baseline value to a higher or lower value and then a rapid return to baseline. Short periods of openness are followed by longer periods when the vocal folds are completely adducted (Childers and Lee 1991). The alternation of these opening and closing gestures (the higher the pressure, the higher the rate) generates the patterns that form the sonic materiality of (human) pulse phonation. These patterns can be made of regular glottal cycles at a low-frequency, irregular impulses, or multiple cycles (Whitehead *et al.* 1984), and six distinct possible realizations were observed (see Chapter 2: 59). The main anatomical characteristic is the presence of thick, short, and lax vocal folds (Cielo *et al.* 2011). The thickening of the vocal folds in conjunction with a flaccid vocal ligament leads to the slow rate and, combined with the creation of difference tones, contributes to the low pitch (Blomgren *et al.* 1998). It is such an essential feature of M0 to produce low frequencies that there is a name for it: "*Strohbass,*" although this term emphasizes the production of extraordinarily stable difference tones through multiple-cycle patterns, rather than basic pulse phonation (see Chapter 11: 327).

The various oscillation regimes thus generated are surprisingly similar across different vocal organs. The way the syringeal labia of birds vibrate through the pulse tone production mechanism, to form a song syllable, for example, is similar to the vibration of mammal vocal folds during pulse phonation, even though birds have two sound sources in the syrinx (see Chapter 5). The vibration mechanism is also analogous in toothed whales (odontocetes), although their phonic lips are located in the nose (see Chapter 12). The similarity extends across the mammal and amphibian larynx. It is worth noting that human pulse phonation can also occur during inhalation (the so-called "inspiratory vocal fry"). While further exploration is required, a preliminary study indicates that this mode of phonation may be physiologically distinct, suggesting its classification as a unique vocal register (Paolillo *et al.* 2019). The same study, however, acknowledges that it shares perceptual characteristics of (expiratory) pulse phonation. Then, regardless of whether these pulse tones are produced by different vocal organs in different patterns (single, double, etc.), whether they are inhaled or exhaled, and whether they acquire any given meaning or function, they all fall within pulse phonation research

(PpR), or the study of creak as a bodily practice and a cultural resource.

The problem here is subtle yet vital. When we examine the physical realities of pulse phonation, we can find striking similarities across different species and cultural contexts. Yet, if we contextualize and compare these voicings, we may find more differences than familiarities. Then, in this sense, the "pulse register" would be little more than a conceptual object: Whether or not we start from an embodied level, we eventually arrive at an element that can be considered a discursive act. However, crucially, if we change our perspective on how we *relate* to pulse phonation—how it is used, what it accomplished, and why—then the pulse register evolves into a larger discourse that cannot be reduced to its physical manifestations. Before being an object of study, as suggested in Diana Sidtis's Foreword, pulse phonation is an aspect of life, and its pulses are part of a soundscape of great complexity.

Humans and nonhuman animals employ the pulse register in a multitude of ways. Diana monkeys in the rainforest, for example, use it to signal alarm (Riede and Zuberbuhler 2003), while in small birds, such as zebra finches or European starlings, it facilitates low-frequency vocalization in song (see Chapter 5). Indeed, PpR suggests that pulse phonation is "a more important component of acoustic communication in nonhuman animals than it is in humans" (Jensen *et al.* 2009). Both crocodilians and members of the Felidae family use pulse phonation in relation to social contact (Peters 1981), while male goitered gazelles use it in mating calls (Frey *et al.* 2011). Pulse phonation patterns have also been observed in many land and sea mammals, such as pigtailed macaques (Grimm 1967), koalas (Smith 1980), and toothed whales (Madsen, Siebert, and Elemans 2023), which use it to echolocate prey at great depths.

Humans may use pulse phonation as a tool for communicating emotion (Loss and Zold 2014), for embodying and shaping identity (Podesva 2013) or with (para)linguistic function (Davidson 2020). The social meanings attached to the pulse register across different social situations include: seductiveness and intimacy (e.g., Shaw and Crocker 2015; Speck 2006), pain (Torres *et al.* 2020; see Chapter 2), negative stance (see Chapter 8), or boredom (Gobl and

Chasaide 2003). The pulse register accommodates the most varied uses in Mandarin, Vietnamese, Zapotec, Montana Salish, English, or Finnish among others (see Davidson 2020 for a review). In Japanese, for example, the pulse phonation manifests as a form of pressed voice, known as "rikimi," which conveys paralinguistic cues related to surprise, admiration, or suffering (see Chapter 4: 141–142).

The identity-related applications of pulse phonation vastly outnumber its linguistic uses. Communities shape the way people use pulse phonation patterns to convey positioning, whether it be a gang persona (Mendoza-Denton 2011) or an educated young woman (Yuasa 2010), only to mention two contrasting cases. Analogous pulse phonation patterns, as another example, have been used to embody identities such as female, male, transgender, or non-binary. Nevertheless, we can consciously use pulse phonation and even develop it into an affectation; however, we more often use it unconsciously (Henton and Bladon 1988), at the end of sentences and words (Redi and Shattuck-Hufnagel 2001), in moans, and in "sad, bored or tired voices" (Ishi, this collection: 141).

In singing and voice work, pulse phonation is used to add "texture" (see Chapter 4), to convey meanings such as "authenticity through struggle" (see Chapter 2), to enhance the bass range (Nix, Emerich, and Titze 2005), or to create vocal effects (Aaen *et al.* 2021), also ingressively (see Chapters 11 and 17). In cinema, sound designers used pulse phonation patterns to voice *fictional* characters (e.g., Gollum, The Grudge, E.T., Predator, The Asset). On the other hand, actors used it to lend *realism* to a performance (see Chapter 8). The use of the pulse register as tool in narrating pain was also reported in medical consultations (Torres *et al.* 2020), which leads us in the field of vocology, or the science of voice habilitation.

Thanks to its physiological characteristics, the pulse register can be used to warm up or cool down the voice (see Chapters 2 and 15), or in cases of organic lesions such as nodules, polyps, cysts, edema (Cielo *et al.* 2011) and glottal hypofunction (Paolillo *et al.* 2019). Its utility has been suggested in the improvement of the pneumophonic coordination and the management of subglottic pressure (Mazzocchi et al. 2002); in hyperkinetic dysphonias (Boone et al. 2010; Cielo *et al.* 2011; in the reduction of hypernasality and improvement of the velopharyngeal sphincter closure (Elias *et al.* 2015); and in the correction of an overly breathy voice (Vennard 1967). Nonetheless, pulse phonation can be part of the cause of specific vocal pathologies, such as contact granuloma (Ylitalo and Hammarberg 2000). Furthermore, it marks the voices of people suffering from cognitive effort in speech planning, such as in schizophrenia or Alzheimer's disease (Cresti *et al.* 2015), or physical effort in voice production, as in the case of cephalargics (Silvestro *et al.* 2021), suggesting that this complex phonatory behavior does not require sophisticated neural control.

As these sound evidences vividly suggest, the pulse register is remarkably versatile. Common sonic materialities (patterns of pulse phonation) can thus serve as a kind of bridge across diverse ways of life (pattern-meaning constructions). However, the direct comparison of these constructions is not suitable due to different conditions, even if similar acoustic situations exist. This sense of *unity in diversity* is an inherent characteristic and a problem of PpR. The umbrella term "pulse register phonation," then, names complex phenomena and not a single, monolithic entity. This is the second position of this collection.

If the pulse register convenes such a complex matter, how can we return complexity to its scrutiny? One solution is to ground our inquiry in the epistemological change established at the outset, abandoning the object "pulse phonation" and instead examining the contexts in which it becomes observable (*in observing*). This approach emphasizes the inextricable link between material and symbolic elements, highlighting that pulse phonation patterns only become perceptible within context-specific practices and knowledge. It allows complexity by incorporating multiple perspectives, each offering particular insights. In practice, this requires a multidisciplinary approach involving different methodologies. These facilitate the juxtaposition of topics such as: The conventional use of creak in Japanese (Chapter 4) and language variation in East Anglia (Chapter 9); infants' production of protophones (Chapter 11) and adult–infant interaction (Chapter 14); singing and voice training (Chapter 2), bird vocal repertoire (Chapter 5), and film studies (Chapter 8). The focus extends beyond similar physiological phenomena, delving specifically into the diverse applications and interpretations of analogous acoustic patterns. In so doing, rather than the pulse register "as such," then, the multidisciplinary field of PpR should be considered the subject of this collection. This is the third position that led to its development.

The study of pulse phonation has yet to be recognized as a unified field. While much of the scholarly exploration has been rooted in sociolinguistics and, to a lesser degree, vocal performance, its relevance in the humanities is underemphasized. Moreover, the role of the pulse register in animal vocalization remains an area of seminal research. Nevertheless, contributions from various disciplines coalesce to create a fragmented yet remarkable subfield of voice studies, speech sciences, and bioacoustics. Merging insights from both the humanities and social sciences can enrich our understanding of pulse phonation and its global significance, while natural sciences can reveal how vocalization both influences and is influenced by behaviors and interactions in an ecosystem. An integrated approach will hopefully lay the ground for the development of PpR and expand our understanding of vocal behavior.

This volume seeks to invigorate and integrate these perspectives. It argues for PpR as an area of multidisciplinary inquiry that requires a combination of approaches, epistemologies, and methods. It proposes such by examining various theories and practices of pulse phonation in three parts: frames (I), narratives (II), and experiences (III).

"Frames" (Part I) alludes to the different places of experience where pulse phonation may become (scholarly) observable, in the context of running speech, vocal performance, or animal vocalization. By offering a multi-context review of the available literature, Part I represents a thorough exploration of the means of studying pulse phonation. In Chapter 1, Nassima Abdelli-Beruh describes the beginnings of PpR and places "the birth of a register" in historical perspective. Focusing on speech-language pathology, the field that granted pulse phonation the status of "normal," she reconstructs the characteristics of pulse phonation as technological advances gradually revealed them. John Nix and Whitney Chappell, in Chapter 2, bring pulse phonation to the surface of contemporary singing pedagogy, vocal therapy, and performance. By examining the use of and attitudes toward pulse phonation among American performers, teachers, and their students, and questioning what are considered stylistically appropriate uses of pulse phonation in style-specific performance and pedagogies, Nix and Chappell lend an attentive ear to the interplay of perception and production.

Alexsandro Meireles's (Chapter 3) and Carlos Ishi's (Chapter 4) contributions then explore in further detail the functioning of pulse phonation in speaking and singing and weave acoustic analyses of spontaneous dialogs and sung excerpts, showing the differences between several possible realizations of pulse phonation. In Chapter 3, Ishi draws on earlier research on the acoustic, aerodynamic, and paralinguistic features of pulse phonation to explore its uses in Japanese and clarify the differences between lax creak and pressed creak. In Chapter 4, Meireles analyzes imitated excerpts from hard rock songs where pulse phonation is used to distort the singing voice. This embodied methodology comes from his experience as a singer and aims at clarifying the difference between mixed voice and unmixed pulse phonation. Concluding Part I, Franz Goller examines the production and use of pulse tones in birds and nonhuman mammals. His writing explores the evolutionary significance of these specialized sounds and what distinguishes them from any other vocalization.

"Narratives" (Part II) refers to the cultural practices in which pulse phonation may embody, perform, and transform lived identities and create social meanings, further exploring the "fascinating intersection between phonetics and language ideologies" (Masters and Strelluf, this collection) where it sits in contemporary societies. In Chapter 6, Madeleine Masters and Christopher Strelluf respond to the "Great Vocal Fry Dustup of 2015" and foreground pulse phonation as a space of linguistic innovation and sociolinguistic change in Suffolk, East Anglia. They highlight pulse phonation as a new feature of innovative English languages *and* a traditional feature of conservative varieties, and zoom in on a local dialect to illuminate intersections and contradictions on a global scale. Robert James Hellyer, in Chapter 7, further explores pulse phonation as a space of change, by examining how the use of pulse phonation evolves as a result of a repositioning of the self. Drawing on Lal Zimman's research on transgender uses of pulse phonation, Hellyer brings one actor's experience—before and after coming out as transgender—to the forefront of pulse phonation as a meaning-making resource.

Míša Hejná and Mark Eaton, in Chapter 8, trace this "meaning-making resource" in the soundscape of attractive female voices in American cinematography. By critically engaging with the "femme fatale" trope in films from 1950s to 2010s, they deconstruct the link between pulse phonation and attractive characters, conveying a counter-image of pulse phonation as a signal of negative stance. In Chapter 9, Joe Pearce further unpacks the complexities of the gendered use of the pulse register by looking at Scottish English and asking to what extent pulse phonation is mediated by sociolinguistic ideologies, and to what extent it can influence gender perception when other cues are ambiguous. In the final chapter of Part II, Tristam Vivian Adams uses Lacanian psychoanalysis and Franco Berardi's thinking on "semiocapitalism" to locate pulse phonation at the limit of speech. By looking at a variety of sources from Neo-Western films to Plato's *Symposium*, Chapter 10 conceptualizes pulse phonation as a symptom of ineradicable excess.

"Experiences" (Part III), therefore, turns to such excess and explores the various ways in which pulse phonation exceeds, excludes, and expands language and meaning, coalesces function and expression, and introduces or reveals "asymmetries" (Edgerton 2015: 22). In Chapter 11, I interrogate the "voices" of early cavemen, modern infants, contemporary musicians, animals, and imaginary beings, to advocate the transformative power that pulse phonation draws from its prelinguistic otherness, and locate in such transformativity its creative force. Bojana Radovanović, in Chapter 12, addresses the space that such "otherness" occupies in contemporary music theory and practice, with a focus on Yugoslav and contemporary Serbian music scenes. By using Jelena Novak's theory of the monstrous voice to discuss the implications of pulse phonation with the standardization of voice, she traces the cultural evolution of pulse phonation from harmful vocal gesture to extended vocal technique. Poe Allphin's contribution further discusses the role of pulse phonation in the formation of the transmasculine voice. From the intentional use of creak as a passing strategy to the unintentional cracks in the voice during hormone replacement therapy, Chapter 13 explores pulse phonation as a method of (re)gaining autonomy in relation to notions of adulthood and perceptions of gender. In Chapter 14, the use of pulse phonation in infant-directed speech is explored by Andrew Cheng, Elise McClay, and H. Henny Yeung. Their analysis focuses on mothers who are unaware of the global cultural meanings of pulse phonation, when communicating with their infants in a structured play session. In Chapter 15, Lisa Quoresimo focuses on the vocal studio. She derives her methodology on her students' experiences and her own experience as a teacher of contemporary singing styles and explores examples of training through pulse phonation and some pedagogical issues this presents. In the final chapter of Part III, Ilona Krawczyk revisits Arnold Mindel's notion of "dreambody" through post-Grotowskian practices of processwork, to trace pulse phonation as revealing of somatic sources.

Transformations

To understand what threads through this breadth of preoccupations, let us go back to the "birth of a register." At that time, Harry Hollien and colleagues initiated with their efforts a tortuous process of normalization (that is still ongoing), which resulted in our ear to be more sensible to pulse phonation. Indeed, an important line of PpR involves exploring and questioning the (de)normalizing processes that may be at play, rather than accepting or reinforcing them. This may involve exploring the ways in which pulse phonation is shaped and influenced by social expectations (e.g., in speech, Chao and Bursten 2020; in singing, Nix *et al.* 2018), as well as the ways in which it resists these norms (e.g., Venturi 2021). It may also be a matter of exploring the power dynamics at play in relation to pulse phonation and how this affects its reterritorialization. In countries such as the United States, New Zealand, or England, the stylized use of pulse phonation evolved into a cultural phenomenon that informed a new generation of speakers, and this phenomenon has reached all corners of the globe through popular media.

Normalizing processes involves developing behavioral standards (e.g., expectations, norms)—for the most varied reasons and not necessarily by someone's will—that can change the way voice quality is used and perceived (see Chapter 1: 27). These processes may also involve the integration or avoidance of previously marginalized voice qualities, among which pulse phonation is at the forefront: The contributions of Abdelli-Beruh, Nix and Chappell, Master and Strelluf, Hellyer, or Radovanović, evidence these processes in several domains, by the means of speech-production literature (Chapter 1), singing pedagogy (Chapter 2), variationist sociolinguistic (Chapter 6), gender theory (Chapter 9), or critical musicology (Chapter 12). These discussions further evidence that the material characteristics of pulse phonation and the cultural meanings that are associated with them (such as ideas about gender, health, etc.) play a joint role in shaping its production. It is precisely in this transformative sense that the problem of PpR lies at the heart of a pattern-meaning construction. This interconnection is indistinguishable.

What are the features that enable this transcendence across cultural boundaries and linguistic paradigms and empower pulse phonation to embody such a wide array of emotional and symbolic connotations? The chapters to follow suggest that retention of a certain "otherness," (Venturi, this collection: 349) or "excess" (Adams, this collection: 310), is perhaps what makes pulse phonation so transformative, and that this is what makes its complexity possible. The fact that the pulse register emerges when other phonations are impaired (Cresti *et al.* 2015) is an indication of its prelinguistic otherness. The fact that it is found also in infants (Oller *et al.* 2019), in primates (Riede and Zuberbuhler 2003), and in languages very different from one another, such as Japanese (Chapter 4) and Scottish English (Chapter 9), is another evidence of nonverbal derivation. The endogenous, presymbolic nature of the pulse register allows it to exist outside or alongside the symbolic systems of language. This allows pulse phonation to convey emotions, meaning, and "force" (Austin 1962).

To borrow the apt words by Katherine Meizel (this collection: 466), pulse phonation "sounds ultra-human and less-than-human, innocent and adult, an ur-voice and a not-voice." For a human, the transformativity of pulse phonation lies in its ability to tap into this prelinguistic otherness and use the possibilities of voice quality to *do things* (Chapter 11). For a nonhuman animal, transformativity lies in the power to develop adaptive functions and communication strategies. Transformativity, then, is a way of thinking about the power of sound to become "quality" and "gesture," to acquire and convey meanings, and to shape our ecological niches and our place within them (Polechovà and Storch 2019). An examination of such power is what brings together the "predator alarm" of a primate in the rainforest (Riede and Zuberbuhler 2003), the "creaking" technique of a human vocalist in a rock concert (Aaen *et al.* 2021), and the "phenomenon of extreme vocal fry" in a country such as New Zealand (Hornibrook, Ormond, and Maclagan 2018). As the pages to follow suggest, this *unity in diversity* is an invitation to approach PpR as the opportunity for new ways of thinking about the sounding voice.

As a further demonstration of such power, Chapters 17 and 18 conclude this volume with a collective exploration of experiences with, uses of, and takes on pulse phonation, by further asking, "what can the pulse register do?" To provide answers, the two chapters bring together the lived experiences of artists, educators, and scholars, seeking to return complexity to a complex phenomenon. In a vivid conflict of voices, feelings, and stories, therefore, it is possible and probable that readers will gain a more nuanced notion of pulse phonation (research).

The complex soundscapes of PpR are an invitation to the thematic explorations of this volume. Readers interested in the acoustic realities of pulse phonation will find insights in Ishi (speech), Meireles (song), and Goller (animals). Historical perspectives run through the contributions of Abdelli-Beruh, Radovanović, and Panzanelli, seeking to retrace the discourse on "vocal fry." The discussions of Adams, Venturi, Masters and Strelluf, and Meizel reflect on the multilayered contradictions introduced by the sounds of pulse phonation. The embodiment of gender is the concern of Hellyer, Hejná and Eaton, Pearce, Adams, and Allphin. A focus on education waves the contributions of Nix and Chappell, Quoresimo, Bana, Pikes, or Cheng, McClay, and Yeung. Finally, readers interested in creative practice will source information in Krawczyk, Nix and Chappell, Venturi, La Barbara, Buccino, Vieira de Oliveira, Kardos, Mozzanega, and

Sitson. There is no unique point of view common to all the voices in this collection, other than an interest in the problem(s) of PpR. My hope and expectation is that these discussions, joined by more voices, will contribute to an emerging field.

References

Aaen, M., McGlashan, J., Christoph, N., and Sadolin, C. (2021). "Extreme vocal effects distortion, growl, grunt, rattle, and creaking as measured by electroglottography and acoustics in 32 healthy professional singers," *Journal of Voice*: S0892-1997(21)00391-X.

Anderson, R. C., Klofstad, C. A., Mayew, W. J., and Venkatachalam, M. (2014). "Vocal fry may undermine the success of young women in the labor market," *PloS ONE*, **9**(5): e97506.

Au, W. W. L. (1993). *The Sonar of the Dolphin Springer*, New York: Verlag.

Austin, J. L. (1962). *How to Do Things with Words*. Oxford: Clarendon Press.

Blomgren, M., Chen, Y., Ng, M. L., and Gilbert, H. R. (1998). "Acoustic, aerodynamic, physiologic, and perceptual properties of modal and vocal fry registers," *J. Acoust. Soc. Am.*, **103**: 2649–2658.

Boone, D. R. and McFarlane, S. C. (2010). *The Voice and Voice Therapy* (7th edition), Boston, MA: Allyn & Bacon.

Buder, E. H., Warlaumont, A. S., and Oller, D. K. (2013). "An acoustic phonetic catalog of prespeech vocalizations from a developmental perspective," in B. Peter and A. N. MacLeod (Eds.), *Comprehensive Perspectives on Child Speech Development and Disorders: Pathways from Linguistic Theory to Clinical Practice*, NOVA.

Chao, M. and Bursten, J. R. S. (2020). "Girl talk: Understanding negative reactions to female vocal fry," *Hypatia*: 1–18.

Childers, D. and Lee, C. (1991). "Vocal qualities factors: Analysis, synthesis, and perception," *J. Acoust. Soc. Am.*, **90**: 23942410.

Cielo, C. A., Elias, V. S., and Brum, D. M. (2011). "Thyroarytenoid muscle and vocal fry: A literature review," *Rev. Sociedade Bras Fonoaudiologia*, **16**: 362–369.

Cresti, C., Dovetto, F. M., and Rocha, B. (2015). "Schizophrenia and prosody. First investigations", in C. Manfredi (Ed.), *IX International Workshop Model and Analysis of Vocal Emissions for Biomedical Applications (MAVEBA), Sept. 2-4, 2015. Proceedings*, Firenze: Firenze University Press: 139–142.

Davidson, L. (2020). "The versatility of creaky phonation: Segmental, prosodic, and sociolinguistic uses in the world's languages," *WIREs Cogn. Sci.*, **12**: e1547.

Edgerton, M. E. (2015). *The 21st-Century Voice: Contemporary and Traditional Extra-Normal Voice*, Lanham: Rowman & Littlefield.

Elias, V. S., Cielo, C. A., Jotz, G. P., and Christmann, M. K. (2015). "Effect of vocal fry on voice and on velopharyngeal sphincter," *Int. Arch Otorhinolaryngol*, (2): 156–162.

Frey, R., Volodin, I., Volodina, E., Soldatova, N. V., and Juldaschev, E. T. (2011). Descended and mobile larynx, vocal tract elongation and rutting roars in male goitred gazelles (*Gazella subgutturosa*, Güldenstaedt, 1780), *J. Anatomy*, **218**: 566–585.

Gobl, C. and Chasaide, A. (2003). "The role of voice quality in communicating emotion, mood and attitude," *Speech Commun.*, **40**(1): 189–212.

Gordon, M. and Ladefoged, P. (2001). "Phonation types: A cross-linguistic overview," *J. Phonetics*, **29**: 383–406.

Grimm, R. J. (1967). "Catalogue of sounds of the pigtailed macaque *Macaca nemestrina*," *J. Zool. Lond.*, **152**: 361–373.

Henton, C. and Bladon, A. (1988). "Creak as a sociophonetic marker," in L. Hyman and C. Li (Eds.), *Language, Speech and Mind: Studies in Honor of Victoria A. Fromkin*, London: Routledge, 3–29.

Hollien, H. (1974). "On vocal registers," *J. Phon.*, **2**(2): 125–143.

Hollien, H., and Michel, J. F. (1968). "Vocal fry as a phonational register." *J. Speech Hear. Res.*, **11**(3): 600–604.

Hollien, H. and Wendahl, R. W. (1968). "Perceptual study of vocal fry," *J. Acoust. Soc. Am.*, **43**: 509.

Hollien, H., Damsté, H., and Murry, T. (1969). "Vocal fold length during vocal fry phonation." *Folia Phoniatr. Logop.*, **21**(4): 257–265.

Hollien, H., Moore, P., Wendahl, R. W., and Michel, J. F. (1966). "On the nature of vocal fry," *J. Speech Lang. Hear. Res.*, **9**(2): 245–247.

Hornibrook, J., Ormond, T., and Maclagan, M. (2018). "Creaky voice or extreme vocal fry in young women," *N. Z. Med. J.*, **131**(1486): 36–40.

Jensen, K. K., Cooper, B. G., Larsen, O. N., and Goller, F. (2007). "Songbirds use pulse tone register in two voices to generate low-frequency sound," *Proc. Biol. Sci.*, **274**(1626): 2703–2710.

Keating, P., Garellek, M., and Kreiman, J. (2015). Acoustic properties of different kinds of creaky voice. In T. S. C. f. I., Ed. (2015), *Proceedings of the 18th International Congress of Phonetic Sciences,* Glasgow, Scotland.

Laver, J. (1980). *The Phonetic Description of Voice Quality,* Cambridge, UK: Cambridge University Press.

Ligon, C., Rountrey, C., Vaidya Rank, N., Hull, M., and Khidr, A. (2019). "Perceived desirability of vocal fry among female speech communication disorders graduate students," *J. Voice,* **33**(5): 805. e21–805.e35.

Loss, S. and Zold, E. (2014). "Use and perception of creaky voice quality in U.S. women's speech," *New Ways of Analyzing Variation,* 43.

MacDougall-Shackleton, S. A. (2011). "The levels of analysis revisited," *Philos. Trans. Royal Soc. Lond.,* **366**(1574): 2076–2085.

Madsen, P. T., Siebert, U., and Elemans, C. P. H. (2023). "Toothed whales use distinct vocal registers for echolocation and communication." *Science,* **379**(6635): 928–933.

Malinowski, B. (1923). "The problem of meaning in primitive languages," in C. K. Ogden and I. A. Richards (Eds.), *The Meaning of Meaning: A Study of the Influence of Language upon Thought and of the Science of Symbolism,* London: Kegan Paul, Trench, Tntbner. (4th revised edition, 1936): 296–336.

Maturana, H. N. (2002). "Autopoiesis, structural coupling and cognition: A history of these and other notions in the biology of cognition," *Cybernetics Human Knowing,* **9**(3–4): 5–34.

Mazzocchi, R., Panzanelli, R., Brizi, S., *et al.* (2002). "L'utilita del vocal fry nell'e-ducazione della voce parlata e cantata: nostra esperienza," *Atti XXXVI Convegno nazionale SIFEL,* Modena.

McLoughlin, D. (2019). "Interview with Vicki Kirby," *Theory Culture Soc.,* **36**(7–8): 261–271.

Mendoza-Denton, N. (2011). "The semiotic hitchhiker's guide to creaky voice: Circulation and gendered hardcore in a Chicana/o gang persona," *J. Linguist. Anthropol.,* **21**(2): 261–280.

Michaud, A. (2004). "Final consonants and glottalization: New perspectives from Hanoi Vietnamese," *Phonetica,* **61**(2–3): 119–146.

Morin, E. (1990). *Introduction à La Pensée Complexe,* Paris: ESF.

Nix, J., Chappell, W., and Parrott, M. (2018). "Social and stylistic correlates of vocal fry in a cappella performances," *J. Voice,* **34**(1): 156.e5–156. e13.

Nix, J., Emerich, K., and Titze, I. (2005). "Application of vocal fry to the training of singers," *J. Singing,* **62**(1): 53–59.

Oller, D. K., Caskey, M., Yoo, H., Bene, E. R., Jhang, Y., Lee C. C., Bowman, D. D., Long, H. L., Buder, E. H., and Vohr, B. (2019). "Preterm and full term infant vocalization and the origin of language," *Sci. Rep.*, **9**(1): 14734.

Paolillo, N. P., Carrozza, L., Oslo, M., Rosa, E., and Scalabrin, M. (2019). "Inspiratory vocal fry: Anatomical and physiological aspects, application in speech therapy, vocal pedagogy and singing. A pilot study," *J. Voice*, **35**(3): 394–399.

Peters, G. (1981). "Das Schnurren der Katzen," *Saugetierkundl, Mitteilungen*, **29**: 30–37.

Podesva, R. (2013). "Gender and the social meaning of non-modal phonation types," *Proceedings of the 37th Annual Meeting of the Berkeley Linguistics Society*, Berkeley: Berkeley Linguistics Society.

Polechová, J. and Storch, D. (2019). "Ecological niche," in *Encyclopedia of Ecology* (2nd edition), Elsevier, 72–80.

Proctor, K., Scherer, R. C., and Perrine, B. L. (2024). "Vocal fry patterns while reading," *J. Voice*, **38**(4): 889–902.

Ramsey, M. (2020). "Vocal fry and how to use it in your voice," Ramsey Voice Studio. https://ramseyvoice.com/vocal-fry/

Redi, L. and Shattuck-Hufnagel, S. (2001). "Variation in the realization of glottalization in normal speakers," *J. Phonetics*, **29**(4): 407–429.

Reynolds, E. (2015). "What's the big deal about vocal fry? An NYU linguist weighs," *New York University* [accessed May 10, 2023].

Riede, T. and Zuberbühler, K. (2003). "Pulse register phonation in Diana monkey alarm calls," *J. Acoust. Soc. Am.*, **113**: 2919–2926.

Roubeau, B., Henrich, N., and Castellengo, M. (2009). "Laryngeal vibratory mechanisms: The notion of vocal register revisited," *J. Voice*, **23**(4): 425–438.

Shaw, F. and Crocker, V. (2015). "Creaky voice as a stylistic feature of young American female speech: An intraspeaker variation study of Scarlett Johansson," *Lifespans and Styles*, **1**: 21–27.

Silvestro, M., Dovetto, F. M., Corvino, V., Apisa, P., Malesci, R., Tessitore, A., Milizia, P., Tedeschi, G., Marciano, E., and Russo, A. (2021). "Enlarging the spectrum of cluster headache: Extracranial autonomic involvement revealed by voice analysis," *Headache. J. Head Face Pain*, **61**(9): 1452–1459.

Smith, M. (1980). "Behavior of the Koala, *Phascolarctos cinereus* Goldfuss, in captivity, III. Vocalizations," *Aust. Wildl. Res.*, **7**: 13–34.

Speck, B. (2006). "Voice and the construction of identity and meaning," in *In-roads of Language: Essays in English Studies*, Castelló, Spain: Universitat Jaume I.

Titze, I. R. (1994). *Principles of Voice Production*, Englewood Cliffs: Prentice Hall.

Torres, P. J., Henry, S. G., and Ramanathan, V. (2020). "Let's talk about pain and opioids: Low pitch and creak in medical consultations," *Discourse Studies*, **22**(2): 174–204.

Van den Berg, J. W. (1958). "Myoelastic-aerodynamic theory of voice production," *J. Speech Hear. Res.*, **1**(3): 227–244.

Van Edwards, V. (2021). "Vocal fry: What it is and how to get rid of it?" *Science of People*, October 25.

Vennard, W. (1967). *Singing: The Mechanism and the Technic*, New York: Carl Fisher.

Venturi, F. (2021). "Creaky voice gender," *J. Interdisciplinary Voice Studies*, **6**(2): 201–218.

Wendahl, R. W., Moore, G. P., and Hollien, H. (1963). "Comments on vocal fry." *Folia Phoniatr. Logop.*, **15**(4): 251–255.

Whitehead, R., Metz, D., and Whitehead, G. (1984). "Vibratory patterns of the vocal folds during pulse register phonation," *J. Acoust. Soc. Am.*, **75**: 1293–1297Ylitalo, R. and Hammarberg, B. (2000). "Voice characteristics, effects of voice therapy, and long-term follow-up of contact granuloma patients," *J. Voice*, **14**(4): 557–566.

Yuasa, I. (2010). "Creaky voice: A new feminine voice quality for young urban-oriented upwardly mobile American women?" *Am. Speech*, **85**(3): 315–337.

Chapter One

From deviance to the birth of a register

Nassima B. Abdelli-Beruh

This chapter focusses on a type of phonatory behavior prevalently called "fry" or "vocal fry" in the field of speech-language pathology (SLP), but for which an array of descriptors have been used in SLP and in other disciplines. This led Moore and von Leden to lament in 1958 that "there is considerable ambiguity in the use of the terms. The semantic problem arises from the fact that various writers have employed one term to designate different qualities, and conversely, a variety of names have been applied to a single quality" (1958: 225). Hence, one of the challenges when reviewing the literature on this laryngeal mode of phonation is the multiplicity of terms used to describe what may or may not be the same vocal behavior: "creak," "creaky voice," "laryngealization," "glottalization," "crackling," "croak," "popping," "frying," "ticking," or "rasping." Given the wealth

Creak: Theories and practices of pulse phonation
Edited by Francesco Venturi

ISBN 978-981-5129-29-8 (Hardcover), 978-1-003-61606-1 (eBook)
www.jennystanford.com

of information on this topic and the "considerable ambiguity in the use of terms" within and across disciplines, the author chose to review selected seminal historical research articles on "fry" in SLP only. This chapter retraces the quest for the term that would, according to the researchers, best encapsulate its distinguishing perceptual, acoustic, aerodynamic, and physiological features, which technological advances gradually revealed. Since the descriptor "fry" is most prevalent in SLP, the author will, therefore, use this term throughout this chapter.

Fry became the focus of scientific interest within SLP, a clinically oriented field, long before it attracted the attention of researchers in other disciplines, became the vocal phenomenon of social media in the early 2000, or a fad adopted by a generation of American English (AE) speakers but reviled by other generations of AE speakers and listeners (Anderson *et al.* 2014; Chao and Bursten 2021; Liu and Xu 2011). The search for the distinguishing features of fry has been accompanied by a flurry of perceptual, acoustic, aerodynamic, and physiological investigations, resulting in the proliferation of new labels that have remained in the province of distinct investigative fields without necessarily crossing over to another. Hence, this review chronicles the scholarly pursuit endeavored by researchers in a single discipline, SLP, and offers a picture of fry as it emerged and gradually evolved in this given field. Therefore, most references used for this review represent the clinical and theoretical perspectives of researchers in this selected discipline. This choice is not random as it is SLP that granted fry its status of register. Experts in other disciplines may avail themselves to this non-exhaustive historical review and decide on whether fry as it is described in SLP is similar or not to what might be called "creaky voice," "laryngealization," "glottalization," "crackling," "croak," "popping," "frying," "ticking," or "rasping" in their own field.

As the review will show, there can be some confusion as to which vocal emission qualifies as fry; this might be attributed in part to the fact that vocal productions are sensitive to variations in eliciting tasks, measurement techniques, perceptual criteria, all of which vary with the expertise of researchers, whose own training and scope of practice/study construe their experimental and theoretical framework (e.g., speech language pathologists, psycholinguists, sociolinguists, physicians, voice scientists, phoneticians, voice pedagogues, and singers, theatre, acting and English teachers, anthropologists and even philosophers) (e.g., Abdelli-Beruh, Wolk, and Slavin 2014; Abdelli-Beruh, Drugman, and Owl 2016; Anderson *et al.* 2014; Chao and Bursten 2021; Shaw and Crocker 2015; Henton and Bladon 1988; Henton 1989; Hollien 1974; Jelle 2017; Gordon and Ladefoged 2001; Moore and von Leden 1958; Roubeau, Heinrich, and Castellengo 2009; van den Berg 1955; 1956; Wolk, Abdelli-Beruh, and Slavin 2012; Yuasa 2010).

Indeed, it is at times dubious whether authors within SLP are referring and examining the same vocal phenomenon: some 60 years ago, Moore and von Leden noted that the "description of the accompanying voice was too general to permit definite identification with the phenomenon under discussion" (1958: 226). Hollien also aptly pointed out that the research cited is limited by the fact that one cannot always "determine the register system utilized in the research" (1974: 127).

Views on vocal fry as defined in this chapter and in SLP have evolved considerably from the nascent days of this field in the United States and over the course of the past 100 years. In the course of a century, fry went from being viewed mainly as deviant (i.e., a sign of vocal pathology and/or affectation) to being viewed as "normal," acquiring along the way the status of register in the late 1960s and therapeutic value in the 1990s. The healing virtues of fry were already put to use by Moser as early as 1942. He utilized the "glottal fry attack" as a particular technique "to establish normal voice" (Allen and Hollien 1973; Cielo *et al.* 2011; Hollien and Michel 1968; Hollien *et al.* 1966; Hollien and Wendahl 1968; Hollien 1974; Moser 1942: 174). Thus, a historical review, by no means comprehensive due to page limitations, is in order to ascertain how its rehabilitation to normal phonation parallels the advent of techniques that revealed its complex and distinctive acoustic, physiological, and aerodynamic traits.

Desirability of vocal emissions

In the United States, the interest in socially acceptable speech productions emerged in the early 19th century as the parameters of what "good" American speech should sound like were being defined (Pierre and Pierre 2018; Ray 2005). In the early part of the twentieth century, norms of speech were being established concurrently as immigrants with speech defects or heavy accents were denied immigration status and sent back to their departing countries. The relationship between protecting the purity of the American speech and the safety of the American nation transpires in the stances of a pledge of allegiance written by the American Speech Committee of the Chicago's Women Club: "I promise that I will say a good American 'yes' and 'no' in place of an Indian grunt 'um-hum' and 'nup-um' or a foreign 'ya' or 'yeh' and 'nope'" (Robbins 1918: 175). Crumpton blamed the "most embarrassing deficiency of speech" on "the influence of negro dialect" (1917: 96). Additionally, McDonald (1916) reported that 90% of delinquent juveniles had speech defects. Robbins (1918) also echoed such sentiment by noting that "among delinquents speech deficiencies are noted with greater frequency than among the normal population" (1916: 586). They believed that if "delinquents" had received adequate speech corrections, they would not have committed acts that warranted incarcerations, thus establishing a link between "good" American speech, morality, mental health, social adequacy, and more. As accents and speech defects begun to be considered a threat to American citizenry and identity, speech betterment programs such as Better Speech Week and Better Babies Week proliferated nationwide (Crumpton 1917). McDonald offered to create a Special Speech Department with the goals to restore proper "tonal and enunciation standards" through the "standardization of English without accent for the foreign born" because "a good speech, unhampered by accent, is a requisite for the highest mental and moral development of the immigrant" (McDonald 1916: 863–864). The desire to help those with speech defects and the wish to purify the so-called "good" American speech became the focus of political, scientific, and medical concerns.

Human behaviors are complex and context-dependent; they may be robust or elusive, inherited and/or acquired. They may be the reflection of an intact normally functioning or disordered physiological system. Vocal emissions participate in verbal as well as nonverbal communications and the nonverbal aspects of the voice can, at will, supersede the verbal ones. When used volitionally, voice quality can serve to communicate a wide array of information about the speaker: emotional states, sexual desirability, power, authority, masculinity, femininity, sociocultural factors, and more (Anderson *et al.* 2014; Jelle 2017; Shaw and Crocker 2015; Sorokowski *et al.* 2019; Yuasa 2010).

Vocal productions are endowed with a quality that does not leave the listeners indifferent and can, when perceived as disordered and/or different than the norm, distract and even hinder communication. Hence, it became necessary for clinicians to classify in order to evaluate such vocal emissions and coincidently assess the integrity of the laryngeal apparatus. The classification of a range of vocal qualities necessitates the use of labels. Even though there is no agreed definition of vocal quality, there are plenty of subjective descriptors used to qualify a voice: "smooth," "breathy," "hoarse," "harsh," "rattling," "rumbling," "crackling," "ticker-like," "popping," "strident," "X factor," "frying," "ticking," and "rasping" (Moore and von Leden 1958: 225, 231; Voelker 1935: 471).

In search of the best descriptors and the importance of titles

In the field of SLP, the terms vocal fry or glottal fry are used interchangeably in the literature. These terms refer to a unique perceptual vocal quality that differs from the perceptual voice quality that is most typically associated with productions in the modal/normal or falsetto/loft phonatory registers. It took a few decades though before the term "fry" became widely used in SLP. Already, Sherman and Linke (1952) indicated that the terms used to describe fry varied immensely from one source to another. They found that in eight out of the nine texts they reviewed, the most recurrent descriptor associated with such perceptual vocal quality was "harshness." A few years later, Moore and von Leden (1958) confirmed that indeed "harshness," not fry, was the term used most frequently among clinicians to describe this voice quality. They reported that the term fry was purposely selected by Moser because it best encapsulated the perception of "the sound of vigorously popping corn" (1958: 174). Interestingly, Moser noted as early as 1942 that it is the "sound produced by many youngsters in imitating a motor boat" (1942: 174), which corroborates recent findings that younger speakers of AE use it more often than older AE speakers, for whom fry is annoying when not abhorrent (Abdelli-Beruh, Wolk, and Slavin 2014; Anderson *et al.* 2014; Chao and Bursten 2021; Ligon *et al.* 2019; Wolk, Abdelli-Beruh, and Salvin 2012).

Early on, dissatisfied with ambiguous descriptors, researchers elected to devise their own terms, creating a plethora of perceptual labels, confounding professionals even more. Asking students to speak sentences "in a careless and indecisive manner," i.e., in a manner incompatible with proper discourse as defined at the times, Voelker (1935) reported the emissions to be an "unpleasant" vocal quality, characterized by a "rattling, rumbling, cracking or ticker-like substitution for phonation… [and] considerable regularity" (1935: 471). He then attributed such vocal emissions to being the product of "dysphonia ventricularis." The use of "ventricularis" was meant to suggest the involvement of ventricular folds (i.e., vestibular folds or false vocal folds), a putative explanation, which would later be disputed by Moore and von Leden but revived by Hollien and colleagues and corroborated by others (Moore and von Leden 1958; Hollien 1974; Stager *et al.* 2000). The involvement of ventricular folds in fry will be discussed later in this chapter. The early use of the clinical term "dysphonia" highlights the already strongly held belief by SLP clinicians that fry productions were signs of impaired laryngeal mechanisms despite the lack of supportive objective measures. Such position was not novel and may have stemmed from the fact that such productions were judged "unpleasant," as noted by Voelker (1935), because they were far too different from the "mellow, vibrant quality," which individuals were expected to exhibit when speaking (1935: 471–472). Hence, rather than being rooted in any physiological evidence, this belief simply reflected degrees of deviance from the standards of speech etiquette of the times. Indeed, Voelker ascribed the "false voice" to personality traits and proposed therapy to restore the "finest of enunciations" considered proper for the times and meeting the standards of good elocution in social settings (1935: 471). Moore and von Leden (1958) admitted that the use of terms such as "harshness," "strident voice," "hoarseness," "X factor," or other terms approximating perceptions of "popping," "frying," "ticking," "rasping" were equally unsatisfactory as they made no reference to the concomitant physiological, acoustic, or aerodynamic idiosyncrasies of the vocal quality labelled fry (1958: 225, 231). As other researchers had done in the past, Moore and von Leden chose to coin their own descriptor after equating "harshness" with "glottal fry," and concurrently admitting to the use of a "confused terminology" (1958: 235). Using high-speed filming and perceptual evaluations, they proposed the term "dicrotic dysphonia" to describe fry; this composite term relied on objective measures for the choice of "dichrotic" and on perceptual/subjective evaluations for the choice of "dysphonia." The term "dysphonia," rooted in the prevalent belief that fry emanates from a dysfunctional phonatory system, perpetuated the idea of "faulty" vocal quality (ibid.). To date, fry remains associated with vocal pathology in SLP. While the term of "dichrotic dysphonia" was used in few articles thereafter (Chen, Robb, and Gilbert 2002; Hedelin and Huber 1990), it was supplanted by the term "fry." Dr. Harry Hollien, a prolific researcher in SLP, together with his colleagues are to be credited with such a shift, when they gave this descriptor a place of choice in the titles of a volley of research papers published in the 1960s, and in which they granted fry the status of register: On the Nature of Vocal Fry, Vocal Fry as a Phonational Register, Perceptual Study of Vocal Fry, Vocal Folds Length During Vocal Fry Phonation, Comments on Vocal Fry, Perceptual Differentiation of Vocal Fry and Harshness, Fundamental Frequency Investigation of Vocal Fry and Harshness (Hollien, Damsté, and Murry 1969; Hollien and Michel 1968; Hollien, Moore *et al.* 1966; Hollien and Wendahl 1968; Michel 1968; Michel and Hollien 1968; Wendahl, Moore, and Hollien 1963).

However, shortly after, Hollien relabeled "fry" as "pulse," just as he favored the term "modal" over "normal" and "loft" over "falsetto" (Hollien 1974: 126). He justified the relabeling by remarking that the term "pulse" was a better fit for the "pulse-like" laryngeal vibratory patterns (ibid.). He also proposed synonymous terms for fry: "glottal fry," "creak," and "strohbass," increasing the numbers of permissible descriptors for this vocal quality newly baptized as a register (ibid.). Just as for vocal quality, "agreement cannot be obtained even with respect to a reasonable definition of the nature of registers themselves" (Hollien 1974: 125; Ferranti 2004; Mörner, Fransson, and Fant 1963; Hollien and Michel 1968; Blomgren *et al.* 2002; Roubeau, Henrich, and Castellengo 2009; Titze 1994). Just as Moore and von Leden (1958) lamented the profusion of descriptors for fry, Mörner, Fransesson, and Fant (1963) deplored that "the terminology with regard to voice pitch levels, i.e., "registers," suffers from the existence of an abundance of terms and an ambiguity of their use" (1963: 18).

A daring proposal and an unforeseen rehabilitation

Early on, a few clinical researchers expressed the need to offer a multifaceted description of registers and vocal qualities. Before the advent of instrumentation capable of documenting aerodynamic, acoustic, physiological, and electromyographic features, registers were only defined in terms of one's perceptions. Manuel Patricio Rodríguez García, a famed voice teacher, singer, and composer, proposed in his New Treatise on the Art of Singing that registers are "a series of consecutive and homogenous tones going from low to high, produced by the development of the same mechanical process" (García 1857: 6). Later, other investigators began to frame their definitions in the context of their specific experimental investigations: for instance, Mörner, Fransson, and Fant (1963) indicated that the only "secure denominator for defining register is by means of its range on the musical scale" (1963: 20). They suggested the use of the international standard pitch scale (A_4 = 440 c/s) and parsed its continuum into five registers "deepest range, deep level, the mid-level, the high level and highest range" (ibid.). Hollien and Michel (1968) defined registers as "a series or range of consecutive (vocal) frequencies of similar vocal quality; in addition, they should be little or no overlap in fundamental frequency between adjacent registers." (1968: 600). Albeit the absence of an agreed definition for register, there were admittedly only two registers in SLP until Hollien and colleagues adamantly added fry to the list, proposing that there should be three distinct registers for the non-pathological voice (Hollien 1974; Hollien, Moore, Wendahl, and Michel 1966). Yet, Hollien (1974) admitted that there might be more than three "voice registers" (1974: 127). Hollien and colleagues stated that fry should be considered a "normal mode of laryngeal operation that results in a distinctive acoustic signal" lying low below the modal/normal voice register (Hollien *et al.* 1966: 246; Hollien and Michel 1968; Hollien and Wendahl 1968; Murry and Brown 1971). Hollien reduced the definition of fry or "pulse" to the lowest part of the fundamental frequency (f_0) continuum partitioned into registers (Hollien 1974; McGlone 1967; Hollien and Wendahl 1968; Hollien and Michel 1968; Blomgren *et al.* 1998).

The "distinctive acoustic signal" to which Hollien and his colleagues refer is f_0 or the rate at which the vocal folds vibrate (i.e., the number of cycles per second) (Hollien *et al.* 1966: 246). It is the main contributor to pitch, the subjective correlate of f_0. It is one of the first objective measures cited in the research literature on fry. As early as 1935, Voelker reported that f_0 in fry could range between 6 Hz and 30 Hz, exhibiting a very restricted range of 24 Hz. He did not provide any details on how he elicited fry or on the instrument used to measure f_0. While mean f_0 and range values in fry may differ from study to study because of the differences in elicitation procedures, instrumentation used, and vocal training of the speakers, f_0 values were consistently reported to remain below that of the modal register: f_0 values reliably do not overlap with that of the modal register and fry f_0 range is also systematically smaller than that for the modal/normal register (Blomgren *et al.* 1998; Chen *et al.* 2002; McGlone 1967; Hollien, Girard, and Coleman 1977; Hollien and Michel 1968; Hollien and Wendahl 1968; Murry 1971; McGlone and Shipp 1971; Voelker, 1935). Strikingly, gender differences in f_0 documented in modal/normal and loft/falsetto registers vanish in fry (Blomgren *et al.* 1998; Chen *et al.* 2002; Hollien and Wendahl 1968; McGlone 1967; McGlone and Shipp 1971; Murry 1971).

But f_0 naturally varies from cycle to cycle: Such variability is known as jitter in the frequency domain and shimmer in the amplitude domain. Additionally, the signal/noise (S/N) ratio provides information on the prevalence of signal energy over noise energy. Together, the measures of periodicity/aperiodicity and variability provide insight into the efficiency of the various laryngeal functions. Voelker noted that while "this timbre resembles more a noise, it has considerable regularity" (Voelker

1935: 471). As had been noted by Voelker, Hollien and Wendahl (1968) also reported that fry is characterized acoustically by periodicity, whereas "hoarseness" and "harshness" are associated with irregular laryngeal vibrations and high perturbation factors (Voelker 1935; Hollien and Wendahl 1968; Moore and Thompson 1965; Lieberman 1963). Blomgren (1998) found that indices of periodicity were affected in fry: Jitter and shimmer values were significantly greater in fry than in the modal register for both males and females, and the S/N ratio was reduced in fry (i.e., higher noise energy component relative to the periodic energy component), possibly indicating a "decreased phonatory stability" (1998: 2655).

In favor of fry as a distinct normal laryngeal behavior

While many clinicians used "harshness" or "hoarseness" interchangeably with "fry," a few clinicians recognized that it existed as a separate vocal entity distinguishable from pathological vocal productions deemed harsh or hoarse (Moore and von Leden 1958; Michel 1968; Timcke, von Leden, and Moore 1959): Van Ripper and Irwin reported that fry could be perceptually distinguished in the context of harsh voices (Hollien and Michel 1968); Timcke, von Leden, and Moore (1959) added that fry is often misdiagnosed as ventricular dysphonia; and Michel (1968) documented experimentally that clinical harshness and vocal fry could be differentiated on the basis of their f_0. Furthermore, Michel reported that mean f_0 values for clinical harshness were higher than that of fry, encompassed a broader range than that of fry, and fell within the f_0 range of the modal register: The mean f_0 for fry was 36.4 Hz (range of 30.9 Hz to 43.7 Hz), and it was distinct from that of harshness for which the mean f_0 was 122.1 Hz (range of 103.7 to 120 Hz). Michel (1968) added that while f_0 could serve to differentiate between fry and clinical harshness, it could not distinguish between clinical harshness and modal phonation, suggesting that f_0 may be key in perceiving fry. Voelker (1935) as well as Hollien and Wendahl (1968) also reported that hoarseness and harshness were associated with greater irregular laryngeal vibrations and higher perturbation factors than that observed in fry (Moore and Thompson 1965; Lieberman 1963).

In addition, several researchers had noted the prevalence of fry at the end of intonational phrases, at syntactic boundaries and at the end of declarative sentences, where f_0 typically drops in the spontaneous speech of speakers devoid of any vocal disorder (Hollien and Michel 1968; Hollien *et al.* 1966; Hollien and Wendahl 1968; Moser 1942; Voelker 1935). As early as 1942, Moser had already noticed its prevalence in the voice of "many youngsters" devoid of laryngeal disorders, leading Hollien to write that "such usage is not considered abnormal" (Moser 1942: 246; Hollien *et al.* 1966: 26; Shipp and McGlone 1970). Much of the research in the field of psycholinguistics and SLP have subsequently confirmed the prevalence of fry at the end of sentences (Abdelli-Beruh, Wolk, and Slavin 2014; Slifka 2006; Redi and Shattuck-Hufnagel 2001; Wolk, Abdelli-Beruh, and Slavin 2012).

The perceptual distinctiveness of this vocal emission, its singular locations at syntactic boundaries, and its volitional use are all evidence in favor of fry being a reflection of normal laryngeal phonation. Yet, in SLP, fry was synonymous with "hoarseness" and "harshness" and viewed as a tell-tale sign of abnormal laryngeal behavior. Hence, the bold character of the proposition that vocal fry is a product of normal laryngeal functions lies in the facts that: (1) vocal fry was no longer regarded as a sign of vocal pathology in SLP and (2) that, soon after, it was promoted to the status of register. While Hollien and colleagues attributed normalcy to fry, they did not completely divorce it from pathology and remarked that the excessive use of fry could possibly be a sign of pathology (Fraser and Campbell 1978; Hollien *et al.* 1966; Ylitalo and Hammarberg 2000). Hollien, Moore, Wendahl, and Michel (1966) stated that "it is probably true that, if an individual phonates only in the vocal fry mode, he has a voice disorder just as an individual who can speak only in falsetto has a voice disorder" (1966: 246). Such statement would ignore the suggestion advanced by Murry and Brown (1971) that there might be two types of vocal fry: one passive and sporadic when it occurs naturally in the terminal breath group or at the end of sentences, and one active and prolonged when speakers adopt it volitionally. Recent literature and casual observations show that speakers have adopted this speech pattern and that it is more prevalent in some categories of AE speakers than in others (Abdelli-Beruh, Wolk, and Slavin 2014; Wolk, Abdelli-Beruh, and Slavin 2012). The consequences of such habitual usage when it extends beyond terminal breath remain a concern for clinicians that it might ultimately damage the vocal folds.

Asserting that fry should be considered a normal register necessitated that it be accompanied by an understanding of how fry, as a product of normal laryngeal functioning, was produced. After relying for so long on subjective evaluations and descriptors that were admittedly confusing, researchers expressed the need to define vocal emissions objectively in order to identify, differentiate, treat, train, and/or retrain vocal behaviors. Hence, the objectification of registers and vocal emissions in terms of their contemporaneous acoustic, physiological, and aerodynamic parameters became a necessity (McGlone and Shipp 1971; McGlone and Brown 1968; Hollien 1974). No longer could reducing registers to a perceptual definition suffice. Anticipating the complexity of the task, Titze (1994) predicted that registers might not be reduced to one or two temporal or spectral measures. Hollien (1974) had already argued in favor of defining registers using a combination of acoustic, physiologic, aerodynamic, and perceptual measures, invaluable in the clinical assessment, treatment, and monitoring during and after treatments. And with the advent of exploratory techniques came the possibility of operationally defining and quantifying register and voice quality. However, one must sift through the abundance of newly unveiled acoustic, physiologic, or aerodynamic features and keep in mind their perceptual, theoretical, and clinical salience, for not all of which necessarily contribute to the percept.

Revolutionary use of an "ingenuous" instrument

Visualization of the laryngeal structures and vibrating vocal folds did not commence until an instrument inserted in the mouth allowed for an indirect view of vocal cords by a means of a set of mirrors. Such an approach has been termed indirect laryngoscopy because the instrument does not allow a direct line of sight (e.g., requires the use of a mirror) (Jahn and Blitzer 1996;

Lapeña 2013; Pieters *et al.* 2015; Ramai *et al.* 2018). Laryngoscopy is an endoscopic procedure that necessitates the insertion of an instrument in the body to offer an unobstructed view of the internal organs concealed from direct examination.

There is archaeological evidence in the ruins of Pompeii that such instruments already existed in 79 BCE. Yet, in spite of its potential clinical importance, the practice could not gain acceptance in the scientific community, presumably because of three major practical issues: its problematic insertion, its lack of adequate lighting, and its cumbersomeness that limited extended examination (Engel 2003; Morgenstern 2005; Lapeña 2013; Ramai *et al.* 2018). While Paul of Egina, a byzantine Greek physician in the seventh century, is credited for being the first to visualize the entrance of the larynx using the glossotrochus, Dr. Philipp von Bozzini, a German physician and inventor, is believed to be the first to recommend the use of an external light source to illuminate the hollow cavities of the human body and to define the principles of endoscopy. He created a device called the Lichtleiter, a complex device composed of four elementary pieces: a viewing piece, a light source (wax candle), a system of mirrors to direct the light, and attachments or specula for insertion into various body cavities, including the larynx. Although there is evidence that it was used successfully in the examination of female body cavities in maternity wards, there is no evidence that it was successfully used to visualize the larynx. The instrument was deemed "ingenuous," but it failed to gain acceptance by the medical community because Dr. Joseph Andreas Stifft, Emperor Franz I's personal physician and the highest authority in military medicine, was openly hostile to innovations (Arregui *et al.* 2012; Ramai *et al.* 2018). That lent a death blow to the Lichleiter, which delayed the usage of an instrument much later considered to have revolutionized medical diagnosis and treatment (Arregui *et al.* 2012; Morgenstern 2005; Pieters *et al.* 2015; Ramai *et al.* 2018). As to whom one should credit the use of the first laryngoscope, the debate goes on: Dr. Benjamin Guy Babington or Manuel Patricio Rodríguez García? Dr. Babington, a physician by training, conceived the "glottoscope" or "glottiscope," which used direct sunlight instead of a wax candle. While his publications describe the instrument, Dr. Babington never published any report on the visualization of the vocal folds. That is where Manuel Patricio Rodríguez García distinguished himself: He provided for posterity the first and most detailed description of the functioning larynx in his seminal article "Observations on the Human Voice," submitted to the *Proceedings of the Royal Society of London* and published in 1856. García wrote that with the aid of a mirror "fixed on a handle suitably bent … against the soft palate and the uvula … the party ought to turn himself towards the sun, so that the luminous rays falling on the little mirror may be reflected on the larynx" (1856: 399).

Even though the instrument allowed only a partial view of the larynx, as the epiglottis concealed the anterior part of the glottis, García described with considerable details the anatomy and laryngeal vibratory patterns of different phonatory modes/registers in a variety of speakers of various age groups. But the desire to provide an anatomo-physiological basis to the art of singing and to apply it to its pedagogical approaches was not readily accepted by all. Among some of his most fierce critics, tenor Gabriel-Louis Duprez wrote: "just as the poet does not need to know the physiology of the brain to write verses, so it is unnecessary to know the anatomy of the vocal organs to sing" (Catellangello 2005: 11; Duprez 1845).

Length and thickness of vocal folds

Since the days of García and Duprez, more sophisticated techniques have been employed in the pursuit of the objectification of vocal emissions. Indirect laryngoscopy, photographic systems, and X-ray techniques were used to measure vocal-fold length. Ultra-speed photography at about 4300 frames per second also provided a novel approach to studying laryngeal structures and their involvement in complex vibratory patterns. A review of the images from ultra-slow speed and frame-by-frame analyses permitted accurate measurements of landmark structures (Moore, White, and von Leden 1962). Hollien (1974) indicated that measures of vocal-fold thickness and length serve to differentiate among the registers.

Chen *et al.* (2002) have argued that the characteristically low f_0 of vocal fry could be attributed to an increase in vocal-fold thickness, a decrease in vocal-fold stiffness (Whitehead, Metz, and Whitehead 1984), a decrease in subglottal pressure (P_S) (McGlone and Shipp 1971), and related physiological and aerodynamic factors (Chen *et al.* 2002; Hollien and Colton 1969). Using vocal-fold photography and lateral radiography, Hollien, Damsté, and Murry (1969) asked speakers to match the frequency of various standards (25 Hz, 44 Hz, 58 Hz). They found that vocal folds are shorter and thicker in fry than in modal register by about 3.5 mm and that their length remained unchanged across changes in f_0 unlike what has been observed in modal register (Allen and Hollien 1973; Hollien, Damsté, and Murry 1969; Hollien 1974; Hollien, Brown, and Hollien 1971; Hollien and Moore 1960; Hollien 1960, 1962; Hollien 1974; Hollien and Moore 1960; Sonninen 1954). Indeed, Hollien and colleagues have shown, by means of vocal-fold photography and lateral radiography, that vocal-fold length varies with f_0 in modal register (Allen and Hollien 1973; Damsté *et al.* 1968; Hollien 1960, 1962; Hollien and Colton 1969; Hollien and Moore 1960; Hollien *et al.* 1966, ; Hollien, Damsté, and Murry 1969).

Relying on the productions of one speaker, Moore and von Leden (1958) reported that during fry the arytenoids come and stay in close contact, the lateral margins of the vocal folds are tense and stiff, but the medial margins of the vocal folds are not as lax as in low-frequency modal register. In fry, the vocal-fold vibrations mainly involve the medial margins of the vocal folds, unlike in the low f_0 end of the modal register (Moore and von Leden 1958). Medial compression in fry also appears lighter than in modal register at low f_0. Although the superior and inferior edges of the vocal folds are well delineated in the low f_0 modal register, they are not in fry (ibid.).

Using electro-myographic (EMG) analysis, McGlone and Shipp (1971) reported that the mean percentage of activity from the cricothyroid muscle was higher in low-frequency modal phonation than in glottal fry, while the mean percentage of activity from the thyroarytenoid was higher in fry than in low-frequency modal phonation. They did not observe any difference in the mean percentage of activity in the posterior cricoarytenoid muscle. Using stroboscopic laminagraphy to obtain X-rays, Allen and Hollien asked 35 volunteers devoid of any vocal pathology to produce "phonation at high and low pulse rates" (Allen and Hollien 1973: 242). They did not find any relationship between vocal-fold thickness, ventricular fold thickness, and f_0. Using high-speed photography, Zemlin (1998) reported that in fry the vocal folds are tightly adducted but relatively slack with air pushing through at about the anterior–posterior portion of the thyroarytenoid muscle. According to Blomgren *et al.* (1998), decreased longitudinal tension along the margins of the folds allows the folds to remain lax, which also requires less subglottal air pressure to part the vocal folds and contribute to a reduction in the rate of vibration characteristic of fry.

Uncovering the complexity of the laryngeal vibratory patterns

Using ultra-high-speed motion photography, and the productions of only one speaker, Moore and von Leden (1958) reported that the vocal folds behave differently during modal/normal register and during fry (Blomgren *et al.* 1998; Timcke, von Leden, and Moore 1959). In modal register, the vocal folds open and approximate for a short period of time (Timcke, von Leden, and Moore 1958). In fry, the glottis opens rapidly, approximately twice per glottal cycle, and remains closed for a long duration "producing a syncopated rhythm" (Moore and von Leden 1958: 227; Timcke, von Leden, and Moore 1959). Based on the existence of two successive rapid glottal openings within a glottal cycle, which they called "pulses," Moore and von Leden thought it most appropriate to relabel fry phonation "dicrotic dysphonia" as they felt that this term best described its "basic laryngeal physiology" (Moore and von Leden 1958: 231). They estimated that the vocal folds remained abducted for about 27–30% of the glottal cycle, briefly closed, reopened for about 18–20% of the cycle, and then remained adducted for the remaining time of the cycle (i.e., about 50% of the duty cycle). They added that the glottal openings first traveled longitudinally along the edges of the vocal folds and then were followed by a brief closure. Additionally, they noted that the initial opening of the vocal folds could occur anteriorly or posteriorly depending on the speaker (i.e., vocal folds may begin by abducting posteriorly, progressing anteriorly for some speakers) and that in the second closure of the duty cycle, an "abrupt and almost simultaneous closure of the glottis" occurred throughout the length of the vocal folds (Moore and von Leden 1958: 229). They also indicated that when the double-opening pattern changed to a single pattern, the vocal quality changed to a "*tense*" but non-fry voice, suggesting that the double glottal opening was necessary to elicit a vocal quality of fry (Moore and von Leden 1958: 231). This raises the issue of the perceptual significance of the number of pulses, later addressed in this chapter. Timcke, von Leden, and Moore (1959) used an ultra-high-speed camera and magnified frame-by-frame selected pictures for measurements. They reported double-opening and closing laryngeal gestures. Additionally, they noted: (1) that the magnitude of the second opening/pulse was greater than the magnitude of the first one by an amount that varied from speaker to speaker and (2) that the second glottal approximation lasted longer than the first one in agreement with subsequent research (Whitehead, Metz, and Whitehead 1984). Hollien (1974) remarked that the closed phase durations were longest in fry and shortest in loft/falsetto register, while the opening phase durations were shortest in fry and longest in the loft/falsetto register.

The so-called "dicrotic" laryngeal pattern has been documented by other researchers using similar and different investigative techniques (Blomgren *et al.* 1998; Chen *et al.* 1996; Chen *et al.* 2002; Childers and Lee 1991; Hollien, Girard, and Coleman 1977; Proctor, Scherer, and Perrine 2024; Whitehead, Metz, and Whitehead 1984). However, using ultra-high-speed motion photography, Wendahl, Moore, and Hollien (1963), who did not provide any information on the number of speakers or any details about the elicitation task, observed that the prevalent vibratory pattern that generated a perception of fry was not the so-called "syncopated rhythm" (i.e., double-opening pattern) but a single pulse followed by a period of no excitation or damping interval occurring between successive excitations (Cavallo, Baken, and Shaiman 1984; Whitehead, Metz, and Whitehead 1984; Chen *et al.* 2002; Murry 1971; Proctor, Scherer, and Perrine 2024). Murry (1971) also reported that pulse-like waveforms were damped almost to zero in between successive glottal excitations when fry was produced. In agreement with Wendahl and colleagues,

Hollien and Wendahl (1968), who asked eight males to match the pitch of prerecorded steady-state samples within the fry register, found that among the eight rigorously selected samples of fry, left after rejecting a substantial number of other productions, only one sample displayed the double-pulse rhythm described earlier by Moore and von Leden (1958).

Is that "true" fry?

Hollien, Girard, and Coleman (1977) questioned the works of Moore and von Leden (1958) and Timcke, von Leden, and Moore (1959) arguing that the "syncopated rhythm" and multiple-opening laryngeal gestures could not be the product of "true" fry phonation, which Hollien and his colleagues restricted to low f_0 and perception of "clean popping" (Hollien, Girard, and Coleman 1977: 201). They added that during "true" fry, the visualization of the vibrations of the vocal folds is obscured by the ventricular folds "loading" onto the true vocal folds, making the scientific investigation of fry difficult if not problematic (Hollien, Girard, and Coleman 1977: 201). Hence, they searched for the speaker "who could expose his or her vocal folds while producing the precise and clean "popping" phonation associated with true pulse register phonation" (ibid.). Using high-speed photography, Hollien, Girard, and Coleman asked one speaker to produce fry at a model f_0 of 25 Hz and 50 Hz, with the f_0 produced being on average 29 Hz and 44 Hz. They only observed single short-lived open, opening and closing gestures and noticeable prolonged closed phases, which was also noted by many investigators before (Coleman 1963; Hollien *et al.* 1966; Hollien 1974; Moore and von Leden 1958; Timcke, von Leden, and Moore 1959; Wehndal, Moore, Hollien 1963). They reported that on average, the vocal folds were closed between 78% and 87% of the time depending on the f_0, and longer for the 44 Hz condition than for the 29 Hz condition. Hollien and his colleagues (1977) also argued that the reduced variability in fry as attested by a restricted range in f_0, smaller than that observed in the normal/modal and loft/falsetto registers, can only allow for single openings within a duty cycle. They further suggested that Moore and von Leden (1958) as well as Timcke, von Leden, and Moore (1959) analyzed a type of phonation that was not "true" fry but rather "mixed" phonation, a hybrid between modal and fry phonations. They speculated that patterns of double and triple pulses are obtained only when speakers use "mixed" phonation. Yet, Hollien and his colleagues admitted that both single and multiple pulses could occur (Hollien, Girard, and Coleman 1977).

Whitehead and his colleagues (1984) disputed the claim of Hollien and his colleagues that multiple pulses could not be the result of "true" fry. They asked a single female speaker, proficient at producing fry, to produce the schwa vowel in the fry register using a laryngeal mirror held in her oropharynx. Prior to analyzing data, they ensured first that all productions met the perceptual (i.e., "popping" or "pulsing quality") and the acoustic (f_0 within the range associated with pulse phonation) criteria for "true" fry as defined by Hollien, Girard, and Coleman (1977: 1296). Whitehead and his colleagues (1984) provided evidence that a single speaker could produce single-, double-, and triple-vibratory laryngeal patterns during "true" vocal fry. However, in their analyses, they noticed a change in f_0 for double- and triple-pulse samples, with double-pulse cycles having a higher f_0 (52 Hz) than triple-pulse cycles (25 Hz). Hence, they concluded that such results suggested "incomplete control of the laryngeal mechanism" during fry, which would concur with the "mixed" phonation argument advanced by Hollien and his colleagues (Hollien, Girard, and Coleman 1977).

Yet, other researchers documented multiple opening and closing phases in what would qualify as "true" fry emissions, attesting to the diversity and the complexity of fry laryngeal vibratory patterns (Blomgren *et al.* 1998; Chen *et al.* 1996; Chen *et al.* 2002; Childers and Lee 1991; Hollien and Wendahl 1968; Whitehead, Metz, and Whitehead 1984). Blomgren and his colleagues asked a large number of speakers (20) to produce vowels in the modal and fry registers (Blomgren *et al.* 1998). They observed that "triplet cycles" always co-occurred with "doublet cycles," but they reported that the "doublet cycles" did not necessarily occur with triplets. Given the diversity of the patterns, the descriptor "dicrotic," which designates the presence of two consecutive glottal openings within a glottal cycle, cannot serve to distinguish fry productions from other vocal emissions since it is only one out of many possible laryngeal patterns for fry. More recently, researchers, who have examined acoustic waveforms only, have documented a variety of acoustic patterns in fry productions (Abdelli-Beruh, Drugman, and Owl 2015; Proctor, Scherer, and Perrine 2024).

Perceptual salience of laryngeal gestures

Decay/damping time

One laryngeal feature that has consistently emerged in all studies, irrespective of the number of opening/closing gestures, is the decay time between glottal cycles (Blomgren *et al.* 1998; Coleman 1963; Hollien, Moore, Wendhal, and Michel 1966; Moore and von Leden 1958; Wehndal, Moore, and Hollien 1963). Moore and von Leden (1958) indicated that the necessary requirements for a fry percept were low f_0 and the presence of a complete decay of a glottal wave before the initiation of the following one. Indeed, based on the perceptual evaluation of fry tokens by highly trained listeners, Wehndal, Moore, and Hollien (1963) claimed that the "primary criterion that must be met for the perception of a voice signal as vocal fry is that there be nearly complete damping of the vocal tract between successive excitations" (1963: 254). When the damping or decay time interval between glottal excitations was reduced, the perception of fry was replaced by the perception of modal/normal phonation. This is in keeping with Coleman's findings (1963) who reported that the experimental manipulation of the time between the maximum energy of a glottal wave and the energy left at the time of a new glottal wave (i.e., decay time) determines the perception of glottal fry. Coleman asked naïve listeners using an ABX design to match the quality of naturally spoken tokens and analog tokens. The decay time between the glottal excitations of the analog tokens was manipulated. He found that when the energy dropped by about "42 to 44 dB of its maximum amplitude for a single pulse, vocal fry was always perceived" and when the decay between pulses dropped to about 30 dB, modal-like phonation was perceived (1963: 262).

Hollien, Moore, Wendhal, and Michel (1966) concurred that in addition to low f_0 "it appears necessary, also, for the damped wave to decay nearly to zero before the next glottal episode excites the supra glottal resonators" to elicit the perception of fry (1966: 247). However, Proctor, Scherer, and Perrine (2024) reported recently that a drop in f_0 by an octave could result in a perception of fry even in the absence of complete damping.

Number of pulses

As for the debate regarding the number of pulses (i.e., single, double, or triple), it might not be so relevant after all as research shows that the number of pulses is perceptually irrelevant. Hollien, Moore, Wendhal, and Michel (1966) reported that fry could be perceived regardless of the number of pulses (1963: 252). Hollien and Wendahl (1968) reported that listeners did not perceive differences among the selected samples of vocal fry with a double and with a single pulse, leading them to conclude that the number of pulses "is immaterial to the perception of phonation as vocal fry" (1968: 509). This is in agreement with follow-up studies, which showed that in a force-choice task, listeners detected fry in all fry samples regardless of the number of pulses within a duty cycle (i.e., both double- and triple-opening) (Blomgren *et al.* 1998).

In spite of this perceptual irrelevance, tentative explanations for the multi-pulse laryngeal patterns were nevertheless proposed. Blomgren *et al.* (1998) surmised that the thickening of the vocal folds, an increased compliance of the vocal ligament, a smaller vibrating portion of the vocal folds paired with a decrease in longitudinal tension and in medial compression would account for the reduction in f_0, and promote vibrations secondary to the primary ones. Yet, Whitehead, Metz, and Whitehead (1984) concluded that "we are unable to delineate the exact physiological underpinnings of the observed multi-phasic pulse register vibratory cycles" (1984: 1296).

Nevertheless, Hollien (1974) and his colleagues hypothesized that the complex patterns of the vocal folds during fry could be accounted for by the descent of ventricular folds onto the vocal folds, which would contribute to damping the glottal wave and lowering f_0 to extreme values, obliterating gender differences in f_0 (Allen and Hollien 1973). Well-documented gender differences in acoustic measures in modal/normal and loft/falsetto have been ascribed to gender-related anatomical differences in the larynx: Laryngeal structures are generally larger and heavier in adult males than in adult females. Among other differences, men have heavier, thicker, and longer vocal folds than women.

Gender differences in objective measures of fry

Yet, in spite of marked anatomical gender difference in laryngeal structures, the most robust results that emerge from the literature remain the fact that the striking gender difference in f_0 evidenced in normal/modal register is not evidenced in fry (Blomgren *et al.* 1998; Chen *et al.* 2002; Hollien and Wendahl 1968; McGlone 1967; McGlone and Shipp 1971; Murry 1971). This implies that women drop their f_0 by a greater amount than men to descend into the fry register. Indeed, Blomgren and his colleagues noted that when producing fry, men lower their modal f_0 by about an octave while women lower their modal f_0 by two octaves (Blomgren *et al.* 1998; Chen *et al.* 1996; Hollien and Wendahl 1968; McGlone and Shipp 1971). Women usually have a broader f_0 range than men in modal register, and yet this difference also disappears in fry (McGlone 1967; Hollien and Wendahl 1968).

Blomgren and his colleagues (1998) noted that all speakers produced one, two, or three peaks per glottal cycle, but they observed some gender differences in other acoustic measures: They reported that jitter was significantly higher in the fry productions of men than that of women and that men had a higher percentage (83%) of complex-repetitive patterns (i.e., more than one peak per cycle) than women (38%), a finding confirmed by Chen and his colleagues (2002). They contended that such gender differences in laryngeal behaviors in fry suggest gender differences in tissue compliance and in the amount of vocal-fold contact. Blomgren and his colleagues (1998) surmised that the lack of gender difference in other reported objective measures of fry may not be so relevant when considering vocal fry. They suggested that the lack of gender differences in f_0 might be due to the fact that speakers do not use fry habitually and, therefore, "gender differences may be obscured" (ibid.: 2654).

Alternatively, Allen and Hollien (1973) ascribed this lack of gender differences in fry to the fact that vocal-fold vibrations only occur over a small portion of the vocal folds for both males and females, adding that the coupling of the true and ventricular folds may explain not only "similarity in f_0 between males and females" but also the damping (1973: 2656).

Ventricular folds: anatomy and their disputed contribution to phonation

Anatomy of ventricular folds

Ventricular folds are also known as vestibular, false vocal folds or plicae ventricularis. Jackson and Jackson (1935) reported that the father of laryngology, Sir Morell McKenzie, favored the term "*ventricular bands*" as he remarked that "the retention of the term false vocal cords perpetuates the memory of a physiological error" (Jackson and Jackson 1935: 157). But his remark fell on deaf ears, and it is common to find this inadequate descriptor in scholarly resources.

The ventricular folds rest above the true vocal folds and make the roof of the laryngeal ventricle, while the true vocal folds constitute the floor of the laryngeal ventricle (Kutta, Steven, Kohla, Tillmann, and Paulsen 2002). The two sets of paired folds are separated by a space, the laryngeal ventricle, larger in males than in females (Agarwal, Scherer, and Hollien 2003). The vestibular folds are composed of muscular fibers, epithelium and lamina propria (adipose, connective, glandular), which enfolds glands with ducts leading to the laryngeal ventricle. These glands aid in lubricating vocal folds and in humidifying the breath by discharging mucus directly on the vocal folds. Reduction or loss of mucus secretion could increase susceptivity to infection, reduce the extent of the glottal closure, increase aperiodicity, and thus adversely affect vocal quality (Kutta *et al.* 2002). Fibers of the ventricular muscle originate from the anterior-lateral border of the arytenoid cartilages and insert in the ventricular folds (Motta 1953). The ventricular muscle splits into two tracts: One inserts in the medial layers of the ventricular folds and another into its lateral margins. Most of the fibers of the ventricularis muscle mass course anteriorly and laterally to the tissues that comprise the bulging portion of the folds (Moon and Alipour

2013). The muscle is composed of three sets of muscular systems: a postero-lateral muscular (PLM) tract, which courses sagitally, allows the folds to approximate; at half or third of the anterior portion of the ventricular folds, the PLM splits into antero-lateral muscular (ALM) fibers, which insert in the thyroid cartilage, and an antero-medial muscular (AMM) tract, whose fibers disperse medially and dorsally in the laryngeal ventricle (Reidenbach 1998). The stimulation of the AMM fibers presumably allows their approximation and descent. Kotby *et al.* (1991) reported that the bulk of the ventricularis muscle is concentrated in the area, where the ventricular folds are widest (i.e., anterior portion of the posterior section) and vertically where the folds are also thickest (Kotby *et al.* 1991; Moore and Alipour 2013). Because the ventricular folds have less muscle mass than the true vocal folds, they are more flaccid and thus are considered inefficient vibrators (Haji *et al.* 1992).

Contribution of ventricular folds to normal/modal phonation

There has been a debate as to whether the ventricular folds have been actively participating in phonation given their muscular configuration, its anatomical vicinity and connections with laryngeal structures. García surmised that "the superior ligaments" (i.e., ventricular vocal folds) were inconsequential to phonation, while von Luschka, a German anatomist, hypothesized that the ventricular folds contribute to phonation by "isolating" the true vocal folds and "facilitating their vibration" (García 1856, also see 1947; von Luschka 1871). But in general, it was assumed that ventricular folds do not participate in the production of speech sounds, unless the vocal folds are unable to function optimally if at all (Jackson and Jackson 1935; Voelker 1935). While the involvement of ventricular folds to phonation still remains unclear, there are clinical evidence that demonstrate that speakers can exploit the adductor muscles of the vestibular folds and develop a compensatory voice when the vocal folds cannot be set into vibration for various reasons such as "underactivity of the true vocal folds," "overactivity of the ventricular bands," or "absence of true vocal folds" (Jackson and Jackson 1935: 157–158). More recent findings have shown that ventricular folds are recruited during the production of glottal stops, pressed phonation, and fry (Allen and Hollien 1973; Jackson and Jackson 1935; van den Berg 1955; von Doersten *et al.* 1992; Pinho *et al.* 1999; Stager *et al.* 2000; Reidenbach 1998; Voelker 1935).

Additional findings show that incidental changes in the local physiological and aerodynamic environment of the ventricular folds and space affect the vibrations of both sets of folds. To prevent the involvement of the ventricular folds during phonation, the ventricular space is maintained wide by means of laryngeal and/or ventricular muscles. When the ventricular space is narrowed, because the muscles fail to keep the ventricular folds apart, the ventricular folds are too large, or because medial compression of the ventricular folds is applied, the resistance to glottal airflow rises, changing the aerodynamic and acoustic features of sounds (Moon and Alipour 2013). A smaller laryngeal space is generally accompanied by an increased glottal resistance, which is conducive to setting the vestibular folds into vibrations, thereby changing the dynamics of true vocal folds (ibid.). Indeed, there may be an optimal gap between the vestibular vocal folds that is ideal for true vocal-fold vibrations (Agarwal, Scherer, and De Witt 2004). Interestingly, van den Berg (1955) reported that the size of the laryngeal ventricle varies with the production of different vowels.

Participation of ventricular folds in fry

In the early part of the 20th century, hoarseness and vocal fry were conceived as the descriptors of the same type of laryngeal emissions. Voelker as well as Jackson and Jackson suggested that ventricular bands were implicated in the production of fry/hoarseness/harshness, also labelled "dysphonia plicae ventricularis" (Voelker 1935; Jackson and Jackson 1935). Voelker stated that "the ventricular bands supply the vibrations rather than the true vocal lips" (1935: 471). The problem is that he provided no substantive evidence to support such claim and no indication as to how he reached this conclusion. Jackson and Jackson (1935) claimed that hoarse phonation was due partly or exclusively to the vibrations of the "ventricular bands" when laryngeal pathologies (removal, paralysis, congenital, or underactivity) affected the true folds or when the ventricular bands were overactive. He contended that such vocal emissions perceived as hoarse could be of the "vicarious" (e.g., due to a pathology) or "usurpative" (e.g., due to "aggressively overactive ventricular bands") types (Jackson and Jackson 1935: 167). But, as Moore and von Leden (1958) aptly remarked, it is not clear from their descriptions of the voice qualities that Jackson and Jackson dealt, strictly speaking, with fry productions: "Their description of the accompanying voice was too general to permit definite identification of the phenomenon under discussion" (1958: 226). On the other hand, Moore and von Leden wrote that Voelker's account of cases of "dysphonia ventricularis" was beyond reasonable doubt cases of glottal fry when he described the "unpleasant vocal quality … characterized by rattling, rumbling, cracking or ticker-like substitution for phonation" (Voelker 1935: 471). But Moore and von Leden (1958) affirmed that the ventricular folds "do not perform any vibratory functions when this voice quality is produced" (1958: 235).

In contrast, and without evidence to substantiate their claim, Hollien, Moore, Wendahl, and Michel (1966) hypothesized that the ventricular folds could adduct during fry and that their inferior portion could "load" onto the true vocal folds and drive f_0 below that of the modal range (Hollien, Girard, and Coleman 1977: 201; Hollien *et al.* 1966). According to Hollien and Wendahl (1968) and Coleman (1963), the descent of the ventricular folds and the "loading" would allow for the damping of the vocal-fold vibrations and the complete decay of the glottal waveform so required for the perception of fry. They further postulated that when the superior surface of the true vocal folds would come in contact with the inferior part of ventricular folds, the latter would be entrained to "vibrate in synchrony with the true folds" (Hollien 1974: 247). If such a hypothesis were true, Hollien suggested that the contributions of each component of the Myoelastic Aerodynamic Theory of Phonation would shift depending on the register. The Myoelastic Aerodynamic Theory of Phonation, first formulated by Johannes Mueller in 1848, and re-tested experimentally a hundred years later by Janwillem van den Berg using current technology, postulates that the frequency of vocal-fold vibrations is determined by a complex interaction of aerodynamic, muscular, and elastic factors (Mueller 1848; van den Berg 1958). This theory repudiates the Neuro Chronaxic Theory of Phonation advocated by Raoul Husson, who posited that the frequency of glottal oscillations was triggered by the stimulation of the vocalis muscle by the recurrent nerve (Husson 1957). Hollien surmised that in the framework of the Myoelastic Aerodynamic Theory of Phonation, the myo- (i.e., muscle) and elastic (i.e., compliant) components would contribute more to the production of fry than the aerodynamic (i.e., movement of gaseous fluids) ones, all the components would contribute equally in the modal register, and the aerodynamic factors would weigh more in the falsetto/loft register (Hollien 1974). Alternatively, Hollien *et al.* (1966) proposed that the position of the ventricular folds during fry might be "the incidental result of basic laryngeal adjustments" if it does not "serve to produce a damping of vocal fold movement" (1966: 247).

A few years later though, Hollien and colleagues provided evidence of the involvement of the ventricular folds with fry using images from conventional and stroboscopic laminagraphy. Allen and Hollien (1973) tested seven male and six female speakers. They noted that the inferior layer of the ventricular folds "impinged" on the superior layers of the true vocal folds during vocal fry phonation consequently reducing the ventricular space (1973: 249). This led Allen and Hollien to suggest that this may account for the complex laryngeal vibratory pattern so characteristics of vocal fry (ibid). Interestingly, Moon and Alipour (2013) have indicated that changes in the size of the ventricular space have cascading effects: They result in changes in the ventricular pressure and in the aerodynamic environment, which in turn affect the dynamics of the ventricular and vocal folds during phonation. Allen and Hollien (1973) ventured to suggest that while the ventricular folds are not contractile (a "non-muscular mass"), they could participate in the vertical traveling motion of the true vocal folds when they descend onto the true vocal folds and add their mass to that of the true vocal folds (1973: 249). This would plausibly account for the prolonged closed time and the damping interval between glottal cycles characteristic of fry laryngeal patterns. But the notion that the ventricular folds are non-muscular and passive has been challenged by anatomical and physiological evidence as well as by clinical data that show speech relying on the ventricular folds vicariously: "The percentage of muscle tissue in the VF [ventricular folds] cross-sections defines the active tension contribution of the VF to its stiffness and maneuverability" (Moon and Alipour 2013: 562; Reidenbach 1998).

Aerodynamic measures

Aside from the acoustic measures such as f_0, jitter, shimmer, and S/N ratio, researchers have relied on other measures such as aerodynamic measures to differentiate vocal emissions as they are central to the prevalent Myoelastic Aerodynamic Theory of Phonation. Such measures allow the objective assessment of physiological systems participating in phonation. Other measures such as electroglottographic (EGG) measures of fry phonation have been examined, but they are not discussed here because of space limitation.

Subglottal air pressure

Aerodynamic measures are highly sensitive to measurement techniques, and thus findings can be contradictory (Isshiki and von Leden 1964; Kunze 1964; Ladefoged 1962; van den Berg 1956). Van den Berg predicted that subglottal air pressure should be higher in fry than in any other registers because of the combination of short-opening laryngeal gestures and long periods of closure, estimated to be about 50% of the duty cycle (Timcke, von Leden, and Moore 1959; Moore and von Leden 1958; van den Berg 1956). In contrast, Hollien and his colleagues predicted that low f_0 vibrations typical of fry could only be sustained by low subglottal air pressure (Hollien and Michel 1968: 247).

Murry and Brown (1971) asked speakers to produce sentences with fry where it could be expected to be produced in the sentences, i.e., at pauses and at the end of declarative sentences. They also asked speakers to utter "Joe ate the soup" entirely in fry. In keeping with Hollien and Michel's prediction (1968), they reported that subglottal air pressure decreased as f_0 dropped from modal to fry registers. It is worth noting that the two male speakers were in a supine position while the subglottal pressure was measured via a hypodermic needle inserted between the first and second tracheal rings. Blomgren and his colleagues (1998) have argued that such an unusual position might have affected the position of the laryngeal structures, the size of the laryngeal ventricle, consequently affecting aerodynamic factors, and the putative involvement of the ventricular folds. Murry and Brown (1971) emphasized that because the findings were only based on two speakers, such results should be considered tentative. Yet, Blomgren and his colleagues, who used a different elicitation task and speakers in erect positions, reported similar results. Among other measures, they calculated glottal resistance (transglottal pressure difference/glottal airflow as per Smitheran and Hixon 1981: 1). They concurrently reported an increase in "glottal resistance" values in vocal fry compared to modal register for both male and female speakers, which they ascribed to a decrease in both subglottal pressure and flow rate, with flow rate decreasing to a greater extent than subglottal pressure (Blomgren *et al.* 1998: 2655). Blomgren and colleagues noted that the decreased longitudinal tension along the edges of the vocal folds allows them to remain flaccid, in which case less air pressure is required to set them into vibration. They suggested that a decreased subglottal pressure indirectly supports the notion that reduced vocal-fold tension may be required to produce fry.

In favor of van den Berg's prediction, Murry (1971), who asked speakers to produce the vowels /a/ and /i/ at different frequencies, reported higher subglottal air pressure in fry than in modal registers. However, Murry suggested that this result might be due to the fact that speakers were phonating in a register they were not accustomed to.

Finally, some researchers reported data that neither supported van den Berg's nor Hollien's hypotheses. McGlone (1967) found subglottal air pressure to be statistically the same in fry and in modal phonation when speakers were asked to match an imposed frequency and sustain it for some time. This is in agreement with Shipp and McGlone (1971), who concluded that subglottal pressure measures, therefore, cannot serve to differentiate between modal and fry registers.

Blomgren *et al.* (1998) proposed an alternative hypothesis: They suggested that subglottal pressure may not be a passive consequence of fry but rather an active aerodynamic maneuver that consists in reducing subglottal pressure considerably to lower f_0 below 60 Hz. This notion of passive versus active strategies had been raised by Murry and Brown (1971) to differentiate between two types of fry: Passive fry is presumably episodic and occurs naturally in the terminal breath group or at the end of sentences, while active fry is prolonged and deliberately adopted by speakers. Interestingly, when contrasting active/voluntary fry with passive/involuntary fry, Murry and Brown found greater air pressure for voluntary fry than for passive fry phonations (Murry and Brown 1971). Such finding suggests that greater respiratory effort may be involved for voluntary than for passive fry. But Shipp and McGlone (1970) tasked one speaker with sustaining fry throughout sentences (instances of active fry) and with producing sentences in which fry occurred at a pause or the end of a sentence (instances of passive fry). The results indicated that the subglottal air pressures for the terminal breath groups were of similar magnitude in the two speaking conditions.

Airflow rate

Van den Berg (1956) predicted lower airflow rates in fry register than in other registers due to the prolonged closed phases and short opening times. Indeed, longer glottal closures allow more subglottal air pressure to accrue below the approximated vocal folds, and airflow is thus expected to be greater than during the shorter period of adduction. However, airflow depends on the size of the glottal opening as well: In fry, the glottal opening is smaller than in modal register, allowing little air to escape. McGlone (1967) recorded the vocal emissions and the associated airflow rates (i.e., volume of air expelled/duration of the fry emission) of five male and five female speakers wearing a mask over their mouth and nose and phonating at three different "pitch" levels. The speakers were asked to sustain vocal fry for at least five seconds, but they were not able to produce fry at distinct pitch levels. Whether these elicitations compared to naturally produced fry was never raised, but the data showed that airflow rate was smaller in fry than in low-frequency normal/modal and falsetto registers. This is in keeping with predictions by Hollien and his colleagues and also in keeping with other published data (Blomgren *et al.* 1998;

Hollien and Michel 1968; Murry 1971; Murry and Brown 1971; McGlone and Shipp 1971).

Conclusion

This review of selected research articles shows that researchers in the field of SLP focused primarily on operationally defining fry in terms of multiple objective parameters in order to differentiate it from other types of vocal emissions, historically considered to be signs of pathology on the basis of subjective/perceptual evaluations. Speech language pathologists, whose essential duties are to diagnose and treat voice-related disorders among other disorders unrelated to voice, have traditionally associated fry with vocal pathology. But as early as the 1960s, Hollien and colleagues began to advocate that sporadic fry is a habitual non-pathological vocal quality, concurrently holding the view that its prolonged use could compromise the integrity of the laryngeal system. Aided by technological advances, researchers have endeavored to uncover the complexity of its unique physiological, acoustic, and aerodynamic characteristics. In the process, fry ascended to the status of normal phonation and to that of register, while concurrently retaining some association with pathology. The search for the distinguishing features of fry has provided ample support in favor of the view that fry is a product of normal laryngeal functioning, which occurs naturally in the terminal breath group or at the end of sentences. This is in agreement with psycholinguists, sociolinguists, and researchers in other fields, for whom such distinctive a vocal emission is utilized by speakers with normal laryngeal function as a marker of paralinguistic, linguistic, and/or sociolinguistic information depending on the language and the dialect.

As transpired from this review, researchers in SLP have not been concerned with the "utility" of fry as have researchers in other disciplines (e.g., psycholinguists). Their scientific investigations have brought to light the complexity of its laryngeal, aerodynamic, and acoustic patterns, some of which are more robust than others to methodological variations: single- and multiple-pulsed vibratory patterns; long decay time between consecutive pulses;

low f_0; restricted f_0 range; perturbation measures higher in fry than in other registers; striking absence of gender differences in some acoustic measures differentiating vocal emissions of male and female speakers in modal/normal and loft/falsetto registers; low airflow rate; variable subglottal air pressures; the controversial contribution of the ventricular folds to the production of fry.

The possibility of different types of fry productions—"true" fry versus "mixed phonation," "active" versus "passive"—emerged from the literature. In the end, several questions arise: Are "fry," "creak," "pulse," "strohbass," and "true fry" described in this review and in the field of SLP the same as or different than what researchers in other fields have called "creaky voice," "laryngealization," "glottalization," "crackling," or "croak"? If different, what are the objective measures that most reliably differentiate them? And, to what extent are these distinctions perceptually meaningful?

References

Abdelli-Beruh, N. B., Drugman, T., and Owl, R. R. (2016). "Occurrence frequencies of acoustic patterns of vocal fry in American English speakers." *J. Voice*, **30**(6), 759-e11.

Abdelli-Beruh, N. B., Wolk, L., and Slavin, D. (2014). "Prevalence of vocal fry in young adult male American English speakers." *J. Voice*, **28**(2), 185–190.

Agarwal, M., Scherer, R. C., and de Witt, K. J. (2004). "The effects of the false vocal folds on translaryngeal airflow resistance." In *Proc. Interna. Conf. Voice Physio. and Biomech.* (pp. 18–20).

Agarwal, M., Scherer, R. C., and Hollien, H. (2003). "The false vocal folds: Shape and size in frontal view during phonation based on laminagraphic tracings." *J. Voice*, **17**(2): 97–113.

Allen, E. L. and Hollien, H. (1973). "A laminagraphic study of pulse (vocal fry) register phonation." *Folia Phoniatr.*, **25**, 241–250.

Anderson, R. C., Klofstad, C. A., Mayew, W. J., and Venkatachalam, M. (2014). "Vocal fry may undermine the success of young women in the labor market." *PloS one*, **9**(5), e97506. Retrieved from https://journals.plos.org/plosone/article?id=10.1371/journal.pone.0097506.

Arregui, M. E., Robert Jr, J., Katkhouda, N., McKernan, J. B., and Reich, H. (Eds.). (2012). *Principles of Laparoscopic Surgery: Basic and Advanced Techniques*. Springer Science and Business Media.

Blomgren, M., Chen, Y., Ng, M. L., and Gilbert, H. R. (1998). "Acoustic, aerodynamic, physiologic, and perceptual properties of modal and vocal fry registers." *J. Acoust. Soc. Am.*, **103**(5), 2649–2658.

Castellengo, M. (2005). "Manuel Garcia Jr: A clear-sighted observer of human voice production." *Logop. Phoniatr. Vocology*, **30**(3–4), 163–170.

Cavallo, S. A., Baken, R. J., and Shaiman, S. (1984). "Frequency perturbation characteristics of pulse register phonation." *J. Commun. Disord.*, **17**(4), 231–243.

Chao, M. and Bursten, J. R. (2021). "Girl talk: Understanding negative reactions to female vocal fry." *Hypatia*, **36**(1), 42–59.

Chen, Y., Ng, M., Jeng, J., and Gilbert, H. (1996). "Aerodynamic, electroglottographic, and acoustic characteristics of modal and vocal fry registers." *Asha*, **38**, 62.

Chen, Y., Robb, M. P., and Gilbert, H. R. (2002). "Electroglottographic evaluation of gender and vowel effects during modal and vocal fry phonation." *JSLHR*, **45**(5), 821–829.

Childers, D. G. and Lee, C. K. (1991). "Vocal quality factors: Analysis, synthesis, and perception." *J. Acoust. Soc. Am.*, **90**(5), 2394–2410.

Cielo, C. A., Elias, V. S., Brum, D. M., and Ferreira, F. V. (2011). "Thyroarytenoid muscle and vocal fry: a literature review." *Revista da Sociedade Brasileira de Fonoaudiologia*, **16**, 362–369.

Coleman, R. F. (1963). "Decay characteristics of vocal fry." *Folia Phoniatr. Logop.*, **15**(4), 256–263.

Crumpton, C. E. (1917). "Speech betterment in Alabama." *Engl. Lang.*, **6**(2), 96–102.

Damsté, P. H., Hollien, H., Moore, P., Murry, T., Koike, Y., Perkins, W. H., ... and Dubois, C. (1968). "Contents, Vol. 20, 1968." *Folia Phoniatr. Logop.*, **20**(6), I–IV.

Duprez, G. (1845). *L'Art du Chant*. Au Ménéstrel, Paris.

Engel, R. M. (2003). "Philipp Bozzini: The father of endoscopy." *J. Endourology*, **17**(10), 859–862.

Ferranti, T. L. (2004). "A historical approach to training the vocal registers: Can ancient practice foster contemporary results?" Louisiana State University and Agricultural and Mechanical College.

Fraser, W. I. and Campbell, B. M. (1978). "A study of six cases of de Lange Amsterdam dwarf syndrome, with special attention to voice, speech and language characteristics." *Dev. Med. Child Neurol.*, **20**(2), 189–198.

García, M. (1847). *Mémoire sur la Voix Humaine présenté à l'Académie des Sciences; réimpression augmentée de quelques observations sur les sons simultanés et suivie du rapport de la commission de l'Académie des Sciences du 12 avril 1841; imprimerie Duverger, Mai 1847, Paris.*

García, M. (1856). "IV. Observations on the Human Voice." *Proc. R. Soc. Lond.*, (7), 399–410.

García, M. (1857). *New Treatise on the Art of Singing–A Compendious Method of Instruction. Part I.* Boston, Oliver Ditson Company.

Gordon, M. and Ladefoged, P. (2001). "Phonation types: A cross-linguistic overview." *J. Phon.*, **29**(4), 383–406.

Haji, T., Mori, K., Omori, K., and Isshiki, N. (1992). "Mechanical properties of the vocal fold." *Acta Otolaryngol.*, **112**(3), 559–565.

Hedelin, P. and Huber, D. (1990). "Pitch period determination of aperiodic speech signals." In *ICASSP* (pp. 361–364). IEEE.

Henton, C. and Bladon, A. (1988). Creak as a sociophonetic marker. In L. M. Hyman, V. Fromkin and C. N. Li (eds.), *Language, Speech and Mind* (pp. 3–29). London: Routledge.

Henton, C. G. (1989). "Sociophonetic aspects of creaky voice." *J. Acoust. Soc. Am.*, **86**(S1), S26–S26.

Hollien, H. (1960). "Vocal pitch variation related to changes in vocal fold length." *J. Speech Hear. Res.*, **3**(2), 150–156.

Hollien, H. (1962). "The relationship of vocal fold length to vocal pitch for female subjects." In *Proc. XII Inter. Congress Logoped, Phoniatr.*, Scoula Tipografica dell'Ente Nationale Sordomuti, Padua (pp. 38–43).

Hollien, H. (1974). "On vocal registers." *J. Phon.*, **2**(2), 125–143.

Hollien, H. and Colton, R. H. (1969). "Four laminagraphic studies of vocal fold thickness." *Folia Phoniatr. Logop.*, **21**(3), 179–198.

Hollien, H. and Moore, G. P. (1960). "Measurements of the vocal folds during changes in pitch." *J. Speech Hear. Res.*, **3**(2), 157–165.

Hollien, H. and Wendahl, R. W. (1968). "Perceptual study of vocal fry." *J. Acoust. Soc. Am.*, **43**(3), 506–509.

Hollien, H., and Michel, J. F. (1968). "Vocal fry as a phonational register." *J. Speech Hear. Res.*, **11**(3), 600–604.

Hollien, H., Brown Jr, W. S., and Hollien, K. (1971). "Vocal fold length associated with modal, falsetto and varying intensity phonations." *Folia Phoniatr. Logop.*, **23**(1), 66–78.

Hollien, H., Damsté, H., and Murry, T. (1969). "Vocal fold length during vocal fry phonation." *Folia Phoniatr. Logop.*, **21**(4), 257–265.

Hollien, H., Girard, G. T., and Coleman, R. F. (1977). "Vocal fold vibratory patterns of pulse register phonation," *Folia Phoniatr.*, **29**, 200–205.

Hollien, H., Moore, P., Wendahl, R. W., and Michel, J. F. (1966). "On the nature of vocal fry." *J. Speech Hear. Res.*, **9**(2), 245–247.

Husson, R. (1957). "Physiologie de La Phonation Exposé de Travaux récents." *Pract. Oto Rhino Laryng.*, **50**(2), 123–139.

Isshiki, N. and von Leden, H. (1964). "Hoarseness: Aerodynamic studies." *Arch. Otolaryngol.*, **80**(2), 206–213.

Jackson, C. and Jackson, C. L. (1935). "Dysphonia plicae ventricularis: Phonation with the ventricular bands." *Arch. Otolaryngol.*, **21**(2), 157–167.

Jahn, A. and Blitzer, A. (1996). "A short history of laryngoscopy." *Logop. Phoniatr. Vocology*, **21**(3–4), 181–185.

Jelle, J. (2017). "A study in Scarlett: Creaky voice and romantic intention in Spike Jonze's Her." *Leviathan: Interdiscip. Eng.*, (1), 35–44.

Kotby, M. N., Kirchner, J. A., Kahane, J. C., Basiouny, S. E., and El-Samaa, M. (1991). "Histo-anatomical structure of the human laryngeal ventricle." *Acta Oto-Laryngologica*, **111**(2), 396–402.

Kunze, L. H. (1964). "Evaluation of methods of estimating sub-glottal air pressure." *J. Speech Hear. Res.*, **7**(2), 151–164.

Kutta, H., Steven, P., Kohla, G., Tillmann, B., and Paulsen, F. (2002). "The human false vocal folds–an analysis of antimicrobial defense mechanisms." *Anat. Embryol.*, **205**(4), 315–323.

Ladefoged, P. (1962). "Subglottal activity during speech." In *Proc. ICPhS* (Vol. 73, p. 91). Mouton The Hague.

Lapeña, J. F. (2013). "Mirrors and reflections: The evolution of indirect laryngoscopy." *Ann. Saudi Med.*, **33**(2), 177–181.

Lieberman, P. (1963). "Some effects of semantic and grammatical context on the production and perception of speech." *Lang. Speech*, **6**(3), 172–187.

Ligon, C., Rountrey, C., Rank, N. V., Hull, M., and Khidr, A. (2019). "Perceived desirability of vocal fry among female speech communication disorders graduate students." *J. Voice*, **33**(5), 805-e21.

Liu, X. and Xu, Y. (2011). "What makes a female voice attractive?" In *ICPhS* (pp. 1274–1277).

McDonald, D. J. (1916). "Speech improvements." *National Education Association Addresses*, **54**, 862–864.

McGlone, R. (1967). "Air flow during vocal fry phonation." *J. Speech Hear. Res.*, **10**, 299–304.

McGlone, R. E. and Brown Jr, W. S. (1968). "Identification of 'break' between vocal registers." *J. Acoust. Soc. Am.*, **44**(1), 397–397.

McGlone, R. E. and Shipp, T. (1971). "Some physiologic correlates of vocal-fry phonation." *J. Speech Hear. Res.*, **14**(4), 769–775.

Michel, J. F. (1968). "Fundamental frequency investigation of vocal fry and harshness." *J. Speech Hear. Res.*, **11**(3), 590–594.

Michel, J. F. and Hollien, H. (1968). "Perceptual differentiation of vocal fry and harshness." *J. Speech Hear. Res.*, **11**(2), 439–443.

Moon, J. and Alipour, F. (2013). "Muscular anatomy of the human ventricular folds." *Ann. Otol. Rhinol. Laryngol.*, **122**(9), 561–567.

Moore, G. P., White, F. D., and von Leden, H. (1962). "Ultra high-speed photography in laryngeal physiology." *J. Speech Hear. Dis.*, **27**(2), 165–171.

Moore, P. and Thompson, C. L. (1965). "Comments on physiology of hoarseness." *Arch. Otolaryngol.*, **81**(1), 97–102.

Moore, P. and von Leden, H. (1958). "Dynamic variations of the vibratory pattern in the normal larynx." *Folia Phoniatr. Logop.*, **10**(4), 205–238.

Morgenstern, L. (2005). "The 200th anniversary of the first endoscope: Philipp Bozzini (1773–1809)." *Surg. Innov.*, **12**(2), 105.

Mörner, M., Fransson, F., and Fant, G. (1963). "Voice register terminology and standard pitch." *STL-QPSR*, **4**(4), 17–23.

Moser, H. M. (1942). "Symposium on unique cases of speech disorders: Presentation of a case." *J. Speech Dis.*, **7**(2), 173–174.

Motta, G. (1953). "The musculature of ventricular fold." *Otorinolaringol. Ital.*, **21**(3), 215–229.

Mueller, J. (1848). The physiology of the senses, voice, and muscular motion with the mental faculties. W. Baly, translator, Taylor, Walton and Maberly, London (1848).

Murry, T. (1971). "Subglottal pressure and airflow measures during vocal fry phonation." *J. Speech Lang. Hear. Res.*, **14**(3), 544–551.

Murry, T. and Brown, W. S. (1971). "Subglottal air pressure during two types of vocal activity: Vocal fry and modal phonation." *Folia Phoniatr.*, **23**, 440–449.

Pierre, J. S. and Pierre, C. S. (2018). "Governing the voice: A critical history of speech-language pathology." *Foucault Studies*, 151–184.

Pieters, B. M., Eindhoven, G. B., Acott, C., and van Zundert, A. A. J. (2015). "Pioneers of laryngoscopy: Indirect, direct and video laryngoscopy." *Anaesth. Intensive Care*, **43**(1_suppl), 4–11.

Pinho, S. M., Pontes, P. A., Gadelha, M. E. C., and Biasi, N. (1999). "Vestibular vocal fold behavior during phonation in unilateral vocal fold paralysis." *J. Voice*, **13**(1), 36–42.

Proctor, K., Scherer, R. C., and Perrine, B. L. (2024). "Vocal fry patterns while reading." *J. Voice*, **38**(4): 889–902.

Ramai, D., Zakhia, K., Etienne, D., and Reddy, M. (2018). "Philipp Bozzini (1773–1809): The earliest description of endoscopy." *J. Med. Biogr.*, **26**(2), 137–141.

Ray, A. G. (2005). *The Lyceum and Public Culture in the Nineteenth-Century United States: Rhetoric and Public Affairs Series*. Michigan State University Press.

Redi, L. and Shattuck-Hufnagel, S. (2001). "Variation in the realization of glottalization in normal speakers." *J. Phon.*, **29**(4), 407–429.

Reidenbach, M. M. (1998). "The muscular tissue of the vestibular folds of the larynx." *Eur. Arch. Oto-Rhino-L.*, **255**(7), 365–367.

Robbins, K. K. (1918). "The work of the American speech committee of the Chicago Woman's Club, and notes upon its school survey." *The English Journal*, **7.3**, 163–176.

Roubeau, B., Henrich, N., and Castellengo, M. (2009). "Laryngeal vibratory mechanisms: The notion of vocal register revisited." *J. Voice*, **23**(4), 425–438.

Shaw, F. and Crocker, V. (2015). "Creaky voice as a stylistic feature of young American female speech: An intraspeaker variation study of Scarlett Johansson." *Lifespans and Styles*, **1**, 21–27.

Sherman, D. and Linke, E. (1952). "The influence of certain vowel types on the degree of harsh voice quality." *J. Speech Hear. Dis.*, **17**(4), 401–408.

Shipp, T. and McGlone, R. E. (1970). "Comparison of sustained vocal fry and terminal breath group phonation." *J. Soc. Amer.*, **48**, 119A.

Shipp, T. and McGlone, R. E. (1971). "Laryngeal dynamics associated with voice frequency change." *J. Speech Hear. Res.*, **14**(4), 761–768.

Slifka, J. (2006). "Some physiological correlates to regular and irregular phonation at the end of an utterance." *J. Voice*, **20**, 171–186.

Smitheran, J. R. and Hixon, T. J. (1981). "A clinical method for estimating laryngeal airway resistance during vowel production." *J. Speech Hear. Dis.*, **46**(2), 138–146.

Sonninen, A. (1954). "Is the length of the vocal cords the same at all different levels of singing?" *Acta Oto-Laryngolo.*, **43**(sup118), 219–231.

Sorokowski, P., Puts, D., Johnson, J., Żółkiewicz, O., Oleszkiewicz, A., Sorokowska, A., Kowal, M., Borkowska, B., and Pisanski, K. (2019). "Voice of authority: Professionals lower their vocal frequencies when giving expert advice." *J. Nonverbal Behav.*, **43**(2), 257–269.

Stager, S. V., Bielamowicz, S. A., Regnell, J. R., Gupta, A., and Barkmeier, J. M. (2000). "Supraglottic activity: Evidence of vocal hyperfunction or laryngeal articulation." *J. Speech Lang. Hear. Res.*, **43**(1), 229–238.

Timcke, R., von Leden, H., and Moore, P. (1958). "Laryngeal vibrations: Measurements of the glottic wave: Part I. The normal vibratory cycle." *AMA Arch. Otolaryngol.*, **68**(1), 1–19.

Timcke, R., von Leden, H., and Moore, P. (1959). "Laryngeal vibrations: Measurements of the glottic wave: Part II—Physiologic variations." *AMA Arch. Otolaryngol.*, **69**(4), 438–444.

Titze, I. R. (1994). "Toward standards in acoustic analysis of voice." *J. Voice*, **8**(1), 1–7.

van den Berg, J. W. (1955). "On the role of the laryngeal ventricle in voice production." *Folia Phoniatr. Logop.*, **7**(2), 57–69.

van den Berg, J. W. (1956). "Physiology and physics of voice production." *Acta Physiol. Pharmacol Neerl.*, **5**, 40–55.

van den Berg, J. W. (1958). "Myoelastic-aerodynamic theory of voice production." *J. Speech Hear. Res.*, **1**(3), 227–244.

Voelker, C. H. (1935). "XLVIII phoniatry in dysphonia ventricularis." *Ann. Otol. Rhinol. Laryngol.*, **44**(2), 471–473.

von Doersten, P. G., Izdebski, K., Ross, J. C., and Cruz, R. M. (1992). "Ventricular dysphonia: A profile of 40 cases." *The Laryngoscope*, **102**(11), 1296–1301.

von Luschka, H. (1871). *Der Kehlkopf des Menschen: Mit zehn Tafeln Abbildungen.* Laupp.

Wendahl, R. W., Moore, G. P., and Hollien, H. (1963). "Comments on vocal fry." *Folia Phoniatr. Logop.*, **15**(4), 251–255.

Whitehead, R. L., Metz, D. E., and Whitehead, B. (1984). "Vibratory patterns of the vocal folds during pulse register phonation." *J. Acoust. Soc. Am.*, **75**, 1293–1297.

Wolk, L., Abdelli-Beruh, N. B., and Slavin, D. (2012). "Habitual use of vocal fry in young adult female speakers." *J. Voice*, **26**(3), e111–e116.

Ylitalo, R. and Hammarberg, B. (2000). "Voice characteristics, effects of voice therapy, and long-term follow-up of contact granuloma patients." *J. Voice*, **14**(4), 557–566.

Yuasa, I. P. (2010). "Creaky voice: A new feminine voice quality for young urban-oriented upwardly mobile American women?" *Amer. Speech*, **85**(3), 315–337.

Zemlin, W. R. (1998). *Speech and Hearing Science: Anatomy and Physiology*. Boston: Allyn and Bacon.

Chapter Two

Vocal fry, creaky voice, and singing voice pedagogy
Singing teacher attitudes and usage

John Nix and Whitney Chappell

> It seems to me that vocal fry has gotten a really bad rep in the music world. Often people associate it with poor technique or assume that its use 'hurts' the voice. In the SLP world, I have seen it used as a tool to help individuals with muscle tension dysphonia actually relax the laryngeal mechanism. All of that to say, fry seems to be a confusing and controversial topic... with strong opinions on either side.
>
> (Survey participant 31)

Introduction

This chapter explores American singing teachers, use of and their attitudes toward vocal fry and creaky voice. More specifically, we

Creak: Theories and practices of pulse phonation
Edited by Francesco Venturi

ISBN 978-981-5129-29-8 (Hardcover), 978-1-003-61606-1 (eBook)
www.jennystanford.com

explore how teachers define vocal fry and creaky voice and how they use them in their practices, pedagogy, and performances. We raise the question of what they regard as stylistically appropriate uses of vocal fry and/or creaky voice in performance. To set the stage and motivate the current project, the literature review introduces the distinction between vocal fry and creaky voice and the issue of their conflation in literature. From here we explore their use in voice therapy and singing pedagogy, detailing the relationship between perceptions of vocal fry and/or creaky voice and their use by performers, teachers, and their students. The following section presents the methodology of a survey the authors designed to gauge the attitudes of singing teachers toward vocal fry and creaky voice as well as how they use these features. Finally, we detail the results before offering a discussion and concluding remarks.

Literature review

Differentiating between vocal fry and creaky voice

Vocal fry—also known as glottal fry, fry, creak, M0, register 1, pulse register, or *Strohbass*—has traditionally been identified as having a very low fundamental frequency, with each flow pulse damping out nearly completely before the subsequent one commences (Hollien and Michel 1968; Hollien and Wendahl 1968). The vocal folds are shorter in length in fry than in modal production and have little significant change in length as frequency varies (Hollien, Damsté, and Murry 1969). Subglottal air pressure and transglottal airflow are lower in fry than in modal production (Murry and Brown 1971; Blomgren *et al.* 1998). Lower muscle activity levels have been observed during vocal fry in the interarytenoid muscles (a vocal fold adductor), posterior cricoarytenoid muscles (a vocal fold abductor), and the cricothyroid muscles (which lengthen the vocal folds) as compared to modal or chest voice. Conversely, greater activity levels have been found in the thyroarytenoid muscle (which shortens and thickens the vocal folds and adducts their inferior medial portion) during vocal fry as compared to modal (chest) voice (McGlone and Shipp 1971). Comparatively, the interarytenoid muscles contracted nearly double in percentage when compared to the thyroarytenoid muscles in both vocal fry and modal register, suggesting that the interarytenoid muscle contraction, rather than the thyroarytenoid muscles, dominated adduction (McGlone and Shipp 1971). In the vocal tract, greater closure of the velopharyngeal opening and reduced nasality has been observed in vocal fry as compared to modal voice (Behlau *et al.* 1998), and constriction of the laryngeal vestibule has been observed in vocal fry (Scherer 1989; Doscher 1994).

The terms vocal fry and creaky voice are used quite broadly and are often conflated by many practitioners and authors in the research literature, including several sources cited in this chapter. Regarding the term vocal fry, it is used to indicate a wide variety of voicing behaviors. A recent publication in the voice science literature categorized vocal fry in great detail:

> Six vocal fry patterns were observed. The patterns were: single pulse fry (FRY1), double pulse fry (FRY2), multiple pulse fry (FRY3), period doubling fry (FRY4), inaudible fry (FRY5), and indeterminate vocal fry (FRY6). Single pulse fry was divided into a single (one) pulse fry cycle (FRY1a), a series of even single pulse fry cycles (FRY1b), and a series of uneven single pulse fry cycles (FRY1c). Double pulse fry was divided into a primary then another primary pulse fry cycle (FRY2a), a secondary then a primary pulse fry cycle (FRY2b), and a primary then a secondary pulse fry cycle (FRY2c). Multiple pulse fry, where a higher frequency was modulated by a lower frequency, was divided into the higher frequency being near the speaking fundamental frequency (FRY3a) and the higher frequency being inconsistent or well below the expected speaking fundamental frequency (FRY3b) (Proctor, Scherer, and Perrine 2024).

Some of the same vocalizations are also labeled as creaky voice, as can be seen in this quote by the linguist Lisa Davidson:

> In terms of articulation, the term 'creaky voice' is usually used to refer to a collection of phonation types, including 'prototypical creaky voice' (low F0, semiregular periodicity, and strong damping between glottal pulses), 'multiply pulsed creak' (glottal openings that alternate in higher and lower amplitude, with a perceived low F0, with the entire duration between higher amplitudes being semiperiodic), and 'nonconstricted creak' (F0 is low and irregular, but these properties are accompanied by glottal spreading and higher airflow, not constriction), along with other examples (Davidson 2020: e22–e23).

Within the voice science community, vocal fry has sometimes been differentiated from creaky voice, which, authors remark, can occur at higher frequencies and is characterized by a perception of roughness and the presence of sub-harmonics in the glottal waveform (Scherer 1989; Titze 2000; Titze and Verdolini-Abbott 2012). This distinction between vocal fry and creaky voice can be perceived by trained listeners, e.g., voice clinicians and linguists (Gerratt and Kreiman 2001).

A number of studies have examined the prevalence of the occurrence, usage, and meaning of vocal fry in American English. Given the wide range of detection methods used in these studies and the variety of phonation types that have been labeled as vocal fry, in citing the literature below, we have retained the language used by the authors. In speech, vocal fry tends to occur at the ends of sentences and paragraphs, at word boundaries, syllable boundaries, and in unstressed syllables (Wolk *et al.* 2011; Gibson 2017). It has been regarded as more prevalent among female speakers than male speakers (Abdelli-Beruh *et al.* 2014; Podesva 2013), although this prevalence has been recently questioned in a systematic review (Dallaston and Docherty 2020). Bilingual English and Spanish speakers use vocal fry more often when speaking English than speaking Spanish, and bilingual speakers whose native language is English use more vocal fry in speaking both languages than do bilingual speakers whose native language is Spanish (Cantor-Cutiva *et al.* 2021). The use of fry in speech has been associated with a variety of social constructs, including toughness in Chicana gang girls (Mendoza-Denton 2011); masculinity, authority, and whiteness (Lefkowitz and Sicoli 2017); and professionalism and upward mobility in young women (Yuasa 2010). Conversely, use of fry in speech has also been associated with evaluations of lower competence, education, trustworthiness, attractiveness, and ability to hire in young women, especially among older listeners (Anderson *et al.* 2014), as well as vanity, apathy, sleepiness, relaxation, and boredom (Ligon *et al.* 2018). Continuous use of fry has been rated as unnatural, requiring greater listener concentration, and associated with lower employability as compared to habitual voice quality (Venkatraman and Sivasankar 2018).

The perception of pitch appears to be altered by the presence of fry and/or creaky voice. A study (Kuang and Liberman 2016) paired synthesized stimuli ranging from 43 to 103 Hz, with five different frequency modulation ratios (0, 2, 8, 15, and 25%). The results showed that untrained (non-voice professional) listeners hear a lower pitch during vocal-fry-like samples than during unmodulated samples, and that the effect of the fry on perception depends on the fundamental frequency and amount of the modulation. This finding was supported by a study also using non-expert listeners, but the samples were of naturally produced statements by four white female podcasters who used modal phonation, the "prototypical" creaky voice, or the "multiply pulsed" creaky voice (Davidson 2020). The author found that for speakers with a higher pitch, the listeners rated samples with partially creaky phonation as lower in pitch than samples of fully modal phonation, even when the partially creaky phonation occurred within a sample of modal speech with a higher fundamental frequency than samples of fully modal speech (2020: e30).

Vocal fry and creaky voice in voice therapy

Speech-language pathology literature includes indications for and contraindications against vocal fry. Buescher (2002), borrowing a singing voice intervention technique from Riggs described below (1998), advocates for the use of upward-inflected glides starting in vocal fry and gliding into modal production as a post-operative rehabilitation exercise. Emerich endorses this approach as being successful in improving vibration amplitude in previously scarred vocal folds (Emerich 2003). In a later publication, Emerich discussed using vocal fry to optimize the glottal closing pattern and decrease the need for lateral or anterior–posterior compression as a compensatory reaction to incomplete glottal closure or laryngeal asymmetry (Nix, Emerich, and Titze 2005). Emerich's approach is cited by Titze and Verdolini-Abbott (2012) as a means to habilitate vocalists who are too breathy in initiating voicing (2012: 268, 273). Interestingly, Titze and Verdolini-Abbott include an exercise that initiates voicing in either vocal fry or creaky voice and glides upward (from fry) or possibly downward (from creaky voice) into modal production. As the authors state: "Creaky voice is similar to fry in terms of adduction, but it can be produced at any pitch and is characterized by an irregular vibration pattern that sounds like a creaky door" (2012: 273). Inspiratory vocal fry, described as "the voice production during inspiration of a sound with vocal fry perceptual characteristics," has also been explored as a possible therapy facilitation technique (Paolillo *et al.* 2019: 394). Paolillo and colleagues found inspiratory vocal fry (IVF) significantly reduced adduction of the false vocal folds and pharyngeal wall contraction values as compared to expiratory vocal fry (EVF) (2019: 397–398); they also found that the length of the vocal folds was increased in IVF as compared to EVF, and that contact quotient values were higher (but not significantly so after Bonferroni correction) in IVF than EVF (2019: 398). They conclude that "...IVF could be an excellent technique to reduce muscular hypertonic tension and to facilitate mucosal elasticity" (2019: 398). However, not all authors are enthusiastic about the use of vocal fry. In her recent book *Manual of Singing Voice Rehabilitation,* speech pathologist and singing voice specialist Leda Scearce does not include fry or creaky voice as a facilitation technique for improving adduction; rather, she places glottal fry and creaky voice as phonation types to be avoided under the heading: "Reducing inefficient vocal behaviors" (Scearce 2016: 244). Subsequently, a study found that continuous use of vocal fry for 30 min significantly increased both PTP-10 (phonation threshold pressure at 10% of the vocal range) and vocal effort ratings (Venkatraman and Sivasankar 2018). The authors of this study urged caution in reading too much into their results but did indicate that their study offered "preliminary evidence that continuous vocal fry can worsen some voice measures, even when used for a relatively short duration" (2018: 1544).

Vocal fry and creaky voice in singing pedagogy

Opinions on vocal fry in singing training also vary substantially. Some writers, such as Ralph Appleman and Meribeth Bunch, make no statements regarding vocal fry, pulse register, or creaky voice in their books; other pedagogues have discussed fry's existence as a register without making any comments about its utility; others have advised positively or negatively on its use in singing. William Leyerle wrote in *Vocal Development Through Organic Imagery*: "This register is tight and crackly. It is not generally a usable part of the baritone's range, but is useful for low basses...." (Leyerle 1986: 65). He does not go on to describe how it might be used or trained. In *Bel Canto: A History of Vocal Pedagogy*, James Stark mentions the writings of Johann Agricola and Mann Mersenne on the practice of using vocal fry to expand the low range of baritones and basses (Stark 1999). Later, Stark states that vocal fry is often heard in Russian choral music. His remarks conclude with a brief description of some of the physiologic, aerodynamic, and acoustic properties of fry (Stark 1999: 89). Barbara Doscher also takes a descriptive approach in her book, detailing characteristics of fry without making any recommendations for its use (Doscher 1994: 187). Doscher was the teacher of one of the authors (Nix), and in his observations of her teaching, while she sometimes slipped into fry in her speech, she neither used it pedagogically in the studio nor advised for or against its use. Cornelius Reid, on the other hand, was very outspoken in his condemnation of not only vocal fry as a phonatory mode but also the terms used to describe this kind of vocalization.

> Pulse register: a term introduced by scientific investigators to refer to vocalized sounds with extremely low frequencies and pulse-like oscillations; a "vocal fry," "glottal fry," "creak," or *Strohbass*. The concept of a "pulse register" is without value to voice culture for the following reasons: 1. It is a designation based upon acoustic rather than physiological data, and there is no distinct mechanical or muscular system responsible for its articulation; 2. The vibratory impulses it describes have no relevance to the "pulse" or vibrato of a legitimate vocal tone; 3. The vibratory impulses it describes are "musical" only in that they possess measureable pitch; they contain no recognizable vowel phoneme and cannot be swelled or diminished; 4. The sounds to which it refers are "freak" tones with no aesthetic value; like squeaks, groans, screams, and squeals, they should be considered forms of affective expression, and not confused with legitimate vocalization, and 5. The sounds to which it refers have no pedagogic value since they could never be incorporated into, or used to influence or improve, the quality of those tones universally accepted as being legitimate. The term "pulse register"

> should be abandoned, since a further proliferation of terminology, even when justified on acoustic terms, simply adds confusion to an already confused area of thinking with respect to the number and origin of vocal registers (Reid 1983: 284–285).

William Vennard from the University of Southern California suggested the use of fry as a corrective technique for an overly breathy voice.

> This exercise is especially good for breathy pupils. A breathy tone can hardly be initiated by the tension required for the rattle. However, it is an ideal tension which adjusts the glottis without tightening the throat, and as such benefits students who are too tense. The voiceless [sic] rattle may rumble into a tone by adding phonation. This approach is the opposite of that described in Par. 182. The "imaginary h" is more suitable for attacking high notes, where tension is a danger, and the "rattle" is better for low notes, indeed it builds the low part of the range. It is especially beneficial for low voices (Vennard 1967: 48).

Vennard furthermore stated that "The glottal rattle, or fry (Par. 195), requires a loose glottis, and is much more difficult to perform on either Ee or Oo than it is while the resonators are forming an Ah" (Vennard 1967: 124). Like Vennard, Oren Brown advocated for an exploration of vocal fry, especially for the development of the low range. In *Discover Your Voice*, the chapter "Range and Registers" includes the following passage:

> To find the lowest notes in Register 1 [in Brown's text, vocal fry is labeled Register 1], the larynx must rest in a very relaxed, low position and then apply an almost breathy flow of air. It takes more air flow and less pressure to produce low notes than high notes because the vocal folds are looser (that is, they do not come together so often) and the amount of air that escapes at each opening pulse is greater. Any attempt to produce these notes by tensing and pushing down into them prevents the muscles from performing their natural function and shuts off the resonance. For this part of the voice, let the tone slide down and let the air flow do the work (Brown 1996: 58).

Brown includes a 5-4-3-2-1 descending vocalise (Exercise XV in his book) for experiencing fry (Brown 1996: 270–271) and includes audio examples of this type of phonation on the CD that accompanies his book. Additionally, Brown employed a fry onset exercise in a lesson with the first author in July 2000.

Seth Riggs recommends the use of creaky voice exercises in his book *Singing for the Stars* as a means to find what he calls Speech-Level Singing (Riggs 1998). He begins with a single upward-inflected glide in a creaky voice hum (Exercise 14 in the book), then moves on to arpeggiated patterns on a hum with creaky type onsets (Exercises 15–19) (1998). The number of creaky onsets is gradually reduced as the singer progresses through the course of exercises. Riggs's book also contains a CD that demonstrates these exercises. Riggs's concepts were used by Randy Buescher in a single-subject case study reported in the *Journal of Singing* (Buescher 2002).

The noted pedagogue and author Richard Miller takes a more measured view of *Strohbass* (which he distinguishes from vocal fry) and fry.

> Just as an occasional falsetto note is intruded in legitimate upper range for some specific coloration, so an occasional *Strohbass* tone may be introduced in the lowest range of the voice... *Strohbass*, if used at all, should be used judiciously. Exercises for the development of this register phenomenon should be undertaken only with a teacher, and never for more than a few brief moments... At times, a moderate use of vocal fry may help a young, low-voiced male develop a "feel" for additional pitches at the lower extension of his range. Vocal fry encourages an imprecise onset and should not be relied on as a standard vocalizing technique (Miller 1986: 125–126).

Miller includes a 5-3-1-3-5 exercise (Exercise 9.3 in 1986: 123) for developing the *Strohbass* quality, with only the lowest note in the pattern being sung in *Strohbass*.

Writing around the same time, Frederick Swanson discusses developing some teenage boys' low range for choral singing through the use of "pulse register":

> One of the most fascinating discoveries was the group of boys who found and developed the "pulse register," otherwise known as "the fry tones" and the "Russian bass." Oddly enough, these were often the boys who were labeled deficient singers. The pulse

> register has been used in Russia, Rumania [sic], and Yugoslavia for centuries (Hayes). Since the Russian Orthodox Church allowed no instruments in their church buildings and all singing was a capella, if a deep bass was wanted for the choral ensemble, it had to be sung. Therefore, choir directors sought out and developed voices that could sing in the octave below the standard bass clef. Some Americans have done research and written on experiments and findings on the pulse register (Hollien and Michel; Hollien, Moore, and Wendell; Moore and Van Leyden; Shipp; Vennard). But for some reason, American voice teachers and choral directors have never been overly interested in discovering and using this vocal resource. Only lately have reports stated that there are adolescent boys who can sing in those pitches (Swanson, The Male Singing Voice), and in a few cases only in those pitches. When developed and used properly, they can add a depth and richness to the sound of a freshman choir that is highly effective. As a result, some boys who had been labeled deficient, considered flawed nonsingers, have turned out to be valuable assets (Swanson 1984: 49).

In a 2005 article, Nix, Emerich, and Titze provided several guidelines for how vocal fry might be used in singing pedagogy (2005: 56). The target behavior they advise is one that is perceptually very relaxed with a steady air stream and a low fundamental frequency. They recommend using vocal fry as a lead-in or pilot to vocalises or phrases, where there is no pause in phonation between the fry and the vocalise/phrase, only an upward glide into modal or other registers. The fry portion should be on the same vowel with the same vocal tract posture the singer is about to use in the vocalise or musical phrase, and the singer should audiate the target pitch of the sung portion of the exercise or phrase while doing the fry. The fry onset should not be used before extremely high pitches; the authors advise an upper boundary of D4 for tenor, baritone, and bass voices, and D5 for treble voice singers. The authors recommend mixing trials with the fry lead-in with trials of the same exercise or phrase without the fry; as the singer becomes more accomplished with the non-fry-initiated version, the fry trials are gradually reduced in frequency, then eliminated. They note that alternating fry onset patterns with staccato or other onset exercises can be very effective. Furthermore, they advise pairing exercises that use the fry onset be performed in alternation with descending patterns. By pairing the bottom-up approach of fry with other patterns that are top down, the authors suggest a better intrinsic muscle balance may be fostered in the larynx.

More recently, American pedagogue Kari Ragan included a sequence of three vocalises with vocal fry in her book *A Systematic Approach to Voice: The Art of Studio Application* (Ragan 2020). Exercise 4-13 is an acclimation exercise, where a singer sustains the vowel [a] in vocal fry for several seconds (2020: 107) to become familiar and comfortable with that mode of production. Exercise 4-14 begins with sustaining vocal fry on [a], then glides upward into chest registration (M1, modal voice, etc.), where a three-note ascending and descending scale is performed on the vowel [a] (2020: 107–108). Exercise 4-15 once again begins in vocal fry on [a], then glides upward into head registration (M2, falsetto, etc.), where a five note descending scale is sung on the vowel [u] (Ragan 2020: 108–109).

Finally, voice teacher and researcher Donald Miller advocated for the use of vocal fry along with real-time spectrographic software in the voice practice room and teaching studio. The very low frequency and non-periodic nature of fry can be useful as a sound source for revealing the approximate location of vowel resonances for a given vocal tract shape. As the singer shifts into singing from vocal fry, it can easily be determined whether or not a harmonic in the sung production is in close proximity to the resonances that were seen in the vocal fry spectrum (Miller *et al.* 1997; Miller 2008).

Previous work on vocal fry perception and priming in singing

Two studies in particular motivate the present work, as they investigate vocal fry perception and priming in singing. As such, they are outlined in detail in this section. In the first study (Chappell, Nix, and Parrott 2018), the authors employed a matched-guise experiment, which was used to measure the listener's evaluation of fry and non-fry guises. Four singers, two male and two female, sang "The Star-Spangled Banner" with onset vocal fry. These baseline recordings were used to create the two guises: (i) an unmodified recording with onset vocal fry on vowel-initial words and (ii) a recording in which the fry had been removed. In total, 253 participants listened to the recordings and evaluated the singers' social and stylistic attributes along a Likert scale (e.g., how confident, sexy, and sincere each singer sounded). A factor analysis was used to conflate correlated variables, and mixed-effects linear regression models (n = 1012) were fitted to each single or joint factor to determine whether vocal fry significantly influenced the listeners' responses to the singers. The results showed that vocal fry significantly altered the listener's evaluation of the sincerity/commitment, maturity/sophistication, naturalness, and confidence of the singers ($p < 0.05$).

The singer's gender and the listener's age and musical training were found to play a crucial role. Unlike male singers, who were rated as significantly less sincere/committed with vocal fry, female singers were seen as more sincere/committed with vocal fry and younger listeners also found them less natural, suggesting vocal fry is associated with emotional intensity in female voices. Younger listeners perceived singers with fry as less mature/sophisticated, suggesting an association with youth. Finally, listeners with more musical training rated singers with fry as less confident, while less trained listeners did not exhibit this difference. Overall, the findings indicated that listeners are highly attuned to vocal fry in music but respond to it differently based on their age, musical training, and the singer's sex. Vocal fry is evaluated more positively among younger, less musically trained listeners, and it is better received in the voices of women, suggesting that the use of fry strategically targets a specific audience—younger and less trained listeners—who interpret fry as a marker of youth and emotional earnestness. These findings indicate that a single stylistic feature like vocal fry can be imbued with multiple meanings depending on the singer and audience, and its use can serve to include or exclude particular listener groups.

The second study (Nix, Chappell, and Powell 2019) explored the relationship between vocal fry usage in speech and subject perceptions of vocal fry in *a cappella* music. The subjects (n = 60) were recorded speaking the Rainbow Passage (Fairbanks 1960) before and after responding to the matched-guise listening task used by the authors in the previous study (Chappell, Nix, and

Parrott 2018). In this second study, all subjects responded to the listening task on the same computer using the same headphones. The two recordings of the subjects speaking the Rainbow Passage before and after responding to the listening task were compared to examine whether hearing the samples where fry was used as a stylism in a musical context had an entrainment effect on the amount of vocal fry in the subject's speech. We also sought to test if habitually using fry in speech affects how a subject evaluates fry in the singing of others. It was hypothesized that listening to singing that features fry would tend to increase its use in speech and that fry usage in speech would be associated with positive evaluations of fry in singing. Looking at the voice recordings alone, no entrainment effect from the mixed guise listening task was found. There were no significant relationships across most of the social attributes. However, a best-fit mixed-effects linear regression model was made for evaluations of vulnerability. Those subjects who used more fry after the listening task tended to give lower evaluations of vulnerability to recordings with fry, while those who used less fry after the listening task tended to give higher evaluations to recordings with fry ($p = 0.02$). The fact that subjects who perceive fry as a marker of greater vulnerability in song decrease their use of fry in speech may be an avoidance strategy. Perhaps those who perceived fry as intensifying vulnerability avoided its use in their second reading, as vulnerability would not be desirable in a formal task (reading aloud). The other striking result was an interaction between fry usage and evaluations of the singer's ethnicity. The results of a best-fit multinomial regression model for perceived race showed that subjects who perceived fry as signifying a Caucasian singer decreased their amount of fry in their own speech pre to post, while subjects who regarded fry as signifying an African–American singer tended to increase their amount of fry pre to post ($p = 0.0166$). In other words, those who perceived fry as a white characteristic in song avoided its use in their second reading, as being identified as such might make one appear less credible as a performer.

Research questions

The aforementioned studies have determined that the listener's attitudes toward vocal fry in singing are conditioned by musical training and that vocal fry use can be primed, but this priming depends on the listener's attitudes. Given the established relationship between attitudes and use, on the one hand, and the fact that more musically trained participants perceive vocal fry differently than others, the present study seeks to specifically shed light on the attitudes of singing teachers toward the use of vocal fry and creaky voice. In particular, we address the following research questions:

1. How do singing teachers define vocal fry and creaky voice? Do they make a distinction between the two?
2. Do U.S. singing teachers use vocal fry and/or creaky voice in their practicing, performing, and teaching, and if so, how do they use it?
3. What do U.S. singing teachers regard as stylistically appropriate uses of vocal fry and/or creaky voice in performance?

Methods

Approval from the University of Texas at San Antonio Human Subjects Research Board was obtained for the study. A survey was created and distributed through the online survey portal Qualtrics (www.qualtrics.com) to explore the attitudes of trained singing teachers toward the use of various phonation types in singing training and in the performance of singing styles. To be eligible, participants had to be at least 22 years of age, hold at least a bachelor's degree in music, speak and read English, and reside in the U.S. To attract participants, a brief recruitment statement was emailed to singing teachers in the U.S. and posted on singing teaching forums on social media (Facebook) and on the American Speech and Hearing Association's Special Interest Group Three list-serv. Of the 124 participants who began the survey, 87 participants completed it and are considered in the discussion that follows. More information is provided about the participants in Table 2.1.

Table 2.1 Basic sociodemographic information of the participants

Age (Years)	Range: 23–73 (Median: 42; Mean: 44)
Gender (*n*)	Man: 29; Woman: 55; Non-binary: 2; No response: 1
Race (*n*)	White: 73; Mixed: 6; Black: 2; Asian: 1; Native American: 1; Prefer not to answer: 4
Hispanic (*n*)	Yes: 12; No: 73; Prefer not to answer: 2
Highest education level (*n*)	Doctoral degree: 30; Post-master's: 2; Master's: 39; Post-bachelor's: 4; Bachelor's: 12
Voice type (*n*)	Bass: 2; Baritone: 9; Tenor: 16; Mezzo: 19; Soprano: 36; Other: 5
Emphasis (*n*)	Voice performance: 35; Voice perf. and pedagogy: 22; Opera performance: 4; Music education: 6 Speech-language pathology: 4; Other: 16
Experience (years)	Range: 1–46
Total number of complete responses	87

The survey included basic sociodemographic questions, such as age, gender, education, race, and ethnicity, as well as information specific to singing teachers, such as voice type, primary musical style performed and taught, emphasis of highest degree, types of teaching (academic, private studio, voice clinic), and years of voice-teaching experience. The participants then answered a series of multiple-choice and open-ended questions about vocal fry and creaky voice to establish what attributes they most associate with these phonation types (low pitch, indeterminate pitch, low airflow, low breath pressure, insufficient airflow, pressed or strained quality, can blend with other registers, separate from other registers, individual clicks or impulses can be heard [non-continuous sound], or other), what term they use to describe the lowest voice register, whether (and how) the voice teachers who trained them used vocal fry and/or creaky voice in lessons, whether they currently use vocal fry

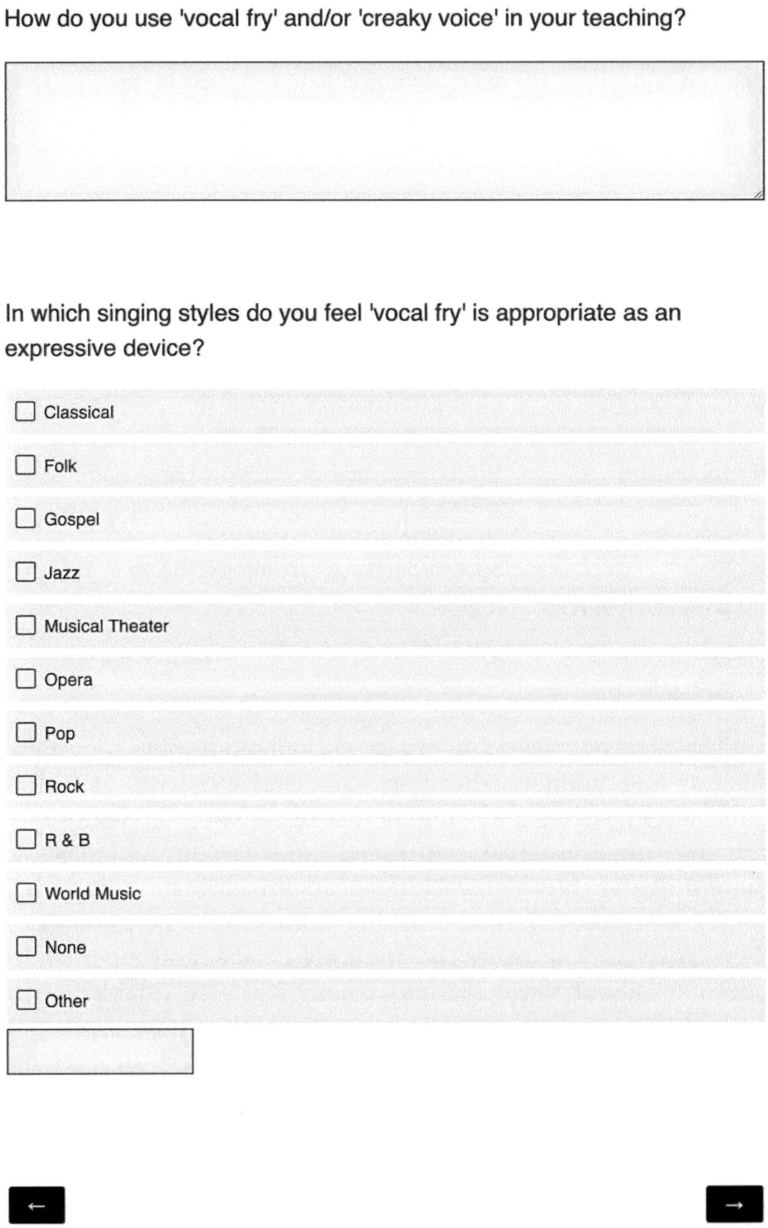
How do you use 'vocal fry' and/or 'creaky voice' in your teaching?

In which singing styles do you feel 'vocal fry' is appropriate as an expressive device?

- [ ] Classical
- [ ] Folk
- [ ] Gospel
- [ ] Jazz
- [ ] Musical Theater
- [ ] Opera
- [ ] Pop
- [ ] Rock
- [ ] R & B
- [ ] World Music
- [ ] None
- [ ] Other

← →

Figure 2.1 Screenshot of one page of the survey.

and/or creaky voice in their practice, performance, and teaching, if (and how) they view vocal fry as synonymous with or distinct from creaky voice, and in what singing styles they find vocal fry and creaky voice to be appropriate to, as an expressive device. Written comments were solicited after multiple-choice questions to gain a better sense of why participants selected the option they did. A screenshot of one page of the survey is provided in Fig. 2.1, which includes one open-ended and one multiple answer question.

Results

This section explores the quantitative and qualitative results of the survey with the goal of better understanding teacher attitudes toward vocal fry, creaky voice, and singing. Below we share a statistical breakdown of quantitative data along with a sampling of written responses, grouping the written responses into categories.

How do singing teachers define vocal fry and creaky voice?

When asked what attributes they associated with the lowest register, participants selected a range of different responses based on production, perception, and acoustic factors. The most widely selected attributes included low pitch (41/87), indeterminate pitch (40/87), low airflow (37/87), low breath pressure (34/87), and the perception of individual clicks or impulses (33/87); additionally, a number of participants found vocal fry and creaky voice to be separate from other registers (28/87). The singing teachers less frequently selected that the phonation types can blend with other registers (17/87), have a pressed or strained quality (16/87), and are associated with insufficient airflow (11/87).

Next, to understand how (and if) teachers distinguish between creaky voice and other pulse-like phonation types, we asked participants whether they found creaky voice to be synonymous with or distinct from vocal fry, providing three possible responses: synonymous, distinct, and unsure. The following question asked participants to elaborate on their response in writing, describing why they categorized them as synonymous, similar or distinct. Figure 2.2 shows the breakdown of participants' multiple-choice responses. This and all following figures were made with the *ggplot2* (Wickham 2013) and *wordcloud* packages (Fellows 2018) in R (R Core Team 2021). As Fig. 2.2 makes clear, many participants were unsure as to whether there was any distinction between creaky voice and vocal fry (37/87), and those who found the terms to be distinct were in the minority (21/87).

In their written comments about the distinction, several respondents indicated familiarity with vocal fry, while lacking exposure to the term "creaky voice," for instance: "I have never heard the term 'creaky voice' until now" (Participant 63, man, 30 years old, tenor). Others were quite familiar with the term and drew a clear distinction, such as this teacher: "From my understanding, I have heard the terms being used interchangeably. I feel like there is a difference in production (at least air pressure wise) between vocal fry in the lowest part of a person's range and the type of vocal fry that is used for stylization purposes in non-classical styles" (Participant 75, woman, 31 years old, soprano).

Another shared this perspective: "I see creaky voice more as a pedagogic tool or vocal stylism that is used in (usually pop or rock) singing styles. It is the same functional mechanism (in my mind) as fry, but it is used more consistently in a singing style to provide a creaky voice effect. I think of it in the same way that a belter predominately uses chest voice, but we refer to it as belting instead of chest" (Participant 49, non-binary, 28 years old, tenor).

Two respondents had opposite opinions about which term denotes a sound that is volitional. One wrote, "I equate creaky voice to a sound someone is making and they are unaware of how and why they are making it, whereas I equate vocal fry or *Strohbass* with a purposeful action by the singer" (Participant 87, woman, 37 years old, soprano). The other teacher felt the opposite, stating: "They create the same sound but, for me, creaky voice is a choice to sound a specific way" (Participant 2, woman, 54 years old, mezzo).

In spite of these disagreements among the singing teachers, a trend could be observed in the data: The educational backgrounds of the participants were correlated with their understanding of creaky voice and vocal fry as synonymous or distinct. More

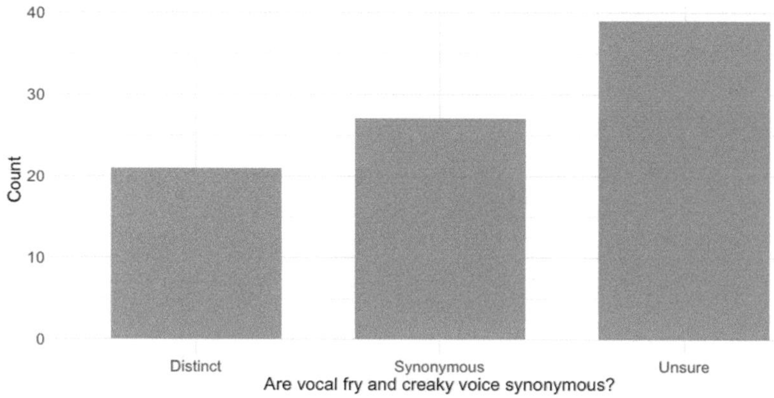

Figure 2.2 Participants' responses to creaky voice being synonymous with vocal fry.

specifically, respondents whose music degrees involved an emphasis in education were more likely to distinguish between the concepts than those with an emphasis in performance. All teachers with an emphasis on speech-language pathology (3/3), most of whom worked with voice performance and pedagogy (9/13), and half with an emphasis on music education (2/4) found creaky voice and vocal fry to be distinct, whereas only five of 30 participants with an emphasis on voice performance thought the terms were distinct.

Exposure of participants to vocal fry and creaky voice in their training

Two questions in the survey sought to establish whether the participants' teachers used vocal fry and/or creaky voice in their lessons, and a written follow-up asked for additional information about how they were used. As shown in Fig. 2.3, the teachers of most participants never used creaky voice and/or vocal fry (47/87), and only four participants claimed it was used frequently in their teachers' lessons.

Written responses regarding how vocal fry and creaky voice were used in their lessons fell into three thematic areas: for relaxation or cool-down, for improved vocal fold closure, and for revealing vocal tract resonances. Here are a few examples.

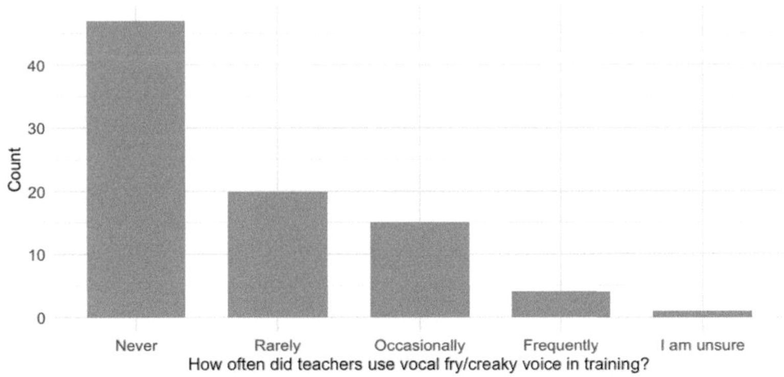

Figure 2.3 Responses of participants to how frequently teachers used vocal fry and/or creaky voice in lessons.

- Relaxation or cool-down:

  "To relax vocal folds as a part of a cool down protocol" (Participant 7, woman, 32 years old, soprano).

  "One teacher used fry as a part of a cool down (just a couple of short bursts) after singing in high and loud coordinations during the lesson. Another teacher used fry with me a few times, in conjunction with twang, as a means of facilitating that coordination with minimal subglottal pressure or squeeze at the level of the folds. He also had me do it with my tongue out so as not to lock the tongue root while I produced the onset" (Participant 29, man, 32 years old, tenor).

  "I had a few use it as a relaxing tool to find loose production in the lower register" (Participant 49, non-binary, 28 years old, tenor).

- Improved closure:

  "As explained by my teacher, vocal fry helps the vocal folds to come together. He would have me phonate in a vocal fry and transition into a vocalise on a neutral vowel" (Participant 69, woman, 25 years old, soprano).

  "To teach closure that wasn't breathy or pressed. To encourage a low/relaxed laryngeal position. To teach closure without external musculature and articulator tension" (Participant 39, woman, 38 years old, soprano).

- Revealing resonances:

  "When using [a] spectrogram to find the natural resonances of specific vowel shapes for formant tuning" (Participant 60, man, 33 years old, tenor).

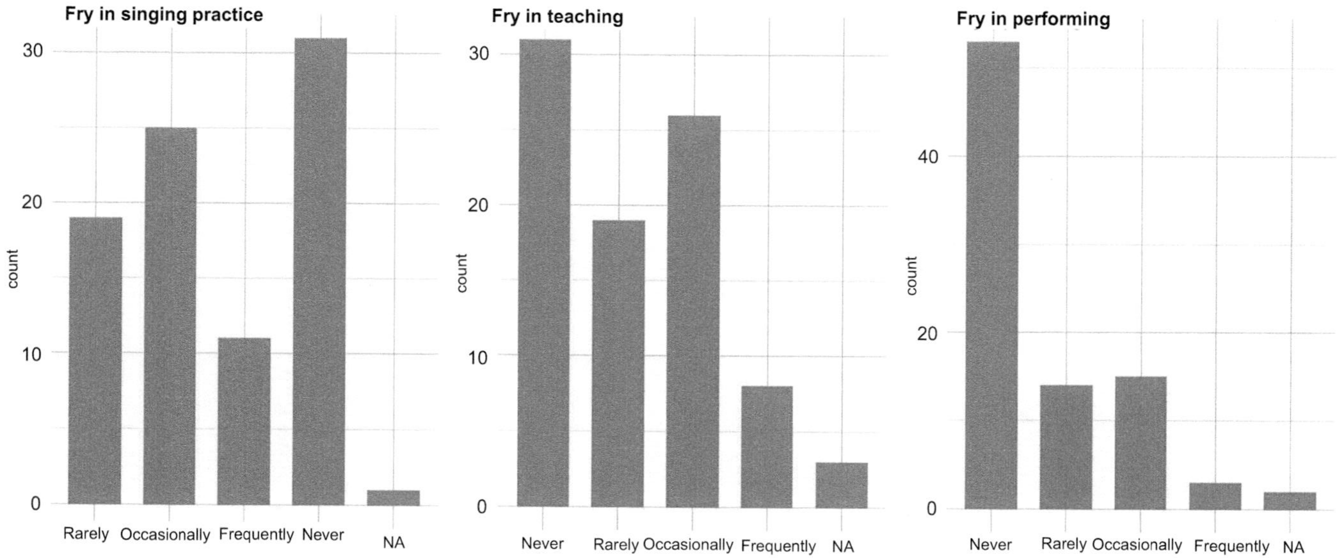

Figure 2.4 Participants' use of creaky voice and/or vocal fry in their practice, teaching, and performance.

> "Only used as a tool to determine formants" (Participant 31, woman, 31 years old, mezzo).

One other respondent mentioned a variety of uses for fry according to their teacher: "One teacher told me to use it as a metric for measuring vocal fatigue. If you can't make fry you should stop singing. Another teacher told me to use it to bring my voice down from my highest register so that I could go from high coloratura back to a musical theatre style of singing within the same performance" (Participant 45, no gender provided, 28 years old, soprano).

Participants' use of vocal fry and creaky voice in their singing practice, teaching, and performance

Participants were also asked to specify how much and in what ways they use creaky voice and/or vocal fry in their own singing practice, teaching, and performance. Figure 2.4 outlines the results in each category. Participants were the least likely to employ creaky voice and/or vocal fry in their performance, but more participants use vocal fry and/or creaky voice frequently or occasionally in their own teaching than their teachers used them, and the participants' use of vocal fry/creaky voice is the most common in their singing practice.

Written responses that elaborated on how creaky voice/vocal fry was used in singing practice centered around relaxation after high or intense singing and working on vocal onset. One teacher wrote, "Sometimes to try and recalibrate and relax into lower sounds after I have done a lot of very high singing" (Participant 42, woman, 42 years old, mezzo). Along these lines, a second wrote, "As a way to relax and check in with my voice after strenuous work" (Participant 45, no gender provided, 28 years old, soprano). As for onset, one provided this extended response:

> "I use it primarily for onset purposes during warmups in order to promote good vocal fold closure without high degrees of sub glottal breath pressure. I will use it if I feel like I'm sending too much air to the glottis, or if I'm trying to access a high note during a warm-up that is not being produced easily. I will sometimes employ it

as a pseudo-onset to help "clean up" sound that indicates a bit of swelling of the folds (eg: if the sound is "hissy"). I like to use it in conjunction with twang exercises in order to find a tonal balance and the appropriate amount of air for that specific pharyngeal shaping. I also use it when I cool-down after laborious, lengthy or high-dynamics singing" (Participant 29, man, 32 years old, tenor).

However, several mentioned stylistic uses:

"The repertoire lends itself to the use of vocal fry stylistically" (Participant 83, man, 37 years old, tenor).

"I don't utilize it in my classical style of singing, but I do in my CCM [Contemporary Commercial Music] styles. I also am not intentionally practicing it or even consciously adding it to my CCM singing, it's sort of reflexive depending on the style of music I'm singing" (Participant 75, woman, 31 years old, soprano).

"I use it stylistically in some styles for expressive purposes" (Participant 55, woman, 41 years old, contemporary female singer).

"It's a favorite of mine when working grit in rock singing or even low growls for metal singing." (Participant 21, man, 27 years old, tenor)

"As an ornament in CCM styles. Occasionally in a vocal exercise." (Participant 18, woman, 45 years old, soprano)

To establish how the teachers used fry and/or creaky voice when working with a student, the participants were asked to elaborate on how they used them in their lessons. Responses fell into four categories: (1) as a stylistic expressive device; (2) for relaxation, especially after high and intense singing; (3) vocal fold closure; (4) to help singers who use fry too much in speech be aware of their habits.

Regarding stylistic usage, one teacher noted, "It helps my students sound more authentic in their pop/musical theatre styles" (Participant 62, woman, 25 years old, mezzo). The idea of fry helping a student rest or relax after higher singing appeared more than once: "To help students access their lower register, to freely produce their lowest tones, and to release vocal folds during strenuous singing sessions" (Participant 24, woman, 59 years old, soprano). One teacher's response covered multiple categories: "To teach closure that isn't pressed or breathy. To encourage low/relaxed laryngeal position. To relax tension in the articulators (mouth/oral cavity space) while also keeping closure intact, to adjust closure in belting, and a myriad of ways to use it stylistically. It's integral to the technique I teach" (Participant 39, woman, 38 years old, soprano).

The final use, that of increasing a student's awareness of speech habits, was interesting. Here are three examples of participant comments:

> "Primarily, to point out how often students are unintentionally doing it, particularly at onset" (Participant 35, woman, 45 years old, contemporary singer).
>
> "For negative practice to sensitize patients to different voice qualities and laryngeal tensions. Non-fry is more desirable" (Participant 20, man, 66 years old, tenor).
>
> "With singers that talk a lot in their fry already. Makes them understand the difference between a fry with flow and a pressed one" (Participant 16, woman, 30 years old, soprano).

The written responses to the question about the use of vocal fry and/or creaky voice while performing were entirely about stylistic responses to different musical genres. A few typical responses included the following:

> "If I'm singing in a pop style that calls for it, I'll use it" (Participant 76, woman, 33 years old, mezzo).
>
> "I don't utilize it in my classical style of singing, but I do in my CCM styles, most commonly at the beginnings of phrases" (Participant 75, woman, 31 years old, soprano).
>
> "In musical theatre it is a common stylization that I use to convey emotion and get a more robust mix" (Participant 62, woman, 25 years old, mezzo).
>
> "It can be used for some elements of growl for affect" (Participant 48, woman, 51 years old, soprano).
>
> "As a slight, passing ornament in quiet singing. More accurately, once in a rare while it happens spontaneously; and because I know that audiences tolerate it in my genre, I do too. Never aim for it, though" (Participant 46, woman, 66 years old, mezzo).

> "Generally as onsets at the beginning or ending of phrases, or preceding a note 'scoop.' Rarely do I use it outside of pop/rock styles of music, although I have occasionally used it when performing musical theatre repertoire, if it is stylistically appropriate" (Participant 29, man, 32 years old, tenor).

Stylistic uses of vocal fry and creaky voice

As the comments above indicate, participants found creaky voice and/or vocal fry to be more appropriate in their teaching, practice, and performance in certain, specific styles. When asked in which styles vocal fry and creaky voice were appropriate, a general consensus emerged among respondents. The word clouds in Fig. 2.5 highlight the most commonly selected styles appropriate for vocal fry and creaky voice. Each style had to be named at least five times to appear in the word cloud, and a larger font represents a more commonly selected style. Overall, pop and rock were believed to be the most appropriate styles for vocal fry and creaky voice, with 62 and 64 of the 87 participants deeming vocal fry appropriate in pop and rock, respectively, and 53 participants deeming creaky voice appropriate in both pop and rock styles. To a lesser extent, vocal fry and creaky voice were frequently said to be appropriate in jazz, musical theater, gospel, folk, and world music, but very few participants found their use appropriate in opera or classical music.

Interestingly, we discovered a significant correlation between the participants' own use of creaky voice and/or vocal fry, on the one hand, and their attitudes toward their use in different styles, on the other. To test this relationship, analysis of variance models were created using the stats package in R (R Core Team 2021). The results showed that an individual's use of vocal fry (frequently/occasionally versus never/rarely) in singing practice ($p = 0.0343$) and in teaching ($p = 0.0162$) is significantly correlated with the number of styles in which they find vocal fry to be acceptable. That is, if the participants practice with and teach vocal fry, they are significantly more likely to evaluate it as appropriate across

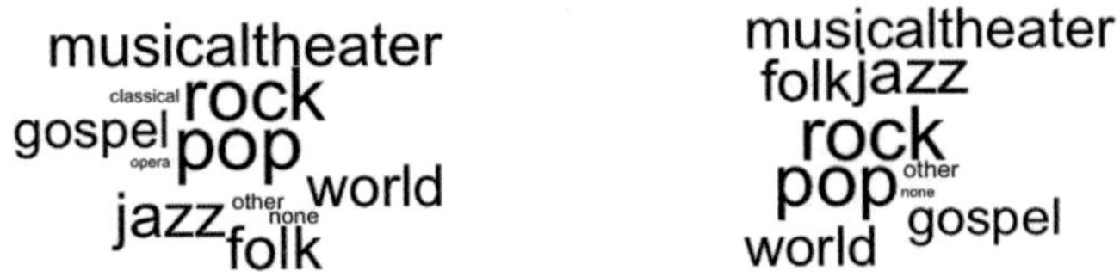

Figure 2.5 Word clouds (made with the wordcloud package in R) showing the styles in which vocal fry (left) and creaky voice (right) are considered most appropriate.

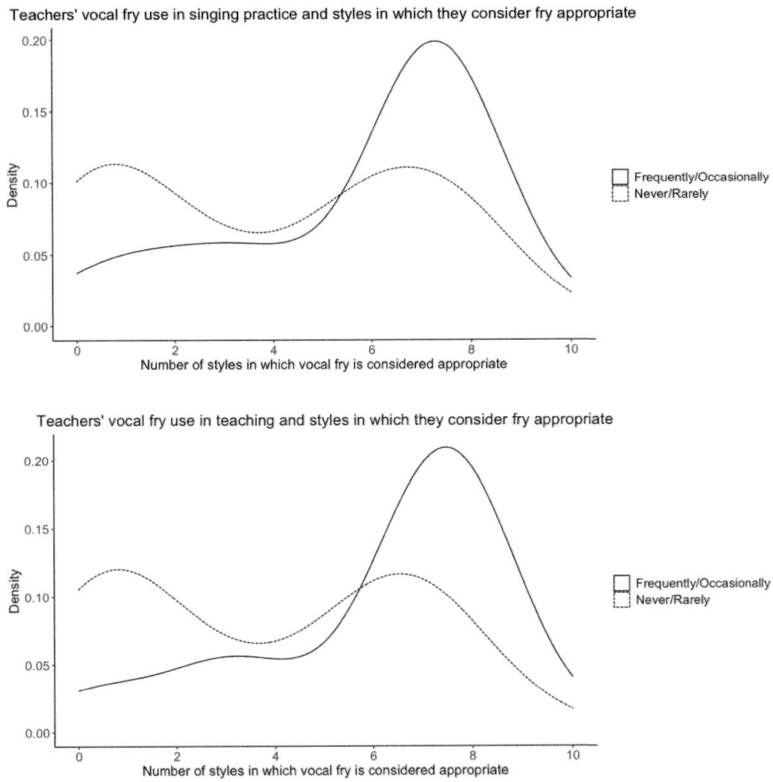

Figure 2.6 Density plots showing the relationship between participants' use of vocal fry in their singing practice (top) and teaching (bottom) and the number of styles in which they found vocal fry to be appropriate.

more styles. Figure 2.6 provides a density plot, or smoothed histogram, to illustrate the distribution of the data. The X-axis shows the number of styles in which the participants found vocal fry and creaky voice to be appropriate (e.g., rock and pop = 2), and the Y-axis shows the probability density, or the probability per unit on the X-axis. The much larger peaks between 6 and 9 styles for teachers who use vocal fry frequently or occasionally demonstrate that these teachers more readily accept the use of vocal fry as appropriate across numerous styles.

The next open-ended question sought to determine how vocal fry is used as an expressive device in singing. Unlike our previous work (Chappell, Nix, and Parrott 2018), in which participants listened to and evaluated singers' voices with and without fry, here we asked singing teachers how fry is used in expression—examining the question of what the fry means from the perspective of the person producing the sounds (or training others to produce the sounds) rather than the listener. Here are several representative responses:

> "In pop singing, it gives a sense of 'sexiness' that subconsciously makes the listener feel more intimate with or closer to the singer" (Participant 83, man, 37 years old, tenor).
>
> "It conveys emotion in a way that other sounds cannot. It's hard to put into words what emotions those might be...could be sadness, anger, frustration, despair. It's a human sound" (Participant 80, woman, 48 years old, soprano).
>
> "Rock singers use it to give the voice an edge for the style. Some world music vocalists use vocal fry as a part of the style" (Participant 78, woman, 25 years old, soprano).
>
> "Commonly, it can be used with onsets in popular styles. Emotionally, it can represent many different things: longing, tiredness, sexiness, etc." (Participant 76, woman, 33, mezzo).
>
> "Particularly used to denote emotional distress, especially when something is difficult to express. It is also stylistically appropriate (almost a requirement) for several singing styles" (Participant 74, woman, 48 years old, mezzo).
>
> "I hear it as a moment of unspeakable emotion" (Participant 70, woman, 64 years old, mezzo).
>
> "As a marker of indecision" (Participant 52, man, 46 years old, bass).
>
> "At the beginning or ends of quiet phrases; sometimes it seems to be used to signal intense intimate emotion, or a holding-back of emotion" (Participant 46, woman, 46 years old, mezzo).
>
> "It conveys heightened emotions to me, specifically frustration, heartache, sadness, relaxation, and a "too-cool-for-school" attitude" (Participant 29, man, 32 years old, tenor).

> "It seems to be used to lend an element of 'authenticity through struggle'" (Participant 17, man, 43 years old, baritone).

As a follow-up, we asked the same question but with respect to creaky voice. There were a limited number of answers, many of them expressing identical sentiments as that found in the earlier responses about vocal fry. There were also these:

> "Is it different from fry?" (Participant 22, woman, 72 years old, soprano).
>
> "Same as fry" (Participant 19, woman, 42 years old, mezzo).
>
> "Seems to indicate a kind of 'false modesty angst'" (Participant 17, man, 43 years old, baritone).
>
> "I think many of the qualities it lends are similar to those that vocal fry lends. It could also give a sense of pain or effort, which is a sought-after aesthetic in some styles" (Participant 15, woman, no age provided, soprano).

Vocal health and general attitudes toward vocal fry and creaky voice

The final open-ended question asked respondents for any additional comments they may have about M0, vocal fry, pulse register, *Strohbass*, creaky voice, glottal fry, or fry. Several mentioned vocal health here:

> "I have always thought it was very unhealthy for the voice" (Participant 85, non-binary, 72 years old, soprano).
>
> "A lot of voice teachers and even SLPs seem to consider fry unhealthy. While I do believe that it doesn't project particularly well, I don't think it's inherently bad for the voice" (Participant 33, woman, 42 years old, soprano).

Others discussed the diversity of opinion on the health ramifications of using vocal fry, including some who expressed they felt vocal fry tended to get an unjust reputation of being unhealthy:

> "I think vocal fry can easily be overused but also feel it gets a bad reputation. I believe it should be applied to the specifics of the singer

and there is no 'one size fits all' with its use. Humans naturally use vocal fry in real life situations when wanting to be held, attract a partner, or are tired mostly. So, it makes sense to use it effectively to portray these emotions in song" (Participant 83, man, 37 years old, tenor).

"It is infrequently used and can be an extremely useful tool both in specific technical areas and also with emotional expressiveness. It has been categorized too often as possibly harmful when it can be utilized to great effect in all genres and styles of singing" (Participant 74, woman, 48 years old, mezzo).

"It would be nice to know if these vocalizations are damaging. For a while they were thought to be, but I have also heard that they aren't. Maybe that means overuse is the issue? As a classically trained singer, I really have no use for these vocal sounds, but pragmatically realize that some singing styles incorporate them. I try to give my students better emotional vocalizations to express themselves, as I believe that vocal inflection can evoke a better result. ..." (Participant 44, woman, 64 years old, soprano).

"Most of the time I dislike fry as a musical element; consider it lazy, and widely over-used by folk/pop singers with inadequate training, especially poor use of breath. More than anything, to me it indicates lack of emotional nuance in the voice, i.e. inability to show emotional range any other way. However, it is used so much that I must accept that younger audiences hear it AS emotional expression, or otherwise find it appealing!" (Participant 46, woman, 66 years old, mezzo).

A few teachers took the middle ground, supporting judicious and intentional usage of fry, creaky voice, etc., in singing:

"I do not avoid fry, but I do make sure my students are aware of when they are using it so that they can make a conscious choice about whether to use it or not use it. I think it can be an appropriate device for many CCM genres" (Participant 80, woman, 48 years old, soprano).

"Underrated and a highly expressive vocal ornament when used sparingly" (Participant 18, woman, 45 years old, soprano).

"To me, they are teaching tools to help solve adduction problems, to add overtones, to add richness to higher pitches, and to help regulate air flow. All these techniques require that the vocal folds adduct fully, both folds coming to the mid-line, so they are often used for singers who have vocal paresis. Too much use of them, however, can result in folds that are too adducted and possibly too stiff/thick" (Participant 30, woman, 73 years old, soprano).

Finally, there were the staunch advocates:

"I love it. It's one of the things that gets the best results with my students and that they immediately feel the difference" (Participant 39, woman, 38 years old, soprano).

Discussion and conclusion

The results of our survey demonstrate that singing teachers exhibit a range of attitudes toward vocal fry. However, several broad generalizations can be drawn. First, the participants associated vocal fry/creaky voice with a range of attributes, including low or indeterminate pitch, low airflow, low breath pressure, and the perception of individual clicks or impulses. Additionally, they were more likely to find vocal fry/creaky voice as separate from other registers than as blending with other registers. This stands somewhat in contrast to the statements of legendary 20th-century pedagogues William Vennard (1967), Oren Brown (1996), and Richard Miller (1986) cited above, who mention the use of vocal fry as a means to extend the low range of male voices. However, unlike Richard Miller and Cornelius Reid (1983), teachers now report using vocal fry and creaky voice to address not only technical issues in singers of all genders and identities, but also to meet the expressive and stylistic elements of musical genres outside of the traditional Western Classical canon. This broader use of vocal fry and creaky voice may be a reflection of larger trends in the field of singing pedagogy, which include the crossing over of intervention strategies from the voice therapy room into the singing studio (and vice versa—see Scearce 2016: 3–12 for a discussion) and the development of style-specific pedagogies based on the functional needs of singers in each genre (Scearce 2016: 157–166; see also the *So You Want To Sing* series [Henderson and Hoch 2014–2022]).

Second, unlike voice scientists, who often distinguish between the low fundamental frequency of vocal fry and creaky voice, which can occur at higher frequencies (Scherer 1989; Titze 2000), most singing teachers do not differentiate between the two. The majority of participants either found the terms to be synonymous or were unsure of the difference between them (66/87). As noted in the literature review, the conflation of these concepts is common among practitioners and researchers from a range of fields; see for example where voice scientists Proctor, Scherer, and Perrine (2024) and linguist Davidson (2020) described identical phonation types with different terms (fry and creaky voice, respectively). It seems unsurprising that singing teachers, whose training is not yet standardized, may or may not have had coursework in voice science, speech pathology, or linguistics, and work in an artistic and perceptually oriented discipline (rather than one based on quantifiable measurements), bear out this confusion/ conflation of terminology. The participants most likely to distinguish between vocal fry and creaky voice came from similar educational backgrounds: speech-language pathology (3/3), voice performance and pedagogy (9/13), and music education (2/4). In other words, those with an emphasis on education tended to differentiate between vocal fry and creaky voice more than performance-oriented participants.

In terms of their use, participants across backgrounds largely agreed on the styles in which creaky voice and vocal fry are most appropriate: pop and rock. Opera and classical were nearly categorically evaluated as inappropriate styles for their use, with musical theater, jazz, folk, gospel, and world music falling between the two poles. When used in styles deemed appropriate, respondents explained that vocal fry can convey authenticity, sexiness or edge, and heightened emotions such as frustration, heartache, sadness, relaxation, anger, and despair, among others. These written observations align with the quantitative results of previous research (Chappell, Nix, and Parrott 2018), which showed that vocal fry in singing increased evaluations of features associated with emotional earnestness and youth. While that study found a difference between the vocal fry evaluations of more musically trained participants and other respondents, the comments of singing teachers in this survey, which bolster the quantitative findings, highlight that this distinction is one of degree, not quality.

Interestingly, the present study found that the participants' attitudes toward the use of vocal fry across styles were significantly conditioned by their own use of fry in their singing practice and teaching. More specifically, the participants who viewed vocal fry as appropriate across the most styles were also the participants most likely to use fry themselves. Similar to Nix, Chappell, and Powell (2019), who found that vocal fry production could be primed, depending on participant attitudes, this study found that, among a musically trained and more homogeneous population, this interplay between perception and production holds true: Self-described vocal fry/creaky voice use is associated with the attitudes of singing teachers toward the phonation type.

Finally, the use of vocal fry and creaky voice in singing practice and teaching appears to be growing to some extent in popularity, based on the results of our survey. Our participants claimed to use vocal fry and creaky voice more than they were exposed to the phonation types in their training, suggesting some changing attitudes toward these vocalizations, especially in particular musical genres. Viewed as an expressive vocal ornament conveying emotion, longing, or distress, participants noted fry and creak voice can add, as Participant 17 in our survey encapsulated so well, "authenticity through struggle" to a performance, especially in popular genres. As these styles comprise a much larger share of the performing arts environment than they did a generation ago, it would appear the acceptance of vocal fry and creaky voice by the teaching community may be in part driven by market forces: Singers are using these stylistic elements, and it has become incumbent on the teaching profession to both understand how to teach their optimal production and their appropriate use in performance situations. Some of the change in attitudes may also be due to greater understanding of the physiology of fry and creaky voice production by the voice-teaching community. Behaviors once considered harmful to vocal health have been studied by voice scientists, laryngologists, as well as speech pathologists, and reconsidered as acceptable methods for the habilitation of voice function. This information has been shared with the voice-teaching community, and, as was seen above in the final open-ended question of our survey, some of the older admonishments about the use of fry and creaky voice appear to be fading away. As the famous song by Danny and the Juniors said about rock and roll, it seems vocal fry and creaky voice are also here to stay (White 1958).

Acknowledgments

A portion of the singing voice literature review was adapted with permission from "Application of Vocal Fry to the Training of Singers," which appeared in *The Journal of Singing* **62**, no. 1 (Sept/Oct 2005), 53–59. The authors thank *The Journal of Singing* (Richard Sjoerdsma, editor), Ingo Titze, and Kate Emerich for their kind permission.

References

Abdelli-Beruh, N. B., Wolk, L., and Slavin, D. (2014). "Prevalence of vocal fry in young adult male American English speakers." *Journal of Voice*, **28**(2): 185–190.

Anderson, R. C., Klofstad, C. A., Mayew, W. J., and Venkatachalam, M. (2014). "Vocal fry may undermine the success of young women in the labor market." *PLoS ONE*, **9**: e97506.

Behlau, M., Machado, L., Guedes, Z., Pontes, P., and Pontes, A. (1998). "Using vocal fry to treat nasality problems" (Abstract). *Phonoscope*, **1**: 4.

Blomgren, M., Chen, Y., Ng, M., and Gilbert, H. (1998). "Acoustic, aerodynamic, physiologic, and perceptual properties of modal and vocal fry registers." *Journal of the Acoustical Society of America*, **103**(5, part 1): 2649.

Brown, O. (1996). *Discover Your Voice: How to Develop Healthy Voice Habits*. San Diego: Singular Publishing.

Buescher, R. (2002). "Postoperative posture memory rehabilitation using speech level singing exercises and balanced onsets." *Journal of Singing*, **58**(3): 223–228.

Cantor-Cutiva, L. C., Jiménez-Chala, E. A., Bottalico, P., and Hunter, E. J. (2021). "Bilingualism and voice production. Differences between bilingual Latin-American Spanish-English female speakers and monolingual Spanish female speakers during Spanish productions." *Journal of Voice*, **37**(5): 716–721.

Chappell, W., Nix, J., and Parrott, M. (2018). "Social and stylistic correlates of vocal fry in a cappella performances." *Journal of Voice*, **34**(1): 156. e5–e13.

Dallaston, K. and Docherty, G. (2020). "The quantitative prevalence of creaky voice (vocal fry) in varieties of English: A systematic review of the literature." *PLoS ONE*, **15**(3): e0229960.

Davidson, L. (2020). "Contributions of modal and creaky voice to the perception of habitual pitch." *Language*, **96**(1): e22–e37.

Doscher, B. (1994). *The Functional Unity of the Singing Voice*, 2nd edn. Metuchen, NJ: Scarecrow.

Emerich, K. A. (2003). "Nontraditional tools helpful in the treatment of certain types of voice disturbances." *Current Opinion in Otolaryngology and Head and Neck Surgery*, **11**: 149–153.

Fairbanks, G. (1960). *Voice and Articulation Drillbook*. New York: Harper & Row, 124–139.

Fellows, I. (2018). *Word Clouds*. Retrieved from https://CRAN.R-project.org/package=wordcloud.

Gerratt, B. R. and Kreiman, J. (2001). "Towards a taxonomy of nonmodal phonation." *Journal of Phonetics*, **29**: 365–381.

Gibson, T. (2017). "The role of lexical stress on the use of vocal fry in young adult female speakers." *Journal of Voice*, **31**(1): 62–66.

Henderson, A. and Hoch, M. (Eds.) (2014–2022). *So You Want To Sing* (series, 21 vols.). Lanham, MD: Rowman and Littlefield.

Hollien H. and Wendahl, R. W. (1968). "Perceptual study of vocal fry." *Journal of the Acoustical Society of America*, **43**: 509.

Hollien, H. and Michel, J. F. (1968). "Vocal fry as a phonational register." *Journal of Speech and Hearing Research*, **11**(3): 600–604.

Hollien, H., Damsté, H., and Murry, T. (1969). "Vocal fold length during vocal fry phonation." *Folia Phoniatrica*, **21**: 264.

Kuang, J. and Liberman, M. (2016). "The effect of vocal fry on pitch perception." Paper presented at the *International Conference on Acoustics, Speech, and Signal Processing* (Shanghai, China), March 25. Retrieved from https://ieeexplore.ieee.org/document/7472681.

Lefkowitz, D. and Sicoli, M. (2017). "Creaky voice: Constructions of gender and authority in American English conversation." Lecture presented at the *106th Meeting of the American Anthropology Association* (Washington, DC), November 28.

Leyerle, W. (1986). *Vocal Development Through Organic Imagery*, 2nd edn. Mount Morris, NY: Leyerle Publications.

Ligon, C., Rountrey, C., Rank, N. V., Hull, M., and Khidr, A. (2018). "Perceived desirability of vocal fry among female speech communication disorders graduate students." *Journal of Voice*, **33**(5): 805.e21–e35.

McGlone, R. E. and Shipp, T. (1971). "Some physiologic correlates of vocal fry phonation." *Journal of Speech and Hearing Research*, **14**: 769–775.

Mendoza-Denton, N. (2011). "The semiotic hitchhiker's guide to creaky voice: Circulation and gendered hardcore in a Chicana/o gang persona." *Journal of Linguistic Anthropology*, **21**: 261–280.

Miller, D. G. (2008). *Resonance in Singing: Voice Building through Acoustic Feedback*. Gahanna, OH: Inside View.

Miller, D. G., Sulter, A., Schutte, H., and Wolf, R. (1997). "Comparison of vocal tract formants in singing and nonperiodic phonation." *Journal of Voice*, **11**(1): 1–11.

Miller, R. (1986). *The Structure of Singing*. New York: Schirmer.

Murry, T. and Brown Jr., W. S. (1971). "Subglottal air pressure during two types of vocal activity: Vocal fry and modal phonation." *Folia Phoniatrica*, **23**: 448.

Nix, J., Chappell, W., and Powell, M. (2019). "The relationship of vocal fry in a speech task with listener ratings of popular vocal music performances." Paper presented at the *48th Annual Symposium on the Care of the Professional Voice* (Philadelphia, PA), June 2. Retrieved from https://voicefoundation.org/wp-content/uploads/2019/05/2019-Abstracts_wTitlePage.pdf, 164.

Nix, J., Emerich, K., and Titze, I. R. (2005). "Application of vocal fry to the training of singers." *Journal of Singing*, **62**(1): 53–59.

Paolillo, N. P., Carrozza, L., Oslo, M., Rosa, E., and Scalabrin, M. (2019). "Inspiratory vocal fry: Anatomical and physiological aspects, application in speech therapy, vocal pedagogy and singing. A pilot study." *Journal of Voice*, **35**(3): 394–399.

Podesva, R. J. (2013). "Gender and the social meaning of non-modal phonation types." In *Proceedings of the 37th Annual Meeting of the Berkeley Linguistic Society*, 427–448.

Proctor, K., Scherer, R. C., and Perrine, B. L. (2024). "Vocal fry patterns while reading." *Journal of Voice*, **38**(4): 889–902.

R Core Team. (2021). *R: A Language and Environment for Statistical Computing*. Vienna, Austria: R Foundation for Statistical Computing. Retrieved from www.R-project.org.

Ragan, K. (2020). *A Systematic Approach to Voice: The Art of Studio Application*. San Diego: Plural Publishing.

Reid, C. (1983). *A Dictionary of Vocal Terminology: An Analysis*. New York: Joseph Patelson.

Riggs, S. (1998). *Singing for the Stars: A Complete Program for Training Your Voice*. Van Nuys, CA: Alfred Publishing.

Scearce, L. (2016). *Manual of Singing Voice Rehabilitation: A Practical Approach to Vocal Health and Wellness*. San Diego: Plural Publishing.

Scherer, R. (1989). "Physiology of creaky voice and vocal fry" (Abstract). *Journal of the Acoustical Society of America*, **86** (Supp. 1): S25.

Stark, J. (1999). *Bel Canto: A History of Vocal Pedagogy*. Toronto: University of Toronto Press.

Swanson, F. (1984). "Changing voices: Don't leave out the boys." *Music Educators Journal*, **70**(5): 47–50.

Titze, I. R. (2000). *Principles of Voice Production*. Iowa City: National Center for Voice and Speech.

Titze, I. R. and Verdolini-Abbott, K. (2012). *Vocology: The Science and Practice of Voice Habilitation*. Salt Lake City, UT: NCVS.

Venkatraman, A. and Sivasankar, M. P. (2018). "Continuous vocal fry simulated in laboratory subjects: A preliminary report on voice production and listener ratings." *American Journal of Speech-Language Pathology*, **27**: 1539–1545.

Vennard, W. (1967). *Singing: The Mechanism and the Technic*, rev. ed. New York: Carl Fischer.

White, D. (1958). "Rock and roll is here to stay" [sound recording], performed by Danny and the Juniors. New York: ABC-Paramount Records, recording number 45-9888.

Wickham, H. (2013). *ggplot2: Elegant Graphics for Data Analysis*. Retrieved from https://CRAN.R-project.org/package=ggplot2.

Wolk, L., Abdelli-Beruh, N. B., and Slavin, D. (2011). "Habitual use of vocal fry in young adult female speakers." *Journal of Voice*, **26**(3): e111–e116.

Yuasa, I. P. (2010). "Creaky voice: A new feminine voice quality for young urban-oriented upwardly mobile American women." *American Speech*, **85**: 315–337.

Chapter Three

Voice creakiness in singing

Alexsandro R. Meireles

Introduction

The sensation of creakiness in voice production is usually described in voice literature by names such as creaky voice, fry, vocal fry, creak, laryngealization, and pulse phonation. Most of the scientific studies related to this topic have been devoted to speech analysis. Little is known about the use of this voice effect in singing. This chapter introduces differences and similarities regarding voice creakiness in speech and then applies this knowledge to the study of singing. The novelty of this study is the imitation of voice creakiness of famous hard rock hits with subsequent acoustic analysis. After recording these imitated voices, I present a detailed acoustic analysis of creaky voice singing, which is based on a previous literature description of this voice phenomenon. The proposal in this chapter is that we need

Creak: Theories and practices of pulse phonation
Edited by Francesco Venturi

ISBN 978-981-5129-29-8 (Hardcover), 978-1-003-61606-1 (eBook)
www.jennystanford.com

to consider "vocal fry" and "creaky voice" as two separate voice phenomena, in order to better describe the types of voices found in singing, which can be used either as vocal effects or as regular singing.

Distinction between vocal fry and creaky voice

Although some authors do not differentiate vocal fry (or creak) from creaky voice in speech and singing (Garellek and Keating 2011; Drugman *et al.* 2014; Miller 2000; Schutte and Miller 1993), the distinction between these two terms is more useful for a deeper comprehension and analysis of this phenomenon in singing.

Laver (1980) gives a scientific definition of creaky voice and vocal fry based on the work of Catford (1964). My interpretation of these voice types is derived from their work. Laver (1980) states that "creak is also called vocal fry or glottal fry in the phonetic literature, particularly by American researchers" (1964: 122), and Catford (1964) describes vocal fry as the "low frequency (down to about 40 cps) periodic vibration of a small section of the vocal folds… The auditory effect is of a rapid series of taps, like a stick being run along a railing" (1964: 32). Laver considers creaky voice to be a compound phonation type formed by the mixture of creak with modal voice and cites Ladefoged (1964) who gives a detailed description of how the larynx behaves in creaky voice as follows: "in this state of the glottis there is a great deal of tension in the intrinsic laryngeal musculature, and the vocal cords no longer vibrate as a whole. The ligamental and arytenoid parts of the vocal cords vibrate separately" (1964: 16). According to Laver, "the creak component in this description would be attributable to the arytenoid location, rather than to the ligamental section" (1980: 139).

The definitions of creaky voice and vocal fry (or creak) in singing are usually derived from the ones found in speech literature. Nevertheless, the Complete Vocal Technique method, developed by Cathrine Sadolin (2021), uses a different terminology to refer to vocal fry and creaky voice. As can be noticed in their website,[1] creak sounds like vocal fry, and creaking, like creaky voice.

[1]https://completevocalinstitute.com

Although they describe these two terms together, the examples they give afterward in the website give us the impression that they consider these two terms to be different. In their words, "Creaks are often used in the beginning or ending of a phrase, often while holding the microphone close to the mouth," and "creaking is when the voice is creaking during singing. It is often used as a delicate version of 'distortion'" (see their website). They define creak/creaking in the following manner:

> Creak and creaking takes place at the vocal cord level, level 1 (see 'Various levels of the vocal tract'). The very delicate creaking sound occurs from vibrations of the vocal cords. The vocal cords vibrations in both creak and creaking have a complex but regular vibratory pattern.
>
> Creaks and Creaking occur when using two slightly opposite ways of using the voice (e.g. with a slightly incorrect use of the modes) in a very controlled manner. Creaks and creaking is performed at a rather low volume, like a consciously failed attempt to sing in a certain amount of metal.

From the acoustic viewpoint, Titze (1994) contrasts vocal fry (pulsed phonation) with creaky voice as follows:

> Pulsed Phonation: Phonation in which temporal gaps are perceived; acoustically, energy 'packets' are perceived below about 70 Hz, where formant energy effectively dies out prior to re-excitation with a new glottal pulse; pulsed phonation or pulse register is also called vocal fry, apparently because of its similarity with popping sounds that are emitted from a hot frying pan.
>
> Creaky Voice: A voice that sounds like a creaking door, like two hard surfaces rubbing against each other; acoustically, a complex pattern of subharmonics and modulations is observed that reflect a complexity of modes of vibration of the vocal folds.

In this paper, Titze also comments that "creaky voice becomes vocal fry (or pulse register) when a subharmonic frequency dominates (perceptually) and is below about 70 Hz" (1994: 7). Recently, Caffier *et al.* (2018) also distinguish these two voice qualities but through physiological evidence. The sound impression of creaky voice—or high vocal fry—is like "a door creaking on its hinges" (2018: 340). They also state that vocal fry and creaky voice are complementary as follows:

> In creaky voice, a pressure with medial compression is created in the arytenoid cartilage area. The vocal folds are tensed, and the effect happens in the glottis. The EGG shows smaller amplitudes, an irregular waveform, and an irregularity of the vocal fold vibration which is perceived as roughness. The closing phase is fast, the opening phase is short, and the oscillation cycles vary in duration.
>
> Vocal fry is a further effect of creaky voice with a comparable mechanism: In contrast to the creaky voice, the vocal folds are maximally shortened and relaxed. In the transition to tonality, the length and tension of the vocal folds increase. As a correlate of hyperadduction, the EGG shows an extended closed phase of the glottis, a fast closing phase, and a short opening phase. There are large amplitudes and long duration of periods due to the low frequency. The vocal folds oscillate relatively evenly four times per period: once large and three times smaller, like swinging out (2018: 342–343).

Five different types of creakiness in speech

Due to the fact that Keating *et al.* (2015) are working with speech data, they do not distinguish creaky voice from vocal fry. Fry is just one of the five possibilities of voice production categorized as creaky voice in the phonetic literature, according to the authors.

Their acoustic descriptions are going to be very useful for the analyses of the next sections. Their five types of creaky voice are:

1. Vocal fry: "the glottis is constricted and F0 is low, but it is not necessarily irregular. Indeed it is often quite periodic... Its special property is high damping of the pulses – this property, due in part to the low F0, makes individual pulses distinct and separately audible (the 'picket fence' effect). Thus the prototypical low-F0 property is enhanced." (2015: 2)
2. Multiply pulsed voice: "A very common form of creak involves a special kind of F0 irregularity: alternating longer and shorter pulses ... In the case of double pulsing (or

period doubling), there are two simultaneous periodicities; higher multiples are also possible. There are thus multiple F0s, usually one quite low and another about (though not exactly) an octave higher, but the resulting percept is usually of an indeterminate pitch, plus roughness" (2015: 2).

3. Aperiodic voice: "Another variant of F0 irregularity is when it is taken to the extreme—vocal fold vibration is so irregular that there is no periodicity and thus no perceived pitch. .... Like multiply pulsed voice, aperiodic voice lacks the prototypical property of low F0; instead, the property of irregular F0 is enhanced, and the voice is therefore noisy." (2015: 2)
4. Nonconstricted creak: "F0 is low and irregular, as in prototypical creak; but the glottis is spreading, not constricted, and therefore airflow through the glottis is higher, not lower. This kind of creak is attested utterance-finally, with the vocal folds beginning to spread before the utterance is over." (2015: 2–3)
5. Tense/pressed voice: "When the glottis is constricted, but the F0 is neither low nor irregular, a tense or pressed voice quality is heard. While not always considered a form of creaky voice, it can function phonologically as such in languages in which a creaky (or laryngealized) phonation can co-occur with high tone. Here the constricted glottis is criterial" (2015: 3).

Examples of vocal fry and creaky voice

We consider vocal fry and creaky voice to be different vocal processes in singing. The main difference is the lack of notes in vocal fry in comparison to creaky voice. Vocal fry, therefore, has a limited use in singing—it is used, mainly, as a vocal effect. On the other hand, creaky voice can be used more extensively to express different melodies. Some examples of these two varieties of creakiness in singing will be presented, but first we are going to present some canonical examples of vocal fry and creaky voice. In order to exemplify the differences between vocal fry, modal voice, and creaky voice (mix of vocal fry and modal voice), I recorded a vowel [a] with a C arpeggio (C2 (65.41 Hz) - E2 (82.41 Hz) - G2 (98 Hz) - E2 - C2) for each of these voice qualities.

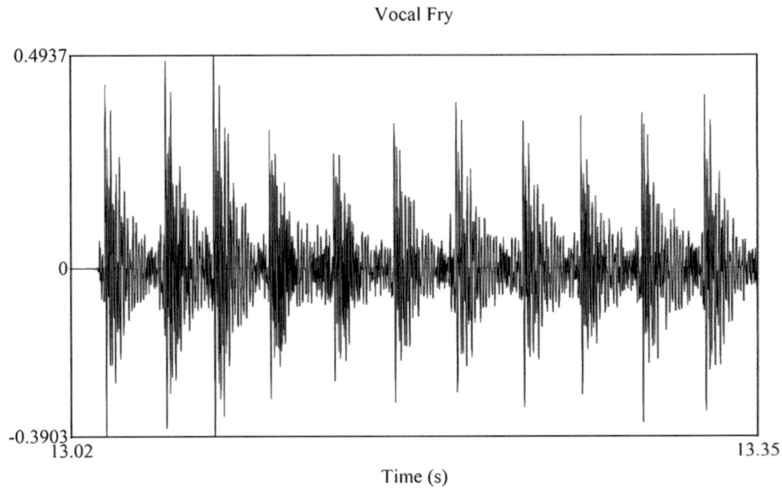

Figure 3.1 Example of pulse phonation. No trackable f0 with Praat's Pitch detector.

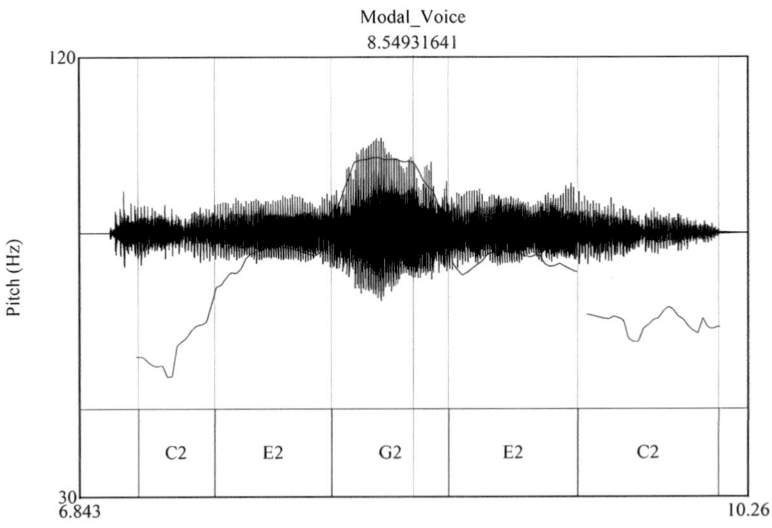

Figure 3.2 Example of modal voice with a clear f0 track (blue line).

Figure 3.1 is a pure vocal fry. As can be seen, no trackable f0 is presented in Praat software (Boersma 2001); therefore, there is no note change. Although it is hard to notice a tune, if we measure the periods (around 0.030s), we have a very low f0 (around 33 Hz). These characteristics are congruent with the definition of vocal fry presented by Keating *et al.* (2015), who consider vocal fry to present low f0, quite periodic pulse, and high damping of the pulses. Next, I recorded the C arpeggio with the modal voice, as can be seen in Fig. 3.2. Unlike vocal fry with no automatic detection of f0, for this voice type, we have a clear f0 trace (blue line in Fig. 3.2), indicating a periodic waveform with variation according to the arpeggio notes. We should take into consideration the fact that some sort of vocal fry occurred before regular phonation if we compare the waveform with the beginning of the f0 contour. We may remember that this frequency is within the frequency range that vocal fry is supposed to occur in speech (below 70 Hz, cf. Titze 1994).

Finally, I recorded the C arpeggio with creaky voice. As we can see in Fig. 3.3, the f0 track (blue line) is not so smooth as for the modal voice. This is, in fact, caused by the irregularities in vocal fold vibration caused by the creak part of the creaky voice. According to Esling *et al.* (2019), this irregularity is caused by supraglottal activity in this voice type. Also, we must take into consideration that the initial C2 also did not have a clear f0 trace. It seems that vocal fry was produced in the beginning of the vocalization as was the case for the modal voice. In addition, if we consider the types of creaky voice proposed by Keating *et al.* (2015), we may consider this voice to be a "multiply pulsed voice."

As can be seen in Fig. 3.4, almost all of their characteristics are present: f0 irregularity, alternation of long and short pulses, and low f0 with another f0 around an octave higher. Nevertheless, the resulting percept is not an indeterminate pitch in the case of singing, as the notes are tracked by Praat's f0 detector.

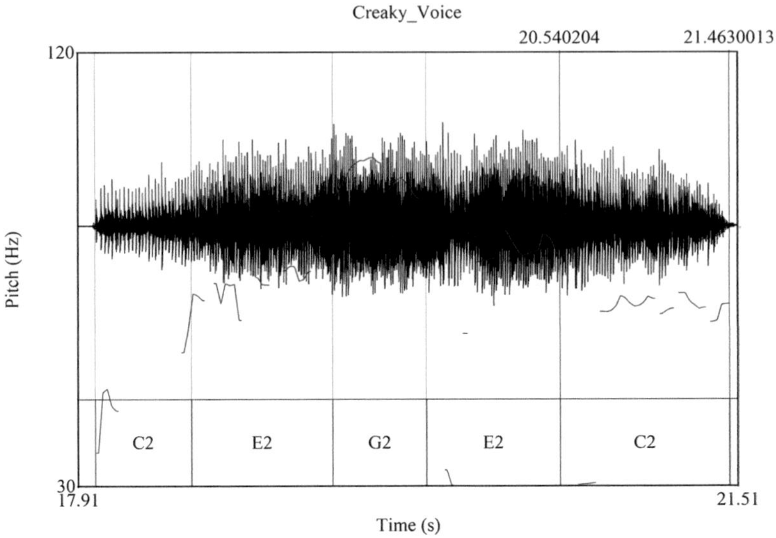

Figure 3.3 Example of a creaky voice with a clear f0 track (blue line).

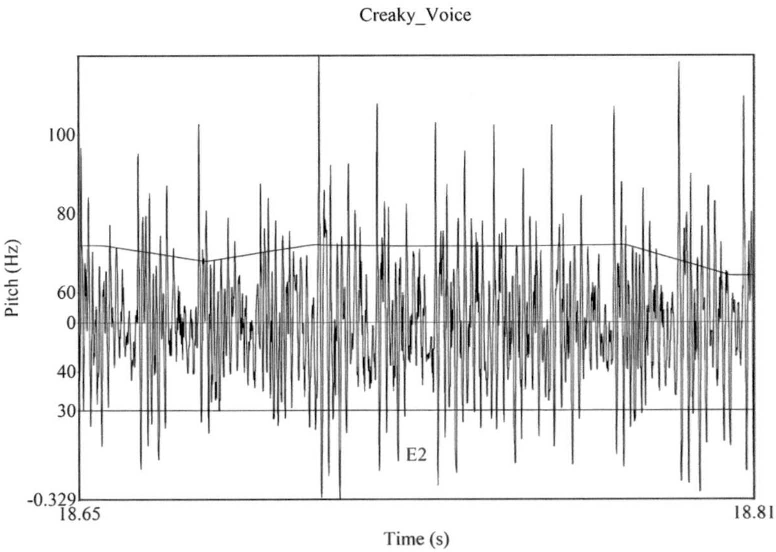

Figure 3.4 Example of a creaky voice with multiple pulses, noted by the alternation of long and short pulses and low f0.

Analysis of creaky voice and vocal fry in singing

Five songs were chosen to represent the use of creaky voice and vocal fry in singing. The references of voices were taken from video clips of some bands, as can be seen in Table 3.1.

The bold words in Table 3.1 were chosen to be recorded by a rock singer—the author of this chapter—trying to mimic the singer's voice as closely as possible. This same methodology was used by Borch *et al.* (2003), so as to have a clear .wav file for acoustic analysis. These representations of creaky voice and vocal fry were selected on the basis of my expertise in studying and singing different voice qualities.

Table 3.1 Song details

Song	Band	Excerpt	YouTube	Singer
1. Janie's got a gun	Aerosmith	Janie's **got a gun**	0:46–0:48	Steven Tyler
2. Lithium	Nirvana	**Yeah, yeah, yeah**	0:37–0:44	Kurt Cobain
3. Born to be my baby	Bon Jovi	We both **got jobs**	0:19–0:20	John Bon Jovi
4. Plush	Stone Temple Pilots	And I **feel it**	0:39–0:42	Scott Weiland
5. I don't want to miss a thing	Aerosmith	**I** could stay awake	0:31–0:31	Steven Tyler

Note: The bold letters represent the excerpts that were analyzed.

Nevertheless, to be more accurate, I asked three speech therapists—who are collaborators in my research with voice quality in singing—to analyze the excerpts and classify the voice qualities as modal, creak (fry), creaky voice, or any other voice quality. They all classified the excerpts from the first four songs as creaky voice and the last one as vocal fry. The voice recordings were made using the free open source software Audacity with a

Blue Yeti microphone (bit rate: 16 bits, polar pattern: cardioid, frequency response: 20 Hz to 20 kHz, maximum sound pressure level: 120 dB). The sample rate of the recording was 44 kHz. After controlling the VU level so as not to have clipped audio, the singer sang the song excerpts through headphones. No audio processing was done afterward.

Figure 3.5 shows the score of the excerpt from song 1 (cf. Table 3.1) that contains the notes F#2 (92.5 Hz), G#2 (103.83 Hz), A#2 (116.54 Hz), A#2 (116.54 Hz), and G#2 (103.83 Hz) sung with modal voice and then with creaky voice (Fig. 3.6). As can be seen, the melodic contours of the notes are similar in both productions, as the notes are presented in both modal voice and creaky voice. The only noticeable difference is the lack of notes in the first syllable for the creaky version, produced with vocal fry instead of creaky voice. Nevertheless, if we take a closer look in the .wav file of the creaky voice (Fig. 3.7), we notice the presence of subharmonics in the creaky voice, which are not present in the modal version of the excerpt.

This creaky voice pattern is similar to our example in the previous section. In order to analyze the differences between the two versions of the song, a spectrum analysis was carried out. Praat's voice report function has shown a consistent difference between the two voice qualities, as expressed in Table 3.2.

For this example, we have analyzed the note A#2 (word "a"). The table shows almost double the frequency variation (jitter), intensity variation (shimmer), as well as more noise added to the signal for the creaky voice in comparison to the modal voice. In order to analyze the differences between the two versions of the song, a spectrum analysis was carried out. Praat's voice report function has shown a consistent difference between the two voice qualities, as expressed in Table 3.2. For this example, we have analyzed the note A#2 (word "a"). The table shows almost double the frequency variation (jitter), intensity variation (shimmer), as well as more noise added to the signal for the creaky voice in comparison to the modal voice.

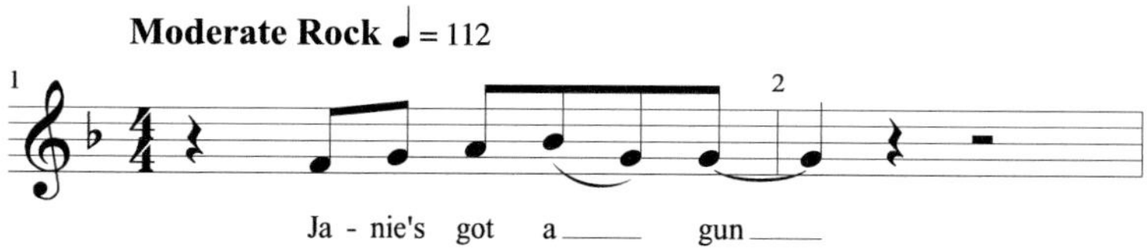

Figure 3.5 Notes of the song *Janie's Got a Gun* by Aerosmith.

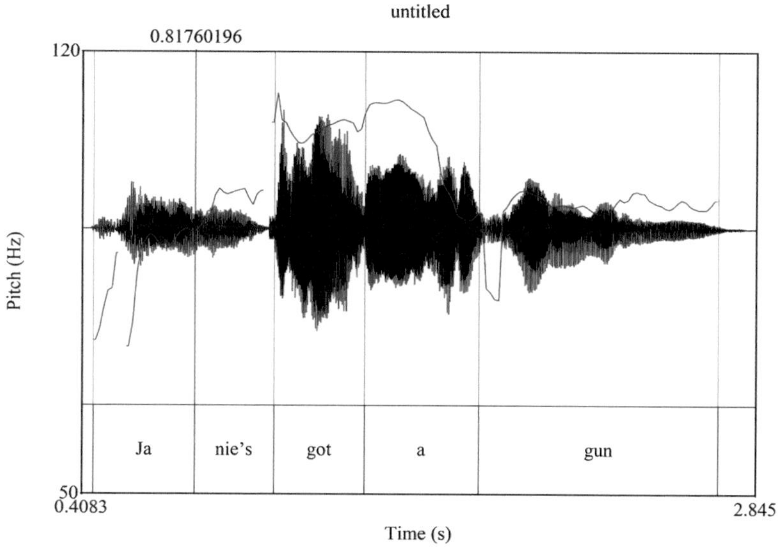

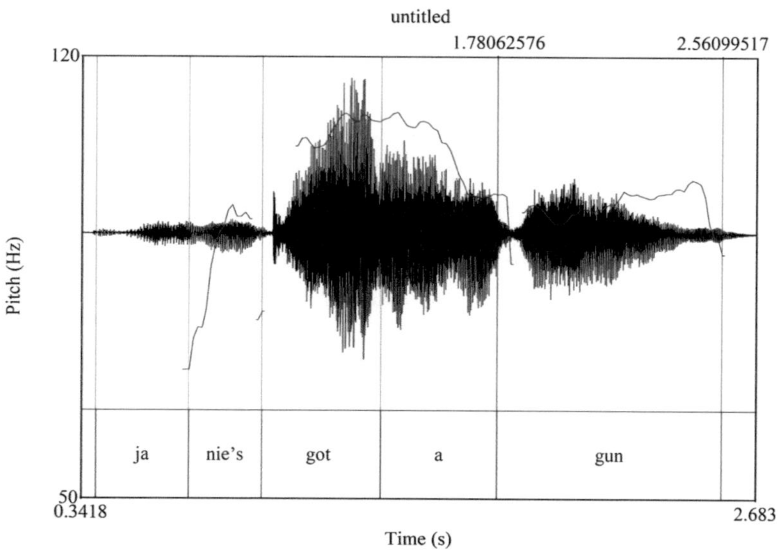

Figure 3.6 Waveform and f0 track (blue) of *Janie's Got a Gun* sung in modal voice (top) and creaky voice (bottom).

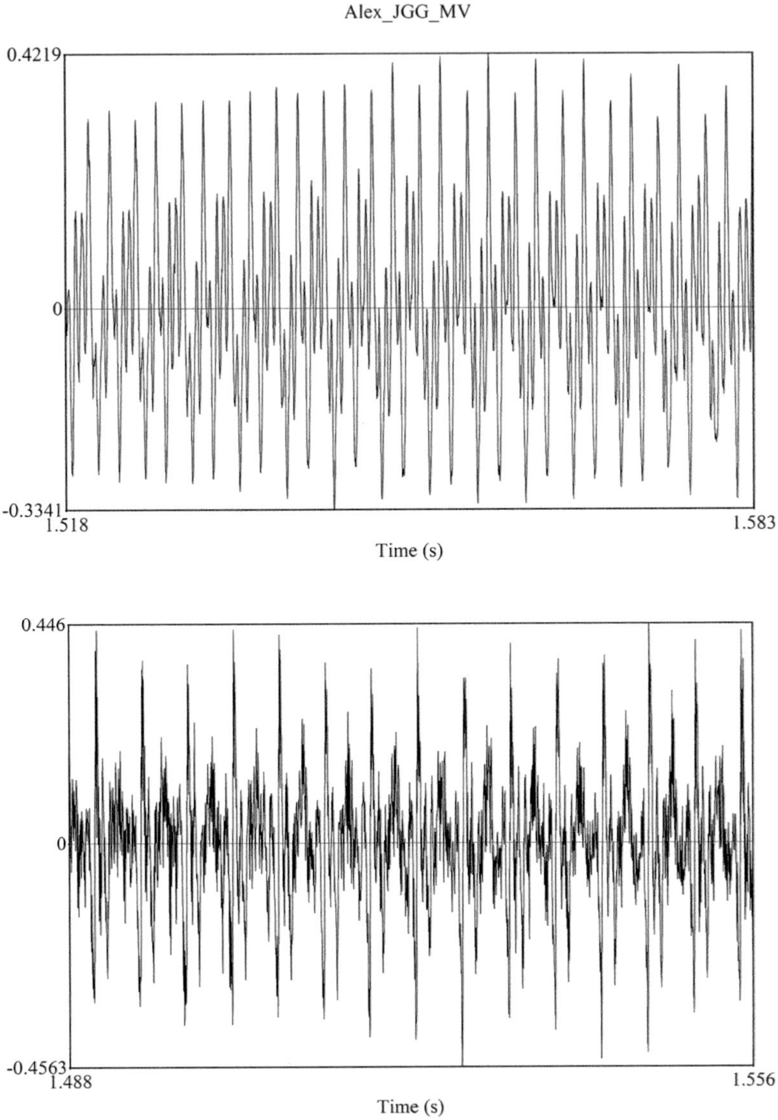

Figure 3.7 Excerpt from the modal voice (top) and creaky voice (bottom).

Figure 3.8 Notes of the song *Lithium* by Nirvana (version in Eb minor, usually played live, transposed a half-step higher than the album version).

Table 3.2 Some spectral differences between modal voice and creaky voice for song 1

Janie's Got a Gun	Jitter	Shimmer	Harmonics-to-Noise Ratio
Modal voice	0.186%	4.611%	15.040 dB
Creaky voice	0.370%	8.742%	7.042 dB

Figure 3.8 shows the score of the excerpt from song two that contains the notes Gb4 (369.99 Hz), F4 (349.23 Hz), Eb4 (311.13 Hz), Db4 (277.18 Hz), and Eb4 (311.13 Hz) sang with modal voice and creaky voice (Fig. 3.9). As for song one, the melodic contour of the notes is similar in both productions, as the notes are presented in both modal voice and creaky voice. As can be noticed in Fig. 3.9., the scale for modal voice is 75 to 500 Hz and for creaky voice is 200 to 400 Hz. As there was greater noise in this high-frequency region for the male voice, the subharmonics messed the f0 detection up, displaying notes that were not intended for the sung parts. In turn, we had to adjust the scale for the higher f0 track to show up. The same creaky voice with multiple pulses was also found here (Fig. 3.10). Moreover, these higher frequencies presented much more irregularities in the spectral parameters than in the lower frequencies of song one, as can be seen in Table 3.3.

Table 3.3 Some spectral differences between modal voice and creaky voice for song two

Lithium	Jitter	Shimmer	Harmonics-to-Noise Ratio
Modal voice	0.063%	1.238%	29.504 dB
Creaky voice	1.303%	17.760%	5.869 dB

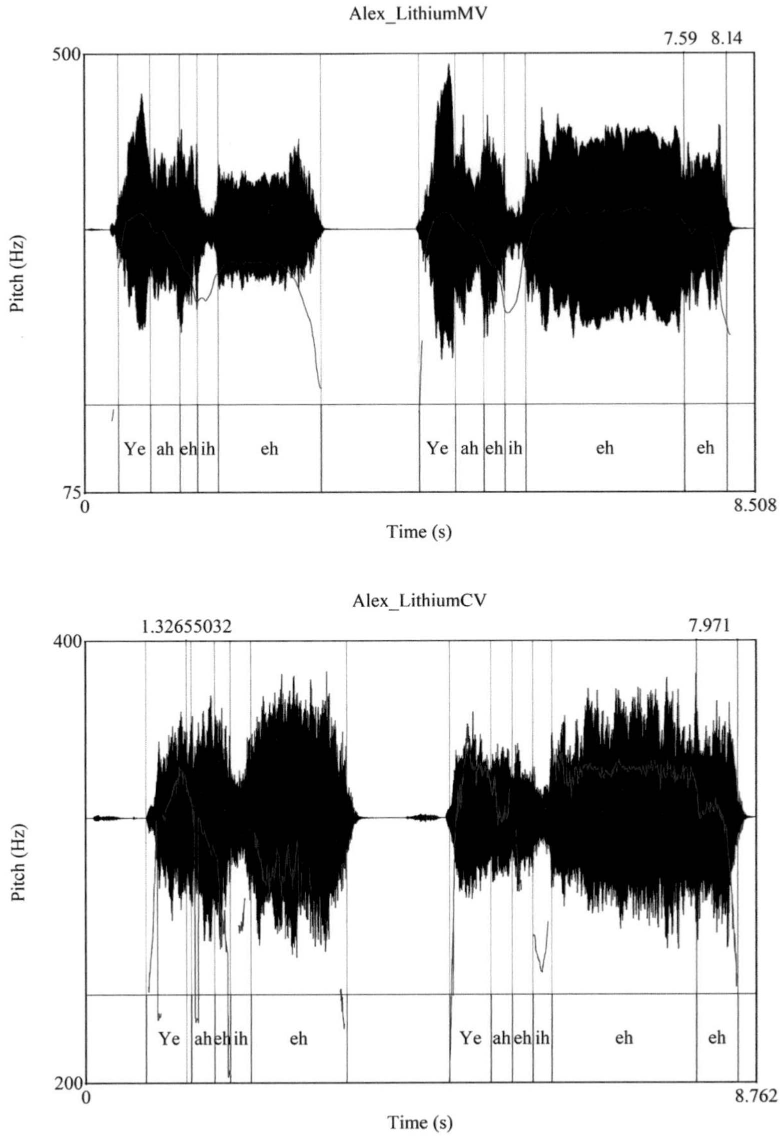

Figure 3.9 Waveform and f0 track (blue) of *Lithium* sung in modal voice (top), and creaky voice (bottom).

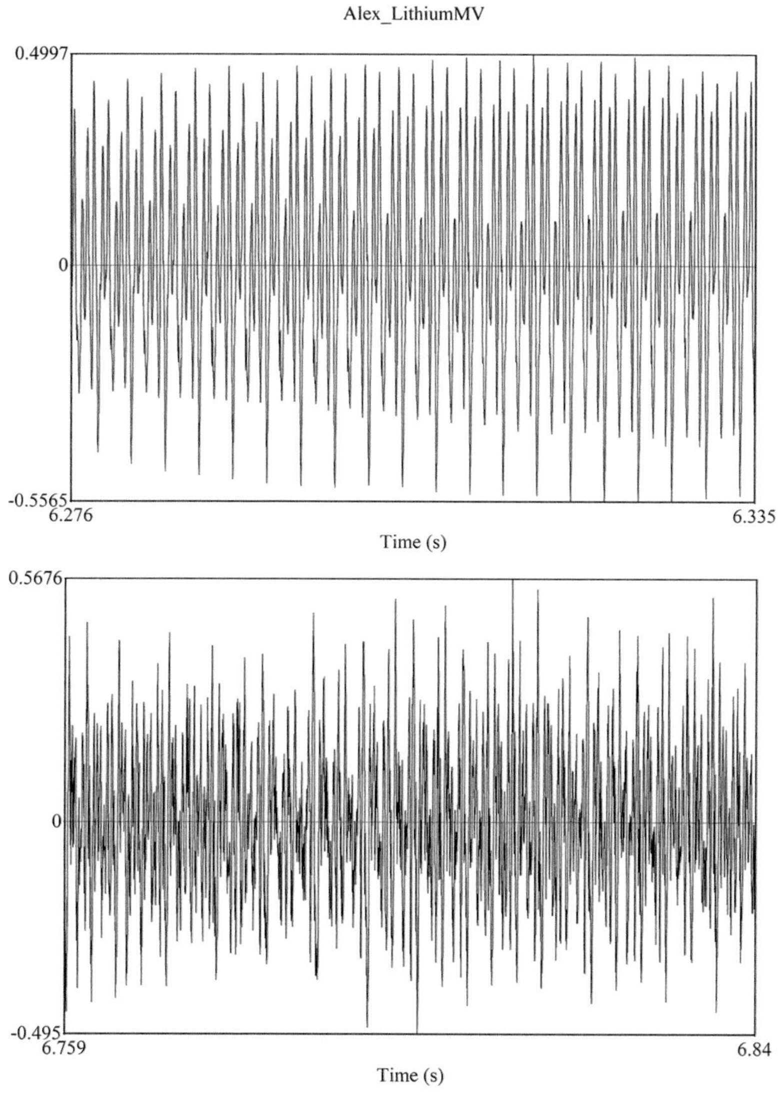

Figure 3.10 Excerpt from the modal voice (top) and creaky voice (bottom) taken from the last note of the excerpt for song two.

Figure 3.11 shows the score of the excerpt from song three that contains the notes F#3 (185 Hz) and A3 (220 Hz) sung with modalvoice and creaky voice (Fig. 3.12). As for songs one and two, the melodic contour of the notes is similar in both productions, as

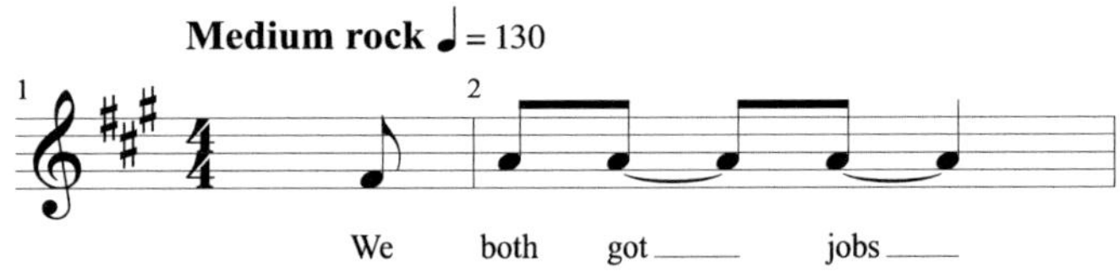

Figure 3.11 Notes of the song *Born to Be My Baby* by Bon Jovi.

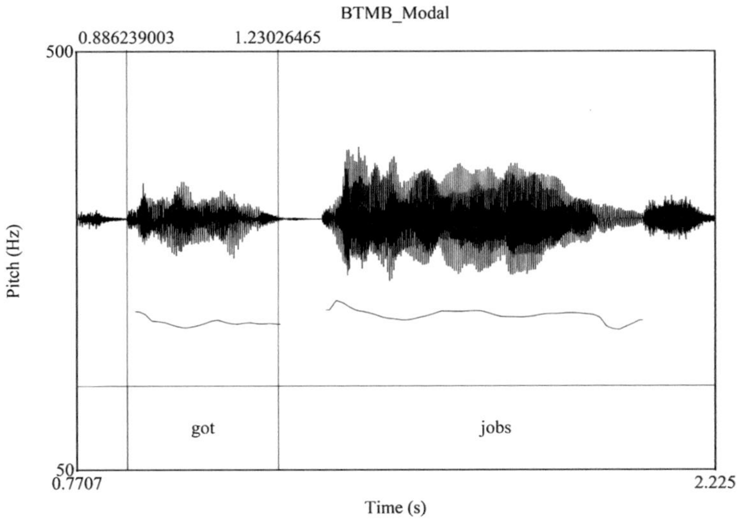

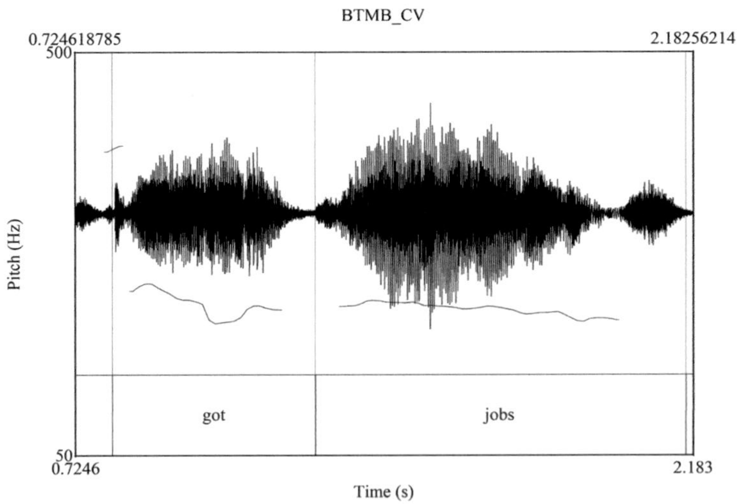

Figure 3.12 Waveform and f0 track (blue) of *Born to Be My Baby* sung in modal voice (top) and creaky voice (bottom).

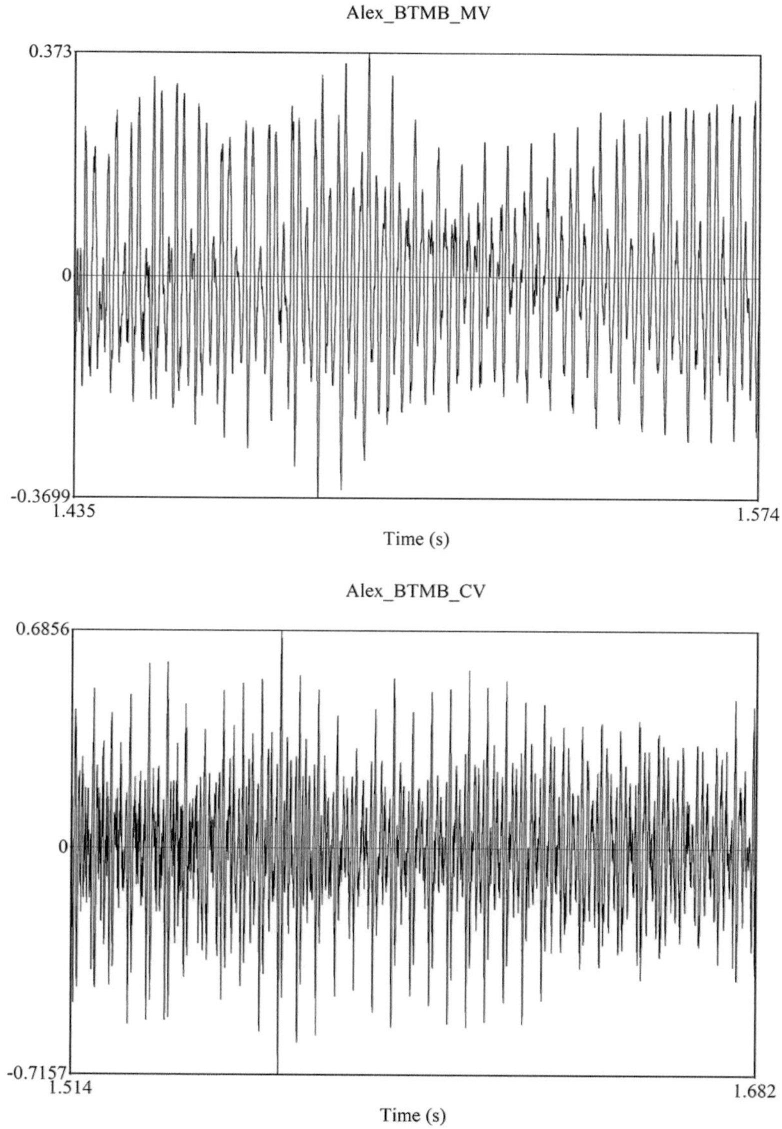

Figure 3.13 Excerpt from the modal voice (top) and creaky voice (bottom) taken from the last note of the excerpt for song three.

the notes are presented in either modal voice or creaky voice. The same creaky voice with multiple pulses was also found here (Fig. 3.13).

Once more the creaky voice presented much more irregularities in spectrum than modal voice, as can be seen in Table 3.4. Nevertheless, there was a different pattern for this song since shimmer was smaller for the creaky voice in this case.

Table 3.4 Some spectral differences between modal voice and creaky voice for song three

Born to be my baby	Jitter	Shimmer	Harmonics-to-noise ratio
Modal voice	0.265%	7.482%	17.834 dB
Creaky voice	0.546%	0.270%	9.173 dB

Figure 3.14 shows the score of the excerpt from song four that contains the notes D4 (293.66 Hz), F#4 (369.99 Hz), and C4 (261.63 Hz) sung with modal voice and creaky voice (Fig. 3.15). As for songs one, two, and three, the melodic contour of the notes is similar in both productions, as the notes are presented in both modal voice and creaky voice. As can be noticed in Fig. 3.15, the scale for modal voice is 50 to 500 Hz and for creaky voice is 200 to 500 Hz. This scale change had to be done because of the same problems with higher frequencies discussed for song two. The same creaky voice with multiple pulses was also found here (Fig. 3.16). Once more, creaky voice presented much more irregularities in spectrum than modal voice, as can be seen in Table 3.5.

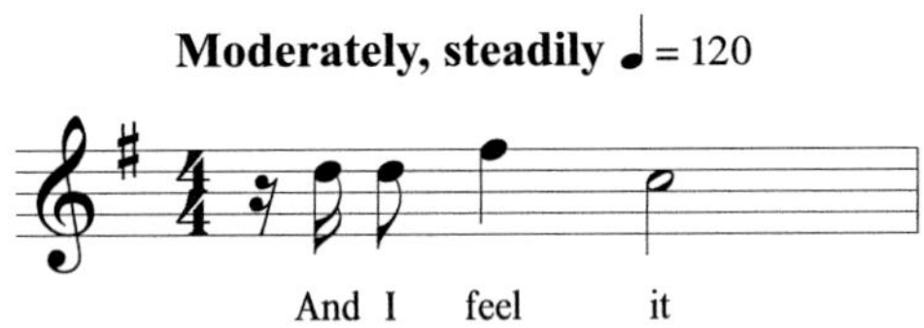

Figure 3.14 Notes of the song *Plush* by Stone Temple Pilots.

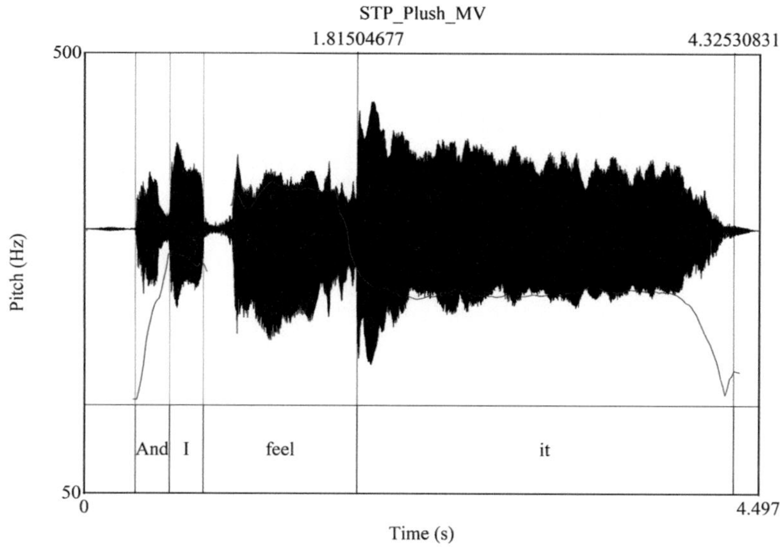

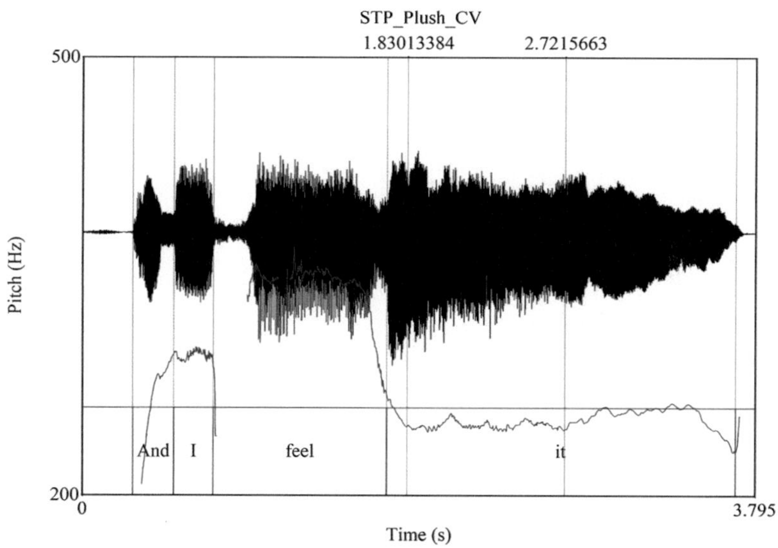

Figure 3.15 Waveform and f0 track (blue) of *Plush* sung in modal voice (top) and creaky voice (bottom).

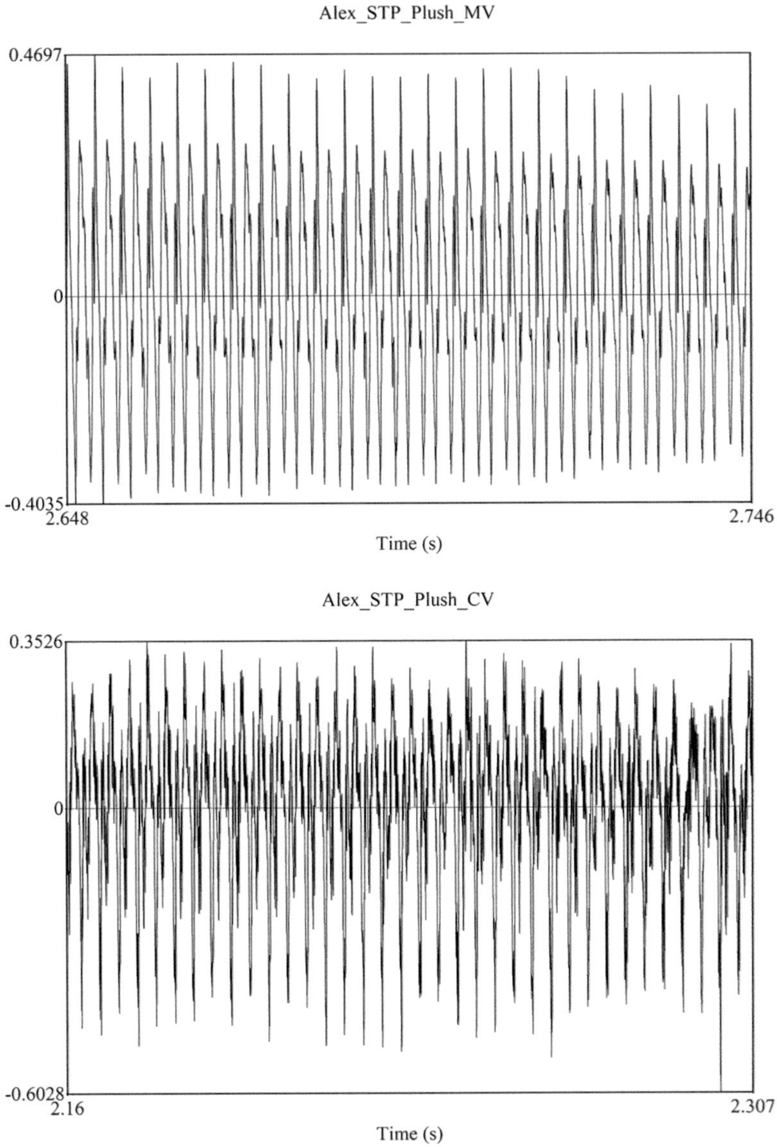

Figure 3.16 Excerpt from the modal voice (top) and creaky voice (bottom) taken from the last note of the excerpt for song four.

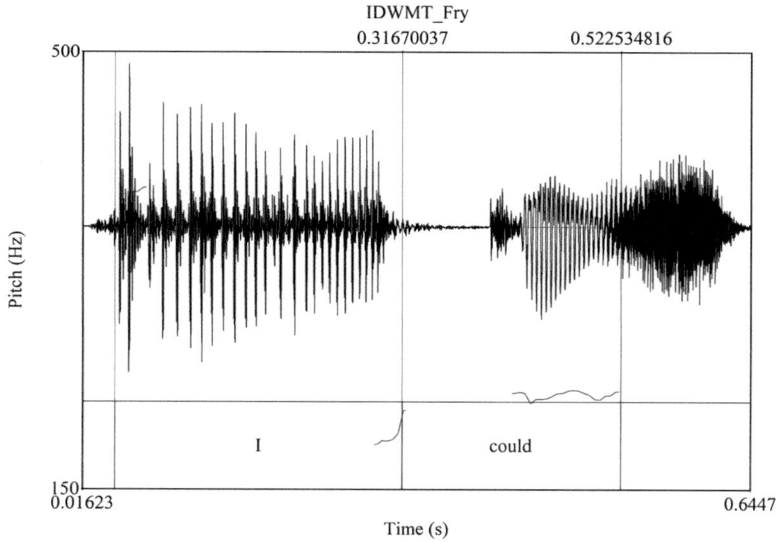

Figure 3.17 Waveform and f0 track (blue) of *I Don't Want to Miss a Thing* by Aerosmith.

Table 3.5 Some spectral differences between modal voice and creaky voice for song four

Plush	Jitter	Shimmer	Harmonics-to-noise ratio
Modal voice	0.240%	2.320%	22.277 dB
Creaky voice	0.618%	5.944%	13.293 dB

Finally, Fig. 3.17 displays an example of vocal fry use in singing. In this case, we did not record the modal voice version, for there was no reference to the note. Remember that vocal fry is supposed to not have musical notes. It is evident that there are no notes for the word "I" in comparison to the word "could" in which there is a blue line representing the f0 trace.

Discussion and final considerations

This chapter has shown that in phonetic linguistics, there is usually no distinction between creaky voice and vocal fry, because these two vocal effects have one single function to distinguish meaning in languages. In other words, as far as is known, there is no language to differentiate words based on variation between vocal fry and creaky voice. Therefore, the speakers produce a vowel categorized as creaky voice with either vocal fry or creaky voice. Nonetheless, once we try to differentiate the vocal processes of these two vocal productions, it is necessary to do so. Moreover, this distinction is even more necessary when studying music, because singing notes are only possible with creaky voice. Vocal fry, because of its atonality, is used just for vocal effects.

The data presented here have shown that vocal fry can also occur naturally in singing when f0 is below 70 Hz. We have seen this phenomenon when reproducing the creaky voice with an arpeggio (see Fig. 3.3). Although we did not have the intention to produce a vocal fry, it naturally happened because of this low C2 note. This is the reason why vocal fry happens in the beginning and ending of sentences in male speech and even female speech. Of course, vocal fry can be controlled as we have done when mimicking the beginning of *I Don't Want to Miss a Thing* by Aerosmith (see Fig. 3.17).

Another difference between creaky voice in speech and singing is the f0 range. In speech, creaky voice usually occurs between 70 and 100 Hz (an extreme low region for singing), but in singing it occurs in low (C2-G2 (65–98 Hz)), medium (G#2-D#3 (104–156 Hz)), and high (E4-B4 (330–494 Hz)) regions of the voice spectrum (see Meireles and Mixdorff 2020, for a detailed description of these voice regions). Our examples show this variation: song one is produced in the low region, songs two and three in the medium region, and song four in the high region (cf. Table 3.6).

The song analysis has shown some similarities when varying from modal voice to creaky voice. As Table 3.6 sums up: jitter (f0 variation) was always higher for the creaky voice; shimmer (intensity variation) was generally higher for the creaky voice; and harmonics-to-noise ratio was always smaller for the creaky voice. These results are in accordance with the voice literature since the irregularities produced in the vocal fold vibration cause more variation within these parameters (jitter and shimmer, among others) and resulted in a signal less harmonical.

Considering the percentage of variation from modal voice to creaky voice (Table 3.6), we notice that songs one and three are more similar between each other, since their variation is about the same for jitter and harmonics-to-noise ratio. The only difference is the decrease in shimmer from modal voice to creaky voice. This needs to be further investigated with additional data, for this is not the expected pattern in this variation. Songs two and four present much more variation from modal voice to creaky voice. We consider this increase in the percentage of variability resulted from the higher subglottal pressure necessary to produce these higher notes. Also, although song two has a lower note than song four, it was produced with much higher subglottal pressure, which can explain its highest percentage of variation.

Table 3.6 Variation in jitter, shimmer, and harmonics-to-noise ratio of the songs used in this study

Song	Note	Type	Jitter	Shimmer	Harmonics-to-noise ratio
1	A#2	Modal voice	0.186%	4.611%	15.040 dB
		Creaky voice	0.370% *(50%+)*	8.742% *(53%+)*	7.042 dB *(47%–)*
2	C#4	Modal voice	0.063%	1.238%	29.504 dB
		Creaky voice	1.303% *(2068%+)*	17.760% *(1434%+)*	5.869 dB *(502%–)*
3	A3	Modal voice	0.265%	7.482%	17.834 dB
		Creaky voice	0.546% *(49%+)*	0.270% **(2770%–)**	9.173 dB *(51%–)*
4	F#4	Modal voice	0.240%	2.320%	22.277 dB
		Creaky voice	0.618% *(257%+)*	5.944% *(256%)*	13.293 dB *(60%–)*

Note: The italic and bold percentages show the percentage of increase or decrease in the parameter from the modal voice to the creaky voice.

Summing up, this chapter presented the definition of creaky voice and vocal fry in voice literature and analyzed five songs that use these voice qualities. The spectral analyses have shown that modal voice is much more irregular than creaky voice in relation to the parameters jitter, shimmer, and harmonics-to-noise ratio. Generally, jitter and shimmer are higher for creaky voice, and harmonics-to-noise ratio is higher for modal voice. In the future, we intend to collect more data from creaky voice in different voice regions with techniques such as videolaringoscopy and/or high-speech camera, in order to further investigate the relations we found here between subglottal pressure and vocal fold variation from modal voice to creaky voice.

References

Boersma, P. (2001). "Praat: A system for doing phonetics by computer." *Glot International*, **5**, 341–345.

Borch, D. Z., Sundberg, J., Lindestad, P. A., and Thalén, M. (2003). "Vocal fold vibration and voice source aperiodicity in phonatorily distorted singing Speech, music and hearing, KTH Stockholm, Sweden TMH-QPRS." *KTH*, **45**, 87–91.

Caffier, P. P., Ibrahim Nasr, A., Ropero Rendon, M. del M., Wienhausen, S., Forbes, E., Seidner, W., and Nawda, T. (2018). "Common vocal effects and partial glottal vibration in professional nonclassical singers." *Journal of Voice*, **32**(3), 340–346.

Catford, J. C. (1964). "Phonation types: The classification of some laryngeal components of speech production." In Abercrombie, D., *et al.* (Eds.), *In Honour of Daniel Jones*, 26–37. London: Longmans.

Drugman, T., Kane, J., and Gobl, C. (2014). "Data-driven detection and analysis of the patterns of creaky voice." *Computer Speech and Language*, **28**, 1233–1253.

Esling, J., Moisik, S. R., Benner, A., and Crevier-Buchman, L. (2019). *Voice Quality: The Laryngeal Articulator Model*. Cambridge: Cambridge University Press.

Garellek, M. and Keating, P. (2011). "The acoustic consequences of phonation and tone interactions in Jalapa Mazatec." *Journal of the IPA*, **41**(2), 185–205.

Keating, P. A., Garellek, M., and Kreiman, J. (2015). "Acoustic properties of different kinds of creaky voice." *Proc. 18th ICPhS*, 0821-4.

Ladefoged, P. (1964). *A Phonetic Study of West African Languages*. Cambridge, England: Cambridge. University Press.

Laver, J. (1980). *The Phonetic Description of Voice Quality*. Cambridge University Press.

Meireles, A. and Mixdorff, H. (2020). "Voice quality in low and high registers in two different styles of singing." *Proc. Speech Prosody 2020*, 601–605.

Miller, D. G. (2000). *Registers in Singing. Empirical and Systematic Studies in the Theory of the Singing Voice*. Wageningen: Ponsen and Looijen BV, Wageningen.

Sadolin, C. (2021). *Complete Vocal Technique*. Copenhagen: Shout Publishing.

Schutte, H. K. and Miller, D. G. (1993). "Belting and pop, nonclassical approaches to the female middle voice: Some preliminary considerations." *Journal of Voice*, **7**(2), 142–150.

Titze, I. R. (1994). "Toward standards in acoustic analysis of voice," *Journal of Voice*, **8**(1), 1–7.

Chapter Four

Acoustic, physiological, and paralinguistic analyses of creaky voice in spontaneous dialogue speech

Carlos T. Ishi

Introduction

Besides the normal (modal) phonation, a diversity of voice qualities can be produced depending on the vibration mode of the vocal folds. A classical categorization of voice qualities includes modal, breathy, whispery, falsetto, creaky, and harsh voices and combinations of these categories (Laver 1980). Part of these voice qualities can be consciously produced, while others are affected by a speaker's mental state being unconsciously produced. Therefore, changes in voice quality may reflect the emotional or attitudinal states of the speaker (e.g., Ball *et al.* 2000; Kreimann et al. 2000; Gordon *et al.* 2001; Gobl and Ní Chasaide 2003; Sadanobu 2004). The terminologies and definitions of voice-quality categories vary depending on the researcher; making it

Creak: Theories and practices of pulse phonation
Edited by Francesco Venturi

ISBN 978-981-5129-29-8 (Hardcover), 978-1-003-61606-1 (eBook)
www.jennystanford.com

important to define them in terms of acoustic, physiological, and perceptual features (Catford 1977; Gerratt *et al.* 2001; Esling 2006).

Creaky voice, creak, vocal fry, glottal fry, pulse register, and laryngealization are terminologies found in the literature for a voice quality characterized by a train of relatively discrete laryngeal (or glottal) excitations (or pulses of brief duration), with almost complete damping of the vocal tract between successive glottal pulses, usually accompanied by extremely low fundamental frequencies, and irregular durations of glottal cycles (Catford 1977; Laver 1980). The auditory perception of creaky voice is rough quality, resembling "rapid series of taps like a stick being run along a railing," or the imitated sound of a motor boat engine, or similar to food cooking in a hot frying pan (Catford 1977; Blomgren *et al.* 1998).

Creaky voice carries important linguistic and paralinguistic information depending on language. It often occurs near morpheme boundaries (Kiessling *et al.* 1995), or near phrase boundaries (Ishi 2004), where pitch is lowered. Paralinguistic correlates of sad, bored, and relaxed voices are reported (Gobl *et al.* 2003; Klasmeyer *et al.* 2000). In Japanese, besides the creaky voices appearing in such low tension voices, it also appears in expressive emphasis utterances as a pressed voice (called "rikimi" in Japanese) that carries paralinguistic information primarily associated with feelings/attitudes of surprise, admiration, and suffering (Sadanobu 2004). Although there are studies on automatic detection of creaky segments based on acoustic features (Ishi *et al.* 2008a; Drugman *et al.* 2014), it is important to additionally distinguish between lax and pressed realizations of creaky voice, since they may convey different types of paralinguistic information.

Furthermore, the irregular characteristics of creaky voice segments can cause severe errors in pitch determination algorithms, which are important for prosodic information extraction. Thus, knowledge about the location of creaky voices could be useful for the indication of paralinguistic information as well as to improve pitch determination.

I have been studying the acoustic, electroglottographic (EGG), and paralinguistic functions of creaky voice appearance in spontaneous dialogues for several years (Ishi 2004; Ishi *et al.* 2007; Ishi *et al.* 2008a; Ishi *et al.* 2008b; Ishi *et al.* 2010; Ishi *et al.* 2018). In this chapter, the features and functions of different realizations of creaky voice appearing in spontaneous dialogue interactions are presented, based on my studies so far. Acoustic and EGG analyses clarify the differences in vibration patterns of the vocal folds from lax to pressed realizations of creaky voice. It is shown that double-pulsed patterns and irregular periodicity patterns can appear in both lax and pressed realizations. While lax creaky tends to show incomplete closure of the vocal folds, pressed creaky shows predominance of closed phase of the vocal folds over the whole segment, regardless of irregularities in periodicity.

This chapter is organized as follows. In Section 4.2, the speech data and the annotation data used for analysis are described. In Section 4.3, acoustic, EGG, and paralinguistic analyses of creaky voice are presented. Analysis results and conclusions are presented in Section 4.4.

Analysis data

Multimodal spontaneous speech data

Most studies dealing with voice quality use the stationary portion of specific voice qualities consciously produced by subjects. However, although both lax and pressed realizations of creaky voice frequently appear in natural conversation, some subjects cannot produce them in a conscious manner. In our study, creaky voice utterances are extracted from natural conversational speech databases, where creaky voice is unconsciously produced. This allows the analyses of the variations on speaking styles in different realizations of creaky voice in spontaneous speech. Furthermore, in part of the database, EGG signals are also available along with the corresponding speech signals, which allows for the physiological analysis of vibratory patterns in the vocal folds during different realizations of creaky voice.

Identification of creaky voice utterances and classification of pressed degree

Candidates of utterances containing lax and pressed realizations of creaky voice were identified by listening to the speech utterances in the dialogue speech database. Note that the term "creaky voice" in this chapter may include a variety of phonatory styles since vocal fry characterized by low fundamental frequencies and harsh voices with higher fundamental frequencies and irregularities in periodicity.

Creaky voice was found in a wide range of ages, regardless of gender (10 female and 9 male speakers from 10s to 60s). Pressed creaky voice was observed in about 70% of the subjects, while lax creaky voice was observed in all subjects.

Then, in the second stage, the degree of perceived pressed quality was rated by four annotators on a three-point scale ("2" when pressed voice is clearly perceived, "1" when pressed voice is weakly perceived, and "0" when pressed voice is not perceived). An additional instruction was given to only choose "2" when they perceive that the vocal folds are clearly more pressed than necessary, i.e., the muscle tensions in the vocal folds are clearly stronger relative to the normal phonation. As a result, many "1" and "0" appeared in the annotations for all subjects. The Crombach's alpha value between subjective scores was 0.77, indicating an acceptable consistency between raters. The scores given by all four subjects were then summed up for each utterance and used in this study as a perceived degree of pressed voice quality. The categories {"p0," "p1," "p2"} were defined to correspond to the summed score ranges of {2~4, 5~6, 7~8} from the subjective pressed voice degrees. The utterances whose summed scores were smaller than 2 were removed from the analysis, resulting in 123 utterances.

Segmentation of creaky voice

In order to evaluate the acoustic features and vibratory patterns in pressed voice segments, a more detailed segmentation of voice quality was conducted at a segmental level, for the utterances containing creaky voice. This is because voice-quality changes usually occur only in part of the voiced portions in a word or a phrase. This type of segmentation and classification requires knowledge about acoustic-phonetics and laryngeal voice qualities. The segmentation was conducted by the chapter author, which has experience in acoustics and voice-quality analysis, based on the visual inspection of spectrograms and waveforms, as well as auditory impression. The segmentation process resulted in 276 segments. For evaluation purposes, modal (normal) phonation intervals were also segmented around the pressed voice portions, and 203 segments were obtained for modal voice.

Analyses of creaky voice in spontaneous dialogue interactions

This section provides acoustic and EGG analyses of creaky voice segments appearing in the spontaneous dialogue data described in Section 4.2, with focus on the features of the pressed realizations of creaky voice.

Periodicity and voice-quality analyses

Irregularity in periodicity is one characteristic of creaky voice. This section aims on clarifying how these irregularities are manifested in creaky voice segments. The different realizations of creaky voice are also discussed based on periodicity information.

Although creaky voice is commonly accompanied by fundamental frequencies (F0) below the modal voice range, we consider that this may not be true in pressed creaky voices. We also consider that a strict categorization in creaky and harsh voice qualities is not consistent due to ambiguities in the acoustic and perceptual spaces. Thus, the discrimination of voice qualities related to creaky and harsh phonations was categorized in three groups, based on periodicity properties, according to the following criterion. The first category is composed of segments with average F0 smaller than 100 Hz, which is more characteristic of vocal fry or creak phonation.

The second category is composed of segments with F0 larger than 100 Hz and with some pulse-to-pulse irregularity—including double periodicity with alternations of small and big glottal pulses, and aperiodic variations between adjacent glottal pulses—which is more characteristic of harsh phonation. Finally, the third category is composed of segments with F0 larger than 100 Hz—i.e., in the range of modal phonations—and no evident irregularity in the inter-pulse intervals. This third category includes segments that are ambiguous for the classical classification of voice qualities. This is because F0 is not as low as in vocal fry, and the auditory impression has a more strident quality compared to normal phonation.

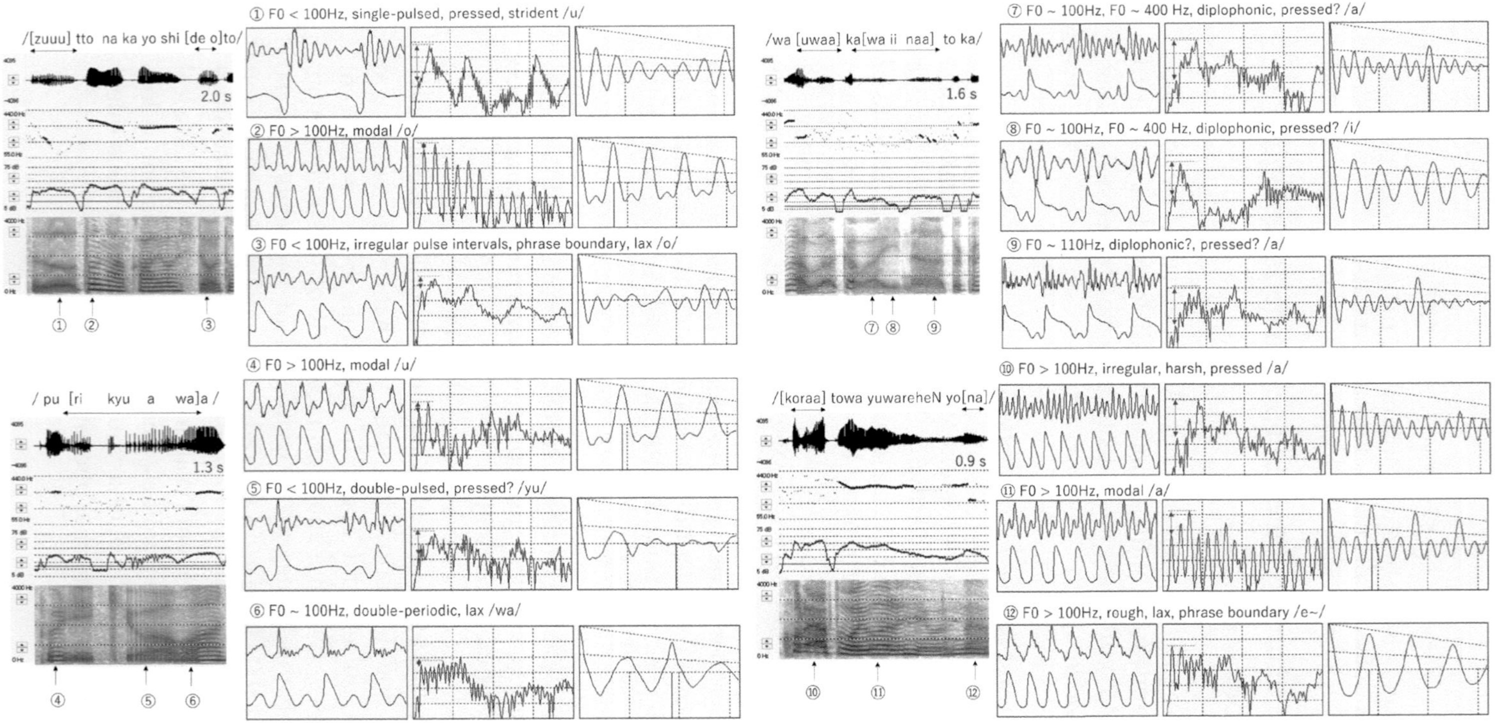

Figure 4.1 Examples of acoustic features of creaky voice segments extracted from spontaneous dialogue speech of a female speaker. Left panels show phonetic transcription, creaky intervals, speech waveform, F0 contour (in log scale from 55 to 440 Hz), power contour, and spectrogram (0 to 4000 Hz). Right panels show speech and EGG (in blue) waveforms (32 ms window length), speech spectrum (0 to 4000 Hz for Y-axis, and 10 dB grids for X-axis), and autocorrelation of the LPC residual signal (0 to 16 ms lag for the Y-axis, and -1 to 1 for X-axis), for the instants indicated in the left panels.

In the present analysis, we estimate F0 based on a classical method of peaking: the peaks on the autocorrelation function of the residual signal after inverse filtering the speech signal by its LPC (linear predictive coding) coefficients (Ishi 2008b). The inverse of the autocorrelation lag of the largest peak corresponds to the fundamental frequency. The autocorrelation of the LPC residual signal will also be used in the present analysis to discuss periodicity patterns of creaky voice segments.

Figure 4.1 shows examples of four utterances including different realizations of creaky voice extracted from spontaneous dialogue speech of a female speaker in her 30s. These examples first show that a diversity of vocal fold vibratory patterns can appear for the same speaker. The left panels in Fig. 4.1 show the phonetic transcription, creaky intervals (indicated by arrows), the speech waveform, the F0 contour (in log/musical scale from 55 to 440 Hz, corresponding to three octaves), the power contour (with grids in 10 dB intervals), and the spectrogram (0 to 4000 Hz). Right panels show speech and EGG (in blue) waveforms (with 32 ms window length, and amplitudes scaled to allow better visualization), speech spectrum (0 to 4000 Hz for Y-axis, and 10 dB grids for X-axis), and autocorrelation of the LPC residual signal (0 to 16 ms lag for the Y-axis, and -1 to 1 for the X-axis), around the instants indicated in the bottom of the left panels. Note that the autocorrelation function decreases as the autocorrelation lag increases, since the number of multiplications decreases. To suitably evaluate the strength of the autocorrelation peaks, dashed downward lines are drawn in the autocorrelation panels of Fig. 4.1 to indicate the maximum autocorrelation values and threshold values for detecting periodicity. Thus, the periodicity of a speech signal is determined by the autocorrelation peak (excluding the one of lag 0) that is closest to the upper dashed line and higher than the threshold dashed line.

In this section, we discuss periodicity patterns in creaky voice. First, the segments around the instants (2), (4), and (11) in Fig. 4.1 are examples of modal voice (i.e., normal phonation) with regular periodicity. It can be observed that the autocorrelation functions in modal voice show clear peaks in the lags corresponding to F0, as well as peaks with similar shapes around the lags that are multiples of the F0 peak. The segments around the instants (1) and (3) in

Fig. 4.1 are typical examples of pressed and lax creaky voices with low F0s (around 60 and 70 Hz, respectively). Lax creaky voice is usually accompanied by some irregularity in the pulse-to-pulse intervals as example (3) shows. Example (5) is a creaky segment with low F0 but is double pulsed and its perceptual impressions are between lax and pressed. The examples in (6) to (9) show segments with F0 around 100 Hz. They are clearly different from modal voice, but the F0s are around or higher than 100 Hz. The example in (6) has a double-periodic pattern and is perceived as lax, while the examples in (7) to (9) are diplophonic with an F0 around 100 Hz and another around 400 Hz and also have a slight pressed quality. These might be categorized as harsh creaky falsetto in Laver's categorization. The example in (10) shows a harsh voice segment with perturbations in periodicity and average F0 around 140 Hz, sounding rasping.

Finally, the example in (12) shows a rough voice quality with slight double-periodic pattern. It can be observed in the autocorrelation function of example (12) that, compared to the first autocorrelation peak, the second peak is slightly thinner and closer to the upper dashed line, which indicates that the periodicity of the second peak (i.e., the subharmonic) is slightly stronger than the first peak (i.e., the F0 peak). An analysis of the spectral and EGG features will be discussed in the subsequent subsections. Table 4.1 shows the distributions of the perceived degrees of pressed quality in segments perceived as creaky or harsh, according to the periodicity-based categories. The distributions in Table 4.1 show that pressed voice utterances are realized with F0 lower than 100 Hz (corresponding to creaky voices) in 54% of the utterances, and with F0 higher than 100 Hz and inter-pulse irregularity (corresponding to harsh voices) in 28% of the utterances. However, it is also shown that 18% of the pressed voice utterances are realized with F0 higher than 100 Hz, without a clear irregularity in the pulse-to-pulse intervals. Further, it can be observed from the table that the majority of the utterances of both creaky and harsh voices appeared in categories with a high degree of pressed voice quality. Note that the term "pressed" used in this chapter is thought to be close to the term "harsh" defined by Laver (1980). However, we consider that part of the realizations of harsh voice could be classified as pressed creaky, from a perceptual viewpoint.

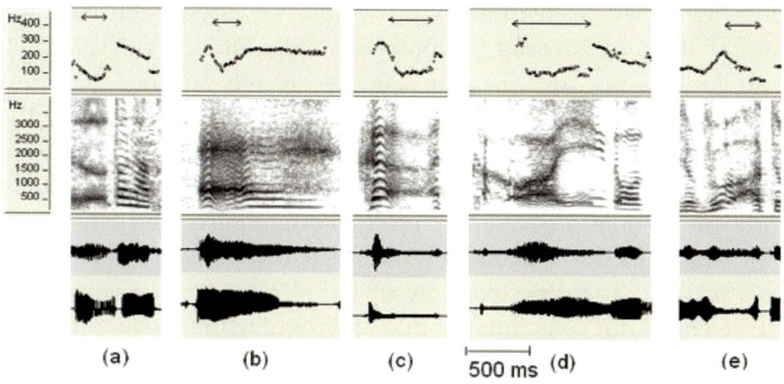

Figure 4.2 F0 contours, spectrograms, speech and EGG waveforms for utterances including pressed creaky voice segments (indicated in the top panels) accompanied by an F0 dip. (a) F0 < 100 Hz; (b) F0 > 100 Hz; (c) F0 ~ 100 Hz; (d) F0 ~ 100 Hz, diplophonic; (e) F0 > 100 Hz.

Table 4.1 Distributions of the perceived degrees of pressed voice quality (p0, p1, p2), according to the periodicity-based categories

	avgF0 < 100 Hz	avgF0 > 100 Hz inter-pulse irregularity	avgF0 > 100 Hz no inter-pulse irregularity	Total (%)
p2	20	15	3	38 (31%)
p1	16	13	11	40 (33%)
p0	30	7	8	45 (36%)
Total (%)	66 (54%)	35 (28%)	22 (18%)	123

The F0 contours in pressed creaky voice utterances were then analyzed. Figure 4.2 shows F0 contours, spectrograms, speech and EGG waveforms for pressed creaky voice segments accompanied by different periodicity categories. In the current sub-section, we discuss the F0 contours, while the spectrogram and EGG features will be discussed in the following subsections.

A common feature found in all utterances containing pressed voice was an F0 dip in the pressed creaky voice portions. Two transition types could be identified between modal to/from pressed creaky voice segments. One is a fast, but continuous, decrease in F0, as shown in the examples of Fig. 4.2(a–c). This pattern was observed in 17% of the pressed creaky voice utterances. The other type is a sudden (discontinuous) F0 jump, as in the diplophonic signal (where multiple phonations can be simultaneously perceived) shown in Fig. 4.2(d). In the remaining 83% of the utterances, the F0 curve was broken, as in the example of Fig. 4.2(d), or a consistent F0 curve could not be estimated due to irregularities in the pulse-to-pulse intervals, as the examples shown in Fig. 4.1.

Spectral tilt analysis: H1′-A1′ measure

Spectral tilt is a commonly used feature for characterizing voice qualities and is reported to be effective for discriminating "tense voice" from "lax voice" (Maddieson *et al.* 1985). As one characteristic of pressed voice is of a tense voice quality, it is expected that spectral tilt can potentially discriminate it from other voice qualities. It can be observed in the spectrograms of Figs. 4.1 and 4.2 that the power of the frequency components of the lower harmonics is reduced in the pressed voice portions. Classical measures for spectral tilt are based on the differences between the amplitudes of the first and second harmonics (H1-H2) or between H1 and the harmonic closest to the first formant (H1-A1) (Gordon *et al.* 2001; Maddieson *et al.* 1985).

The H1-A1 measure was considered to be a good alternative for characterizing pressed voice segments, since it reflects the effects of loss of energy in the low frequencies, due to high vocal fold tensions. However, a problem in applying the above measures for pressed voice is that the harmonic structure is disturbed or sometimes inexistent when irregularities in periodicity are present.

In such cases, in place of H1 and A1, we use the maximum peak amplitude in the range of 100 to 200 Hz (H1′), and the maximum peak amplitude in the range of 200 to 1200 Hz (A1′), where the first formant is likely to occur. For periodic signals, H1′ = H1 and A1′ = A1.

For the computation of the power spectrum (prior to the computation of the H1′ and A1′ values), pre-emphasis processing is

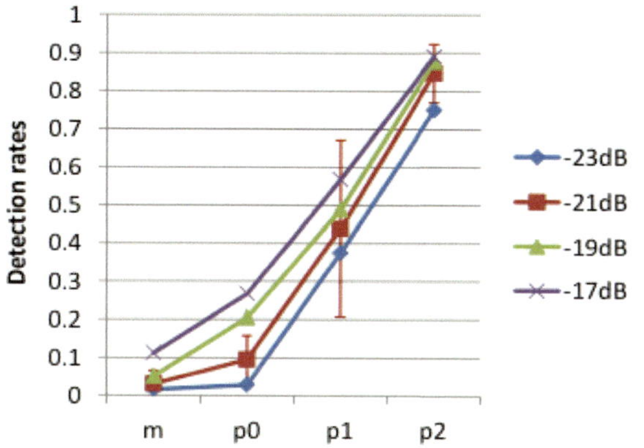

Figure 4.3 Discrimination performances of pressed voice segments by using different thresholds for H1′-A1′.

conducted to emphasize the vocal tract formant peaks and reduce the effects of the glottal formant. A high-pass filter with the transfer function ($1–0.96\ z^{-1}$) is used for implementing the pre-emphasis processing. Segments with H1′-A1′ values smaller than a threshold are judged as pressed voice. Figure 4.3 shows the discrimination performances of pressed voice segments by using different thresholds for H1′-A1′.

Results in Fig. 4.3 show high detection rates in pressed segments ("p2") (true positives), low detection rates in modal segments ("m") (false negatives), and intermediate detection rates for the "p0" and "p1" categories. For example, for a threshold of –21 dB for H1′-A1′, a correct detection rate of 80% is achieved in pressed segments, with a false detection rate of 3% in modal segments. Detailed analysis was also conducted on the pressed segments "p2," which could not be identified as pressed by H1′-A1′.

Figure 4.4 shows examples of spectrums of pressed and non-pressed intervals extracted from our dataset. These examples were selected in a way to illustrate the cases where true/false positive/negative detections occur in the pressed segment detection by H1′-A1′. Note that the figures show the spectrums of both with and without pre-emphasis, which is a common pre-processing in speech spectral analysis. Figures 4.4a and 4.4b show typical examples of pressed creaky and harsh segments, where H1′-A1′ is lower than –20 dB, leading to true positive detections.

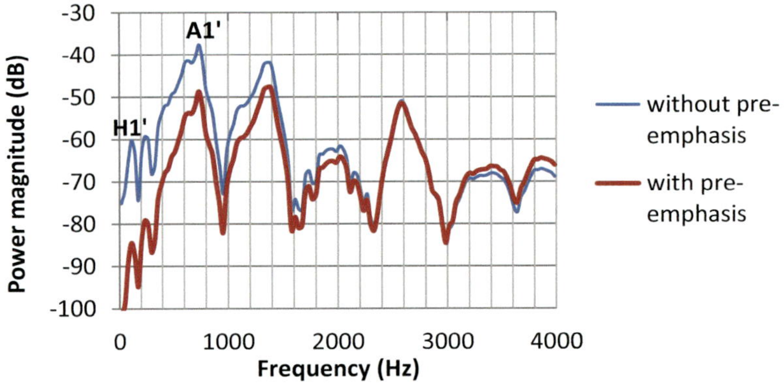

(a) Pressed creaky (F0 < 100 Hz) /a/ (true positive)

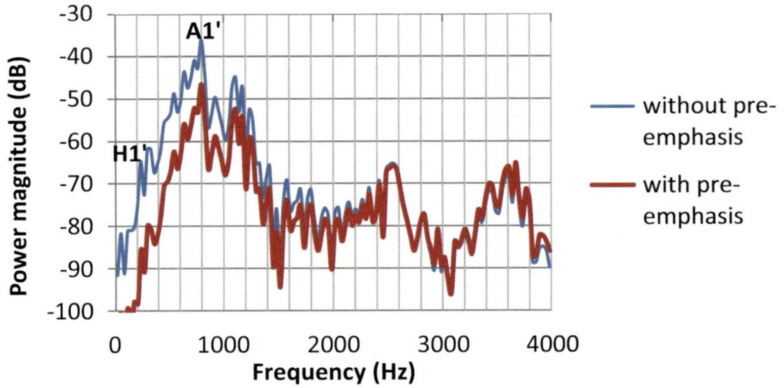

(b) Pressed harsh (F0 > 100 Hz) /a/ (true positive)

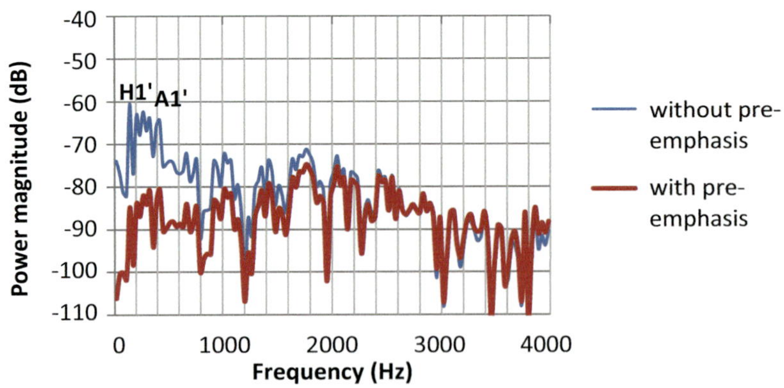

(c) Pressed creaky nasal /n/ (false negative)

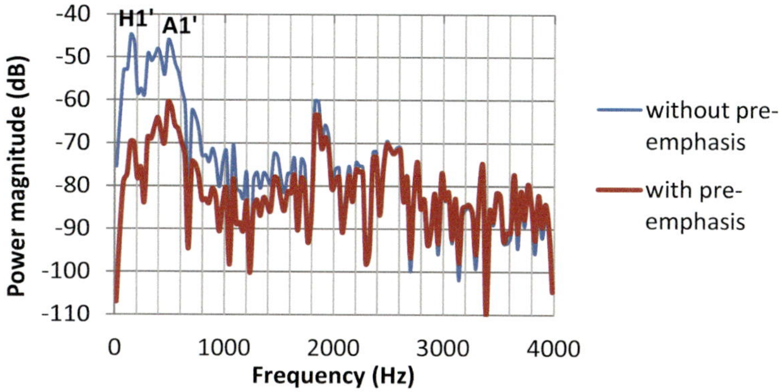

(d) Pressed creaky nasalized /e/ (false negative)

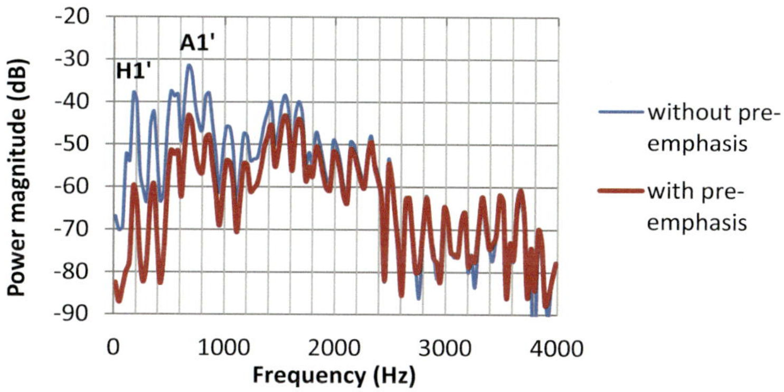

(e) Modal voice /a/ (true negative)

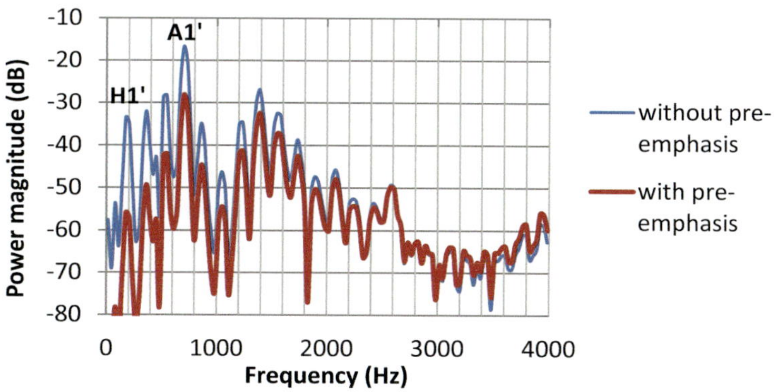

(f) Modal voice /a/ (false positive)

Figure 4.4 Examples of spectrums of pressed and non-pressed intervals, with true/false positive/negative detections by H1′-A1′.

The problem of H1′-A1′ occurs for nasals (Fig. 4.4c) or nasalized vowels (Fig. 4.4d), where a strong spectral component appears in the low-frequency bands around 100 to 300 Hz, due to a nasal formant. This causes an increase in H1′-A1′ values and consequent false negative detections of pressed nasal or nasalized segments. Figure 4.4e shows an example of modal segment, where H1′-A1′ is larger than −20 dB, leading to a true negative detection. Finally, Fig. 4.4f shows an example of false positive detection in a non-pressed segment. Although these segments have a more "tense" voice quality compared with other modal segments, they are not perceptually judged as a pressed voice. Such tense voice segments have intermediate features between modal and pressed.

Regarding the spectrums shown in Fig. 4.1, it can be observed that the examples of segments (1), (7), (8), (9), and (10) can be correctly classified as "pressed" (true positives), and (2), (3), (4), (6), (11), and (12), which are either modal or lax creaky, can be correctly classified as "non-pressed" (true negatives), according to the H1′-A1′ measures. The example in (5) can be considered as a false negative, where the increase in the H1′ value is due to the double pulse pattern of the vocal folds.

How to deal with false negatives in pressed nasalized or double-pulsed segments and false positives in non-pressed segments remains a topic for future investigations.

Vocal fold vibratory patterns: EGG analysis

Vocal fold vibratory patterns were analyzed based on the EGG signals, for supporting the understanding of the different realizations of creaky voice.

Figure 4.5 shows speech, EGG, and DEGG (derivative of the EGG) waveforms of representative samples, including lax and pressed realizations of creaky voice found in our dataset. These samples were selected in a way to illustrate the different vibratory patterns, including single/multiple glottal pulses in different realizations of creaky voice segments. Modal segment is also included for comparison. The length of the segments is 100 ms for all plots. The amplitudes of the waveforms are re-scaled to allow better visualization.

Regarding the interpretation of the EGG waveforms, the peaks represent high vocal fold contact (or glottal closure), while the valleys represent low vocal fold contact (or glottal opening). The DEGG waveforms are computed as the negative of the derivative waveforms of the EGG signals and provide a better visualization of the approximated instants of opening (positive peaks) and closing (negative peaks) of the vocal folds (Childers *et al.* 1991). This way, open intervals can be visualized as the intervals between positive and successive negative peaks in the DEGG waveform, while closed intervals can be visualized as pulse cycle intervals minus the open intervals.

Upward/downward arrows indicate approximate instances of glottal opening/closing in Fig. 4.5. These instances were hand-annotated by visual inspection of the waveforms. The closed and opened intervals are drawn (dark gray for closed intervals and light gray for opened intervals) based on the glottal opening/closed instances. However, incomplete closures and intervals where glottal opening instance is not clear are shown as striped segments.

Figure 4.5a shows an example of modal voice (normal phonation) of a female speaker, for reference. Figure 4.5b shows a typical example of a single-pulsed pressed creaky voice of the same speaker, where F0 is very low (about 80 Hz). It can be observed in the EGG and DEGG waveforms that the open intervals are much shorter than the closed intervals in the pressed segment of Fig. 4.5b. Figure 4.5c shows an example of double-pulsed creaky segment for the same speaker. Small pulses, corresponding to incomplete closures of the vocal folds, can be observed before bigger and longer pulses with complete closures. Figures 4.5d and 4.5e show examples of pressed and non-pressed creaky segments of a male speaker. Both signals have small pulses between large pulses in the EGG waveforms, corresponding to incomplete closures between complete closures. The difference between these signals is that the overall open intervals are shorter than the closed intervals in the pressed creaky segment. Similar behavior was observed in other speakers. Figure 4.5f shows an example of pressed segment with no pulse-to-pulse irregularity in periodicity, and having F0 ranges above 100 Hz, so that individual pulses cannot be perceived, but a strident quality can be perceived. Similar to the previous pressed examples, the open intervals are relatively shorter than the closed intervals. Figure 4.5g shows an example of a pressed segment

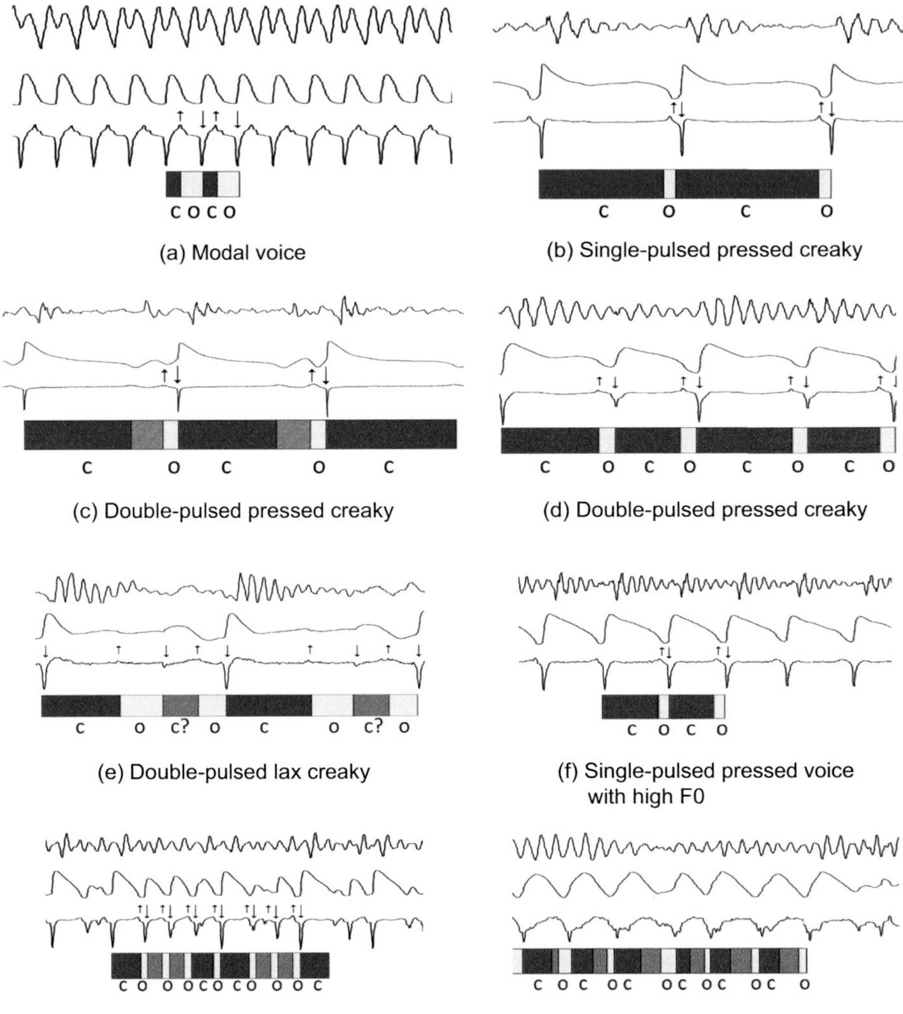

Figure 4.5 Speech, EGG, and DEGG waveforms for several types of vibratory patterns, including lax and pressed creaky segments. Below the DEGG waveforms, closed intervals ("c") are indicated by dark gray regions, open intervals ("o") are indicated by light gray regions, and incomplete closure is indicated by striped regions. Segment lengths are 100 ms.

with a diplophonic quality (where multiple phonations can be simultaneously perceived). It can be noted that larger pulses with stronger negative DEGG peaks occur with larger inter-pulse intervals (lower F0, around 110 Hz), while smaller pulses occur with smaller inter-pulse intervals (higher F0, around 330 Hz).

As shown in the examples of Fig. 4.5, single- and double-pulsed patterns were observed in the glottal pulses of pressed segments. However, a common feature found in most of the pressed segments was that the overall open intervals are much shorter in duration than the overall closed intervals, regardless of periodicity—as shown in the examples of Figs. 4.5b, 4.5c, 4.5d, and 4.5f. Nonetheless, this feature is not clear in the example shown in Fig. 4.5h. Positive peaks in DEGG waveforms are broader and less sharp than the other pressed examples, so that it is difficult to identify a precise glottal opening instance. Correspondence between speech and EGG waveforms is also unclear, due to insufficient closure over the whole glottal pulses.

Regarding vocal fold vibratory analysis, it is also worth to mention that there are measures like open quotient (OQ) and speed quotient (SQ) for characterizing glottal waveform shapes (Childers *et al.* 1991). However, there is not a clear consensus on how to deal with incomplete closures. Other hints may be provided a method for OQ in incomplete closures (Yokonishi *et al.* 2016).

Linguistic and paralinguistic analyses

As stated in the introduction, lax creaky voice is commonly found in morpheme, phrase boundaries, or alongside sad, bored, or tired voices, where the vocal tension is low and F0 is lowered. On the other hand, pressed creaky voice is commonly found in the middle of emphasized words or phrases, including accented syllables.

The pressed creaky segments found in the analyzed Japanese conversational database can be categorized in the following items (the utterance meanings are shown within brackets):

- Sincere expressions of emotions or attitudes (like surprise, admiration, and disgust) through interjections: "*waa*" ("I am really impressed."), "*eee*" ("I am really hesitated; I am really dissatisfied."), "*hee*" ("I am really surprised; I am really impressed."), "*nnn*" ("I am really embarrassed."), "*yaa*" ("I cannot really accept."), "*waa,* kimochi warui" ("That is really disgusting!)

- Expressions of sincere feelings through adjectives and adverbs: "*kawaii*" ("really cute!"), "*zutto*" ("for a really long time!"), "*taihen*" ("really serious!"), "*mottainai*" ("really wasteful!"), "*itai*" ("really painful!"), "*kirei*" ("really beautiful!"), "*umainaa*" ("really delicious!"), "*kimochiwarii*" ("really disgusting!"), "*tsumetaai*" ("really cold!"), "*kyoufu yawaa*" ("I am really scared!")
- Expressions of vivification through onomatopoeia: "*uwaa*"/"*bwaa*" ("a lot, fast"), "*gaan*" ("with vigor"), "*kaaa*" ("with vigor, energetically"), "*pyun*" ("jumping up and down"), "*nyee*" ("baby's strident crying sound"), "*aaa*" ("excited crying sound expressing strong sadness")
- Expressions of hesitation: "*nnn…* mayoimasuwa" ("I am really uncertain…"), "*fukuwa…* kirarenai" ("I really cannot wear it…")
- Expressions of modesty: "*yaa korewa chotto yappari yoku nai to omoimasu*" ("I really think that is not good…"), "*sumimasenga, chotto yurushite kudasaiyo*" ("I am really sorry, please forgive me…")
- Utterance quoting: Pressed voice frequently appeared when the speaker quoted his/her past utterance, or an utterance of another person. The expressivity conveyed by pressed quoted utterances was of excitement or hesitation. "*kora!*" (Hey!), "*moo iikagen ni shii!*" ("*Stop it!*").

Note that in the English translations of the examples above, the word "really" (which conveys sincere feelings) is not explicitly verbally expressed, but paralinguistically expressed—i.e., accompanied with a pressed creaky voice quality.

Another observation is that creaky voice usually appears in the end of phrases and sentences, where F0 is lowered, while pressed creaky voice portions usually appeared in the accented syllables of the words.

Discussion and final remarks

Acoustic, physiological, and paralinguistic analyses were conducted on speech segments, including different realizations of creaky voice, extracted from spontaneous dialogue speech data of several speakers.

Periodicity analysis first showed that creaky voices usually have irregularities in periodicity, but a pressed creaky voice can also be accompanied by periodicity with F0s in the range of modal phonations. It was also found that transitions between modal to creaky voice are accompanied by a continuous or discontinuous F0 dip. A sudden F0 halving tends to occur in lax creaky voice, as a consequence of insufficient vocal fold tensions, while a continuous F0 dip commonly appears in pressed creaky voice segments, as a consequence of the high vocal fold tensions during its production.

From a perceptual viewpoint, higher pitch is commonly perceived in several pressed voice portions even though the acoustic F0 is lower than the neighboring modal portions. Further, the pressed voice portion (which is accompanied by F0 dips) appeared in the accented syllables of the words, where F0 is raised in normal phonation. Therefore, the measured F0 curve may not be an appropriate representation of accent/intonation in both lax and pressed creaky voice utterances, so that special care has to be taken in the interpretation of F0 curves of creaky voice segments.

Spectral analysis indicated reduction in power in low-frequency components during pressed creaky voice. The spectral measure H1′-A1′—which is closely related with tense/lax voice properties—is effective for identifying 80% of the pressed voice utterances with high degree of pressed quality; with a 3% insertion error rate in modal segments. However, H1′-A1′ fails to identify pressed segments when nasalization occurs.

Analysis of vocal fold vibratory patterns from the EGG signals revealed that in most cases of pressed creaky voice segments—including nasalized segments—the completely closed intervals of the vocal folds are longer than the open intervals on average, regardless of periodicity, while lax creaky voice is usually accompanied by incomplete vocal fold closure, often resulting in double-periodic patterns. Finding acoustic parameters that better reflect the closed interval properties could solve the problems of the current spectral features in nasalized pressed segments.

Acknowledgments

This research was supported by JSPS KAKENHI Grant Number JP22H04875 and JST ERATO Grant Number JPMJER1401.

References

Ball, M. J.; Esling, J.; and Dickson, G. J. (2000). "The transcription of voice quality." In: Kent, R. D. and Ball, M. J. (Eds), *Voice Quality Measurement*. San Diego: Singular Publishing Group, 49–58.

Blomgren, M.; Chen, Y.; Ng, M.; and Gilbert, M. (1998). "Acoustic, aerodynamic, physiologic, and perceptual properties of modal and vocal fry registers." *J. Acoust. Soc. Am.*, **103**(5): 2649–2658.

Catford, J. (1977). *Fundamental Problems in Phonetics*. Edinburgh: Edinburgh University Press.

Childers, D. G. and Lee, C. K. (1991). "Voice quality factors: Analysis, synthesis, and perception." *J. Acoust. Soc. Am.*, **90**(5): 2394–2410.

Drugman, T.; Kane, J.; and Gobl, C. (2014). "Data-driven detection and analysis of the patterns of creaky voice." *Computer Speech Language*, **28**(5), 1233–1253.

Esling, J. H. (2006). "Voice quality." In K. Brown (Ed.), *Encyclopedia of Language and Linguistics* (2nd edn.), 13: 470–474. Oxford: Elsevier.

Gerratt, B. R. and Kreiman, J. (2001). "Toward a taxonomy of nonmodal phonation." *J. Phonetics*, **29**: 365–381.

Gobl, C. and Ní Chasaide, A. (2003). "The role of voice quality in communicating emotion, mood and attitude." *Speech Communication*, **40**: 189–212.

Gordon, M. and Ladefoged, P. (2001). "Phonation types: A cross-linguistic overview." *J. Phonetics*, **29**, 383–406.

Ishi, C. T. (2004). "Analysis of autocorrelation-based parameters for creaky voice detection." *Proc. 2nd International Conference on Speech Prosody*, 643–646.

Ishi, C. T. and Arai, J. (2018). "Periodicity, spectral and electroglottographic analyses of pressed voice in expressive speech." *Acoust. Sci. Technol.*, **39**(2): 101–108.

Ishi, C. T.; Ishiguro, H.; and Hagita, N. (2007). "Acoustic and EGG analysis of pressed phonation." *Proc. International Conference on Phonetic Sciences (ICPhS'2007)*: 2057–2060.

Ishi, C. T.; Ishiguro, H.; and Hagita, N. (2008a). "Automatic extraction of paralinguistic information using prosodic features related to F0, duration and voice quality." *Speech Commun.*, **50**(6): 531–543.

Ishi, C.; Ishiguro, H.; and Hagita, N. (2008b). "A method for automatic detection of vocal fry." *IEEE Trans. Audio Speech Language Proc.*, **16**(1): 47–56.

Ishi, C.; Ishiguro, H.; and Hagita, N. (2010). "Acoustic, electroglottographic and paralinguistic analyses of "rikimi" in expressive speech." *Proc. Speech Prosody 2010 (SP2010)*, ID 100139: 1–4.

Kiessling, A.; Kompe, R.; Niemann, H.; Nöth, E.; and Batliner, A. (1995) "Voice source state as a source of information in speech recognition: Detection of laryngealizations," in *Speech Recognition and Coding: New Advances and Trends*, pp. 329–332, Berlin, Heidelberg, New York.

Klasmeyer, G. and Sendlmeier, W. F. (2000). "Voice and emotional states," In *Voice Quality Measurement*, pp. 339–358. Singular Thomson Learning.

Kreiman, J. and Gerrat, B. (2000). "Measuring vocal quality," In Kent, R. D., Ball, M. J., (eds), *Voice Quality Measurement*, pp. 73–101. San Diego: Singular Publishing Group.

Laver, J. (1980). "Phonatory settings." *The Phonetic Description of Voice Quality*. Cambridge: Cambridge University Press, 93–135.

Maddieson, I. and Ladefoged, P. (1985). "'Tense' and 'lax' in four minority languages of China." *J. Phonetics*, **13**: 433–454.

Sadanobu, T. (2004). "A natural history of Japanese pressed voice." *J. Phonetic Soc. Japan*, **8**(1): 29–44.

Yokonishi, H.; Imagawa, H.; Sakakibara, K.-I.; Yamauchi, A.; Nito, T.; Yamasoba, T.; and Tayama, N. (2016). "Relationship of various open quotients with acoustic property, phonation types, fundamental frequency, and intensity." *J. Voice*, **30**(2): 145–157.

Chapter Five

Generation, acoustic properties, and behavioral relevance of pulse tone vocalizations in bird vocal repertoires

Franz Goller

Introduction

Sound generators, displaying vocal-fold-like tissue properties, give rise to a highly diverse range of sounds that may differ in fundamental frequency, spectral properties, and timbre. For the human voice, these differences have been thoroughly described acoustically and linked to the vibratory behavior of the vocal folds (e.g., Hirano 1988; Titze 1988; Vilkman *et al.* 1995; Ŝvec and Schutte 1996; Henrich 2006; Roubeau *et al.* 2009; Echternacht *et al.* 2017). The different oscillation regimes (labeled M0–M3) generate different frequencies. The lowest frequencies emanate from a pulse tone mechanism (M0, also called creak, vocal fry, or *Strohbass*), followed by the modal regime (M1, chest register)

Creak: Theories and practices of pulse phonation
Edited by Francesco Venturi

ISBN 978-981-5129-29-8 (Hardcover), 978-1-003-61606-1 (eBook)
www.jennystanford.com

and above that mechanisms M2 (falsetto) and M3 (whistle and flageolet) (Roubeau *et al.* 2009). Although there is some overlap in frequency between the different modes, the broad frequency range is achieved by switching from mode to mode. Several classifications of modes based on oscillatory dynamics have been attempted (e.g., Roubeau *et al.* 2009).

Vocal organs in terrestrial vertebrates differ in morphological complexity, yet the main mechanism of oscillation is similar across this range. For example, among the terrestrial vertebrates, amphibians, non-avian reptiles, and mammals phonate with vocal cords or vocal folds in the larynx, whereas birds have evolved a unique vocal organ, the syrinx. Nevertheless, oscillating tissues across all these groups share strong similarity in histological composition and use a shared mechanism of sustaining oscillations—as accounted for in myoelastic aerodynamic theory (e.g., Titze 2006; Elemans *et al.* 2015). We can, therefore, assume that the possibility of using the different oscillation regimes described for the human sound source must exist in vocal organs of non-human animals as well.

A detailed characterization of oscillatory regimes, similar to that in humans, has not been established for vocalizations of non-human mammals or birds. We, therefore, do not know to what degree the mechanisms (M1–M3) contribute to the vocal repertoires of most terrestrial vertebrates. In contrast, acoustic analyses have shown that some non-human mammals and birds also use the pulse tone mechanism for some of their vocalizations. Additionally, functional analyses revealed specific information about the generation of pulse tones in birds. Pulse tone generation and use in non-human animals will be the topic of this review. I will address how this pulse tone mechanism gives rise to specific sounds, how these sounds differ from those generated with other mechanisms, and how the specialized vocal organs of birds facilitate additional complexity in pulse tone sounds. Finally, we will explore the evolutionary significance these specialized sounds may have.

Functional morphology of pulse tone generation

Across vertebrates, the sound sources in either the larynx (terrestrial mammals, non-avian reptiles, and amphibians) or the syrinx (birds) are soft flapping structures, consisting of epithelia-encased loose connective tissue. These vocal cords or vocal folds can be induced to vibrate by a passing airstream. Typically, animals phonate during expiration, but some sounds are also produced during inspiration. Depending on the specific composition, vocal folds can engage in different oscillatory regimes with specific nonlinear dynamics.

Typically, airflow draws the vocal folds together and elastic restitution forces pull them apart. This closing and opening of the airway—the aspect of vibration perpendicular to the airflow direction—is accompanied by an additional wave-like motion of the vocal folds along the longitudinal axis, which causes a change in profile during the oscillation. A switch from a divergent to a convergent profile is thought to maintain oscillations once they have started. These two main oscillatory components are accompanied by even more complex tissue movements, all of which give rise to pressure oscillations, which deviate from a sinusoidal pattern and, therefore, generate a complex energy distribution across the harmonic spectrum. This oscillatory mode is typically referred to as modal vibration regime and in human voice is referred to as M1, based on details of the laryngeal mechanisms (Roubeau *et al.* 2009).

The oscillatory patterns of the vocal folds can be studied with electroglottography, and these recordings show distinct patterns when the voice shifts to higher frequencies as it transitions between registers (Roubeau *et al.* 2009). A striking pattern accompanies pulse tone generation with a highly skewed distribution of closed versus open periods during an oscillation cycle. Long periods in which the vocal folds are in contact and thus close the aperture alternate with very brief opening motions. The brief open periods release pressure pulses into the area above the vocal folds, and these pulses of high pressure alternate with long periods of low pressure. This pressure oscillation pattern is far from sinusoidal, thus giving rise to a rich and broad distribution of energy across the upper harmonic frequency spectrum.

In the modal mechanism, stress and tissue density of the vocal folds determine their oscillation frequency (Titze 2011). Interestingly, in pulse tone vibratory mode, the vocal folds are short and flaccid, such that sound frequency is largely determined by the driving pressure. This is illustrated by the fact that pulse tones generated by human males and females show similar sound frequency ranges (e.g., Blomgren *et al.* 1998). Despite differences in vocal fold mass, this mass is distributed such that density ranges overlap. In contrast, the smaller mass of female vocal folds compared to males drives oscillations of higher frequency in the modal regime. This difference in frequency control constitutes an important aspect of vocal behavior, which will be discussed below.

Details of the vibratory behavior of the vocal folds during pulse tone generation are mainly known from investigations of the human larynx. In this vibratory mode, the vocal folds are very short, thick, and lack tautness. High-speed film documentation of the vocal folds and electroglottographic evidence show that vocal fry phonation can involve either singlet opening events or doublet and triplet opening events in rapid succession, followed in all cases by long closed periods (e.g., Blomgren *et al.* 1998). Different phase lags in the longitudinal wave of the vocal folds may explain the occurrence of multiple opening events (Birkholz *et al.* 2011).

Low-dimensional mathematical modeling has been used to describe the behavior of oscillating vocal folds in mammals and birds (e.g., Titze and Story 2002; Cataldo *et al.* 2006; Mindlin 2017). These approaches capture many aspects of the vibratory behavior despite the low dimensionality. The two dynamically different vibratory modes can be modeled as well and point toward a qualitative difference in how the oscillations arise. Bifurcations give rise to self-sustained oscillations. While modal oscillations are born out of Hopf bifurcations, pulse tone sounds arise from saddle-node in limit cycle (SNILC) bifurcations (e.g., Amador and Mindlin 2008; Sitt *et al.* 2008; Mindlin 2017). The two oscillatory regimes have implications for the acoustic quality of the sound. As stated earlier, one main aspect is the energy distribution across the harmonic spectrum. Pulse tones are characterized by high-energy content in the upper harmonics, and this upper harmonic content decreases with increasing pulse frequency (Sitt *et al.* 2008).

Details of pulse tone production in songbirds

The avian sound-generating organ, the syrinx, is located at the tracheobronchial junction, and this location allowed for the evolution of two sound sources, one situated in each bronchus (Fig. 5.1A). The ancestral avian vocal organ may have been one with two sound sources (Kingsley *et al.* 2018), but the syrinx of some extant taxa is situated above the tracheobronchial junction and, therefore, comprises only one source. The most extensively studied

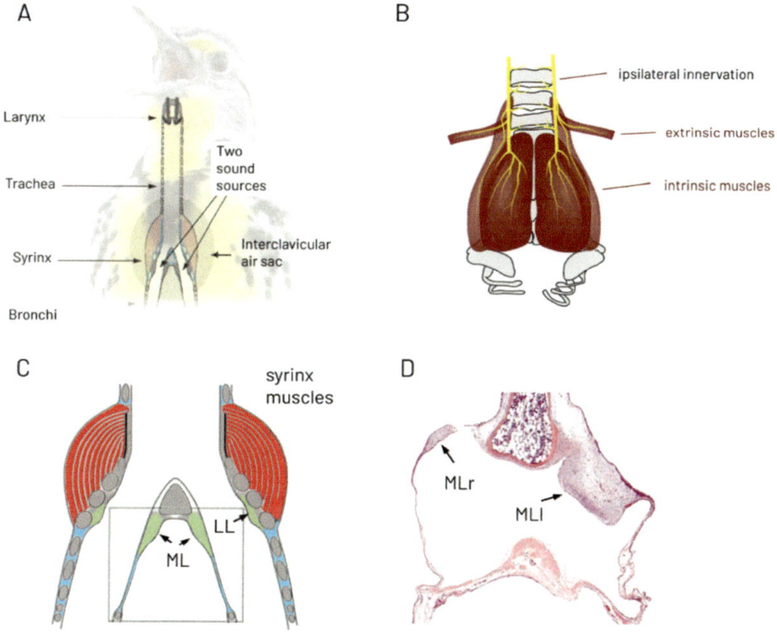

Figure 5.1 (A) The syrinx is the sound-generating organ of birds and is situated at the tracheobronchial junction. In songbirds, it contains two independently controlled sound sources (modified from Kingsley *et al.* 2018). (B) The songbird syrinx is controlled by extrinsic (one insertion site outside syrinx area) and intrinsic (insertion sites are in syringeal area) musculature. Motor control of sound production is affected by syringeal musculature, which is predominantly ipsilaterally innervated from motoneurons whose cell bodies are located in the tracheosyringeal portion of the hypoglossal nucleus (modified from Faunes *et al.* 2017). (C) Schematic depiction of a frontal section of the syringeal area, showing the labial pairs (medial labia (ML) and lateral labia (LL)—in green color), which are present on each side of the syrinx. The gray box illustrates the approximate section shown in (D). Histological section of syringeal section showing the medial labia of a European starling. The left medial labium (MLl) is much larger than the right medial labium (MLr) (modified after Prince *et al.* 2011).

syrinx with two sources is the one of oscine songbirds. In this group, each sound source is independently controlled by ipsilateral syringeal muscles, which regulate airflow and actively control acoustic parameters such as fundamental frequency and sound amplitude (Fig. 5.1B). In some species, the two sources are tuned to different frequency ranges, which is facilitated by an up to eight-fold difference in labial volume (see Figs. 5.1C and 5.1D; for review, see Suthers and Zollinger 2008; Goller 2016).

Additionally, the elastic proteins (collagen and elastin) are arranged in a layered structure, which likely allows recruitment of different portions into vibration and thus facilitates the generation of a broad frequency range (Riede and Goller 2014).

Recordings of airflow through each bronchus allowed determination of the respective sound frequencies contributed by the two syringeal sound generators. In the European starling (*Sturnus vulgaris*), the left sound source contributes sounds from approximately 1.2–3.8 kHz and the right side contributes 3–12 kHz to song. In the song, these contributions can be delivered sequentially or both sides can generate independent tones simultaneously (Goller 2022). If the two frequencies are close, the interaction of the two sounds will generate an amplitude modulation with a rate that is equal to the frequency difference between the two sounds: difference tone or beat note. In addition, starlings also use the pulse tone mechanism to generate specific low-frequency song syllables (Jensen *et al.* 2007). The fundamental frequency of these syllables ranges from 150 to 800 Hz. A striking difference in the use of the two sound sources is notable between the modal and pulse tone mechanisms. During modal phonation, the two sources contribute unilaterally or, if bilaterally, the sounds are harmonically unrelated. In contrast, during pulse tone, typically both sides contribute the same frequency simultaneously.

How the two sides are operating during pulse tone vocalizations was determined through high-speed filming and recording of airflow through two sound sources in crows (*Corvus cornix*), European starlings, and zebra finches (*Taeniopygia guttata*) (Jensen *et al.* 2007). Irrespective of the different frequency ranges during modal sound production, for pulse tones both sides contribute the same low frequency (Fig. 5.2). The short opening events, creating the pulse, can occur simultaneously in both generators or they can be phase shifted relative to each other. The shift induces double pulses and, therefore, influences the harmonic spectrum. If the phase shift is 180 degrees, the sound frequency doubles (Fig. 5.2). In addition to possible coupling effects, it is likely that the specific bilateral pattern is, in part, regulated through neural control.

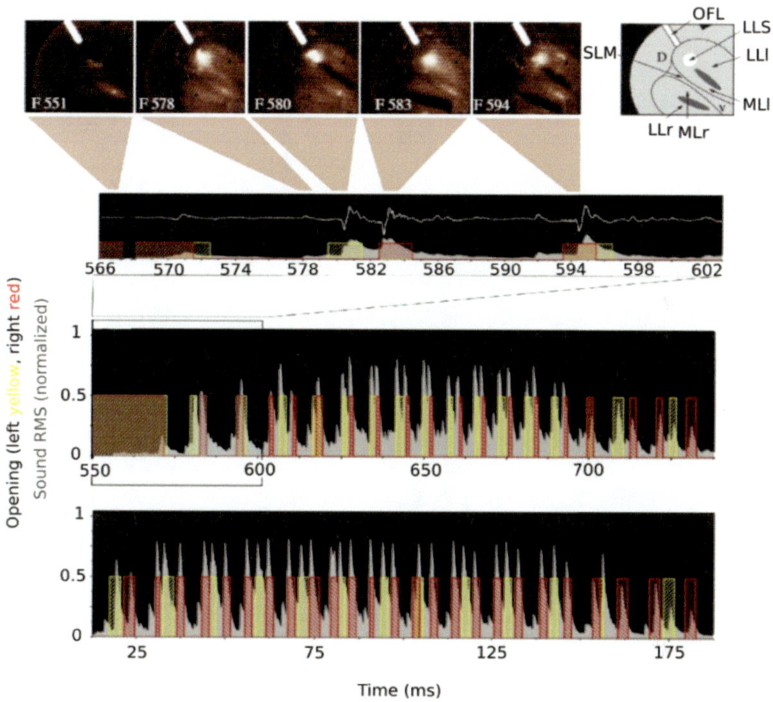

Figure 5.2 (Top) High-speed imaging of the crow syrinx during induced phonation shows the opening events of the two sound sources. A schematic on the right provides explanation for the field of view through the endoscope. SLM, semilunar membrane; OFL, optic fiber layer; LLS, laser light spot; MLl, MLr, left and right medial labium; LLl, LLr, left and right lateral labium; D, dorsal; V, ventral. The next panel illustrates the segment of sound corresponding to the images in the top panel. The microphone trace is depicted in white, and the opening and closing events of the left and right syrinx are depicted in yellow and red, respectively. Bottom panels illustrate two longer segments with different phase relationships of the opening events in the left and right syrinx, thus generating different pulse sequences (modified after Jensen *et al.* 2007).

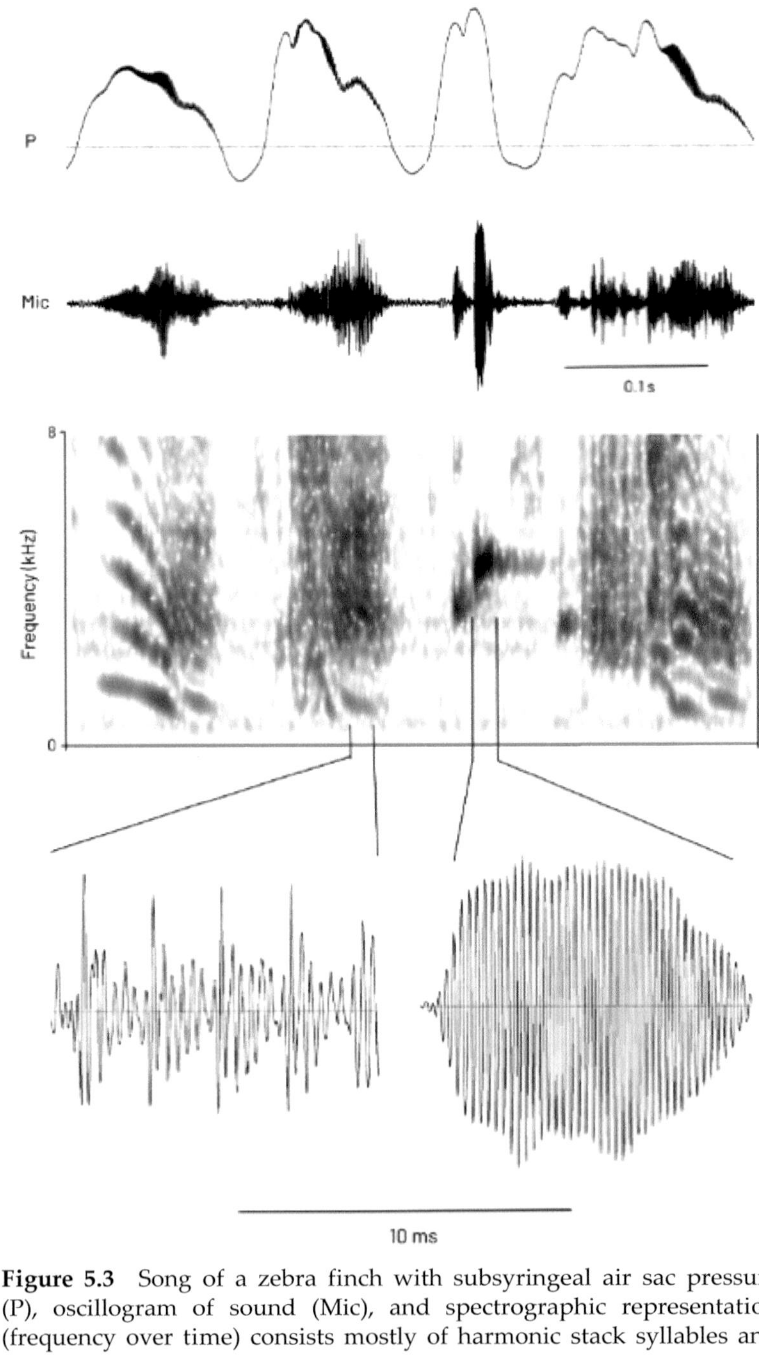

Figure 5.3 Song of a zebra finch with subsyringeal air sac pressure (P), oscillogram of sound (Mic), and spectrographic representation (frequency over time) consists mostly of harmonic stack syllables and

can include high-frequency elements (third syllable). In this example, all syllables are generated during expiration. The silent elements between syllables are used for mini-breaths—i.e., short inspirations indicated by pressure below the horizontal line, which indicates atmospheric pressure. Segments of a harmonic stack syllable and a high-frequency syllable are enlarged to reveal the detailed oscillation patterns (bottom panels). The pulsed pattern of the harmonic stack is distinctly different from that of the high-frequency syllable, which is likely generated by the modal mechanism.

European starlings generate a variety of calls with the pulse tone mechanism, and pulse tone syllables are also incorporated into their warbling songs. These song syllables are instructive as they allow comparison between pulse tone syllables and surrounding tonal syllables. The energy content in upper harmonics is much higher in pulse tone syllables. At the same time, their amplitude is much lower. It can be more than 40 dB lower (Goller unpublished) than that of the surrounding tonal syllables.

Although pulse tone syllables comprise only a small portion of the warbling song of starlings, they are the predominant syllable type in the zebra finch. Zebra finch song consists of a series of harmonic stack syllables (Fig. 5.3), which are generated with the pulse tone mechanism (Sitt *et al.* 2008) and occasional high-frequency elements that are likely generated with the modal mechanism. The fundamental frequency of harmonic stack syllables ranges from 500 to 1500 Hz, and they are typically produced by both sides, as indicated by bilateral airflow (Goller and Cooper 2004; Jensen *et al.* 2007). As in starlings, the opening events in the two sound generators can be in phase or out of phase, giving rise to complex pulse sequences and complex energy distribution across the spectrum. Through upper vocal tract filtering, upper harmonic content is enhanced, such that the dominant frequency ranges between 2 and 6 kHz. Interestingly, the labia of the zebra finch syrinx—large relative to their body size—are not strongly laterally asymmetric and display a uniform composition of elastic fiber proteins (collagen and elastin) (Riede and Goller 2014).

Vocalizations in pulse tone mode across vertebrates

The use of pulse tone vocalizations is probably widespread among mammals and birds. Individual calls may be part of a large vocal repertoire and, therefore, may not have received specific attention. This applies particularly to descriptions of bird vocal repertoires, which may be quite extensive. To appreciate this aspect, we need to briefly review vocal repertoires of birds. All birds produce vocalizations, which are termed calls. Calls are defined as vocalizations with typically a singular function, such as alarm calls, contact calls, begging calls, mobbing calls, etc. Often these calls are short and simple in their acoustic and temporal structure, but many exceptions exist. The part of the vocal repertoire that has received most attention is birdsong. Song may consist of simple vocalizations or be composed of thousands of different sound elements. Again, it is most useful to define song functionally. Song is thought to serve multiple functions, with substantial variation in use observed across species. Singing individuals communicate with conspecific males in a territorial context, which frequently involves maintenance of territorial boundaries and access to females (for review, see Catchpole and Slater 2010). During the breeding season, when most song is produced, it is thought to attract females and serve in courtship. In addition, song may serve functions in the synchronization of pairs for reproduction and population-wide synchronization of the breeding effort.

Most research efforts have focused on song because it is more elaborate than calls—its complexity has attracted attention for centuries. Furthermore, some birds exhibit vocal production learning, which refers to the nature of song development. Songs can be developed either innately or through a process of learning. In the latter mechanism, species-specific sounds need to be acquired after hatching and are then used to shape songs through a sensorimotor learning period. This song-learning process is similar to speech learning in humans and has, therefore, been the focus of most research (e.g., Brainard and Doupe 2002;

Jarvis 2013; Nowicki and Searcy 2014). Consequently, relatively little attention has been given to the rest of a vocal repertoire. It can, therefore, be expected that pulse tone vocalizations are more widely used than is known. In mammals, pulse tone-like vocalizations are generated by species in different taxa (e.g., Riede and Zuberbühler 2003; Shapiro 2006) and appear to be restricted to mid-size to large species. In contrast, in birds we find pulse tone vocalizations in very small—e.g., zebra finch—to very large species (Table 5.1). Especially in small birds, pulse tones facilitate the production of low-frequency vocalizations. The species listed in Table 5.1 illustrate that this vocal production mode allows small birds to generate remarkably low fundamental frequencies.

Table 5.1 Pulse tones in vocal repertoires of small and large birds

Species	**Pulse tone fundamental frequency (kHz)**	**Peak frequency (kHz)**	**Vocal repertoire**	**Body mass (g)**	**Refs.**
Zebra finch (*Taeniopygia guttata*)	0.48–1.6	1.5–3.8	Song syllables, calls	11.9	Jensen *et al.* 2007
White-tipped plantcutter (*Phytotoma rutila*)	0.05–0.3	3–4.3	Song	40.5	Uribarri *et al.* 2020
European starling (*Sturnus vulgaris*)	0.15–0.8		Song syllables	87	Jensen *et al.* 2007
Hooded crow (*Corvus cornix*)	0.3	1.6	Calls	500	Jensen *et al.* 2007
Dwarf cassowary (*Casuarius bennetti*)	0.025	0.04–0.06	Call	17–26	Mack and Jones 2003

Pulse tones contribute unique sounds to vocal repertoires

Pulse tone generation facilitates the production of very low frequencies. In mammals, this mechanism is widely used in middle-sized to larger species, but it is not known to be a typical sound-production mechanism in the vocal repertoire of small species. In part, this may be explained because the smallest mammals communicate with very-high-frequency vocalizations that reach into the ultrasound. However, the vocal repertoires of very few species have been thoroughly described, or vocalizations have not been investigated in respect to their oscillatory dynamic origins. We can, therefore, expect to find occasional use of pulse tone sounds also in small mammals.

In birds, it is also not known how broadly pulse tone vocalizations are used in diverse vocal repertoires. Instead of discussing phylogenetic patterns, I will, therefore, illuminate some functional perspectives based on case examples. As pointed out above, the use of the pulse tone dynamics allows birds to generate low-frequency sounds. However, this mechanism is not the only one (Uribarri *et al.* 2020). Simultaneous use of the two sound generators can lead to interactions of the two harmonically unrelated sounds. This and additional nonlinear interactions can account for many low-frequency calls—e.g., ruby-crowned kinglet, *Regulus calendula*; common white-throat, *Sylvia communis*, etc.—and pulse-like appearing song syllables—e.g., red-winged blackbird, *Agelaius phoeniceus*; saltmarsh sparrow, *Ammospiza caudacuta*—(Goller, unpublished). It has not been systematically investigated how sounds generated by these two mechanisms differ in their acoustic structure. It is, therefore, also unclear why some species use one mechanism whereas others use the other. One possible reason for these patterns could be found in the specific morphology of the vocal folds (labia) of each species and the range of stress–strain conditions that are possible for each labium. If size and attachment to the cartilaginous framework are such that the vocal folds cannot shorten sufficiently to engage in a pulse tone mechanism, the interactive mechanism may be used for generation of pulse-like oscillations. While this explanation indicates why we do not find pulse tone sounds in the vocal repertoires of some species, it cannot explain the use of both mechanisms within the vocal repertoire of individual species. Perhaps acoustic differences between sounds generated by the two mechanisms may provide an explanation in this case. Although specific comparative information is not available to elucidate how different production mechanisms may yield different acoustic features, an instructive example can shed some light on this question.

As we discussed above, European starlings use the pulse tone mechanism to generate low-frequency calls and song syllables. We raised young starlings with brown-headed cowbirds (*Molothrus ater*) to induce them to imitate their song (Cooper and Goller, unpublished). Although brown-headed cowbirds sing very short songs (<1.5 s), their songs are remarkably complex. They span a very broad frequency range. Songs are composed of introductory note clusters of low-frequency components (450–1500 Hz) and high-frequency whistles (up to 12 kHz). All song syllables are tonal—the energy of upper harmonics is much lower than that of the fundamental frequency—and, therefore, are likely generated by a modal oscillatory mechanism (Fig. 5.4). This is especially remarkable for the lowest frequency elements of the introductory note clusters. In these, left-side generated elements alternate rapidly with right-side generated elements, creating a series of elements with stepwise changes in fundamental frequency (Allen and Suthers 1994). Because starlings also generate low-frequency sounds and incorporate them into their songs but do so using the pulse tone mechanism, we were interested whether they copied the introductory note clusters and if so, how they generated the lowest frequencies.

The results demonstrate that the acoustic differences between sounds emanating from different oscillatory regimes appear to matter and that the spectral differences appear perceptually relevant to the birds. Starlings do not seem to be capable of imitating the low-frequency introductory notes with the modal mechanism (M1), which is consistent with the observation that no such sounds appear in their highly variable and large song repertoires. However, starlings do not imitate the fundamental frequency of the lowest cowbird syllables by using a pulse tone mechanism, yet they do copy introductory note clusters

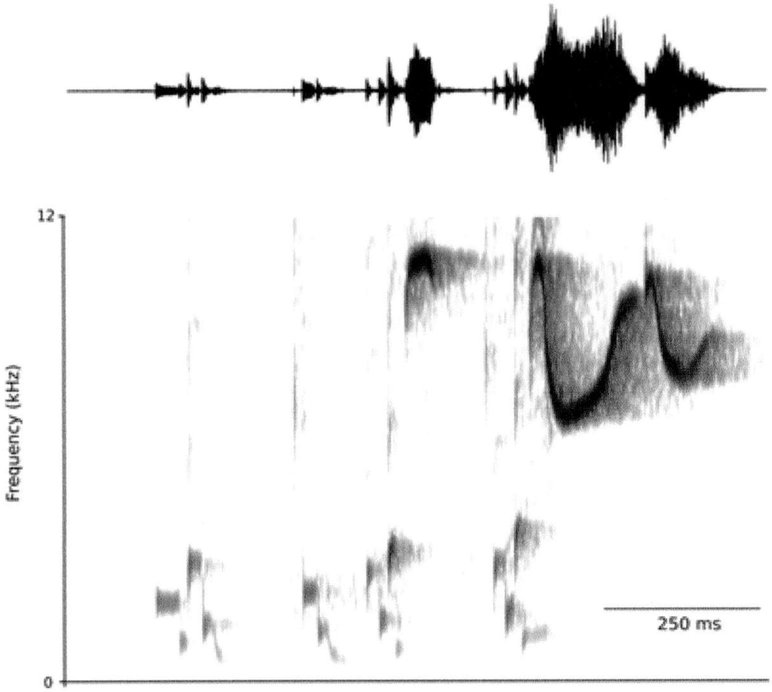

Figure 5.4 Example of brown-headed cowbird song (oscillogram, top; spectrogram, bottom) illustrates the complexity of the short song. Low-frequency introductory notes are highly tonal but significantly softer than the high-frequency whistles.

with some resemblance to those of cowbirds to the human ear. They predominantly use their two independent sound sources, vibrating within their natural frequency range, and interactions of the two independent tones of different frequencies give rise to the low-frequency components (Fig. 5.5). As would be true for a pulse tone mechanism, sounds generated by interaction also do not imitate the tonal quality of the cowbird tutor model. However, the different distribution of energy across the spectrum between interactive sounds and pulse tone sounds (Figs. 5.5 and 5.6) yields a more faithful copy to the human ear. The difference in spectral complexity is illustrated by an example from one individual, which in addition to the interactive mechanism also generated very short segments in pulse tone mode (Fig. 5.6).

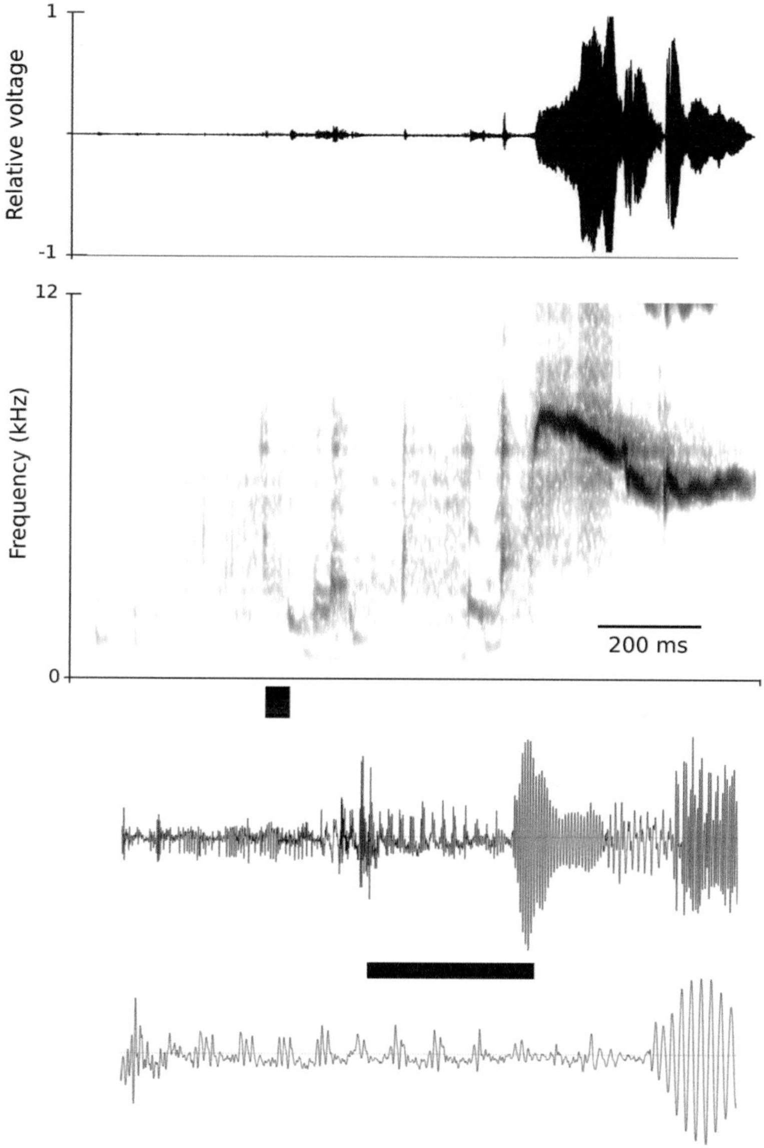

Figure 5.5 The imitation of a brown-headed cowbird song by a European starling (top oscillogram, below spectrogram) documents that low-frequency syllables are not generated with the modal mechanism as occurs in cowbirds. The main mechanism used for the introductory note clusters is interaction of simultaneous bilateral sounds. In this individual, a short segment generated with the pulse-tone mechanism is highlighted in the lower panels with increasing expansion. It is also notable that all syllables of the introduction are of low intensity compared to the loud high-frequency terminal element.

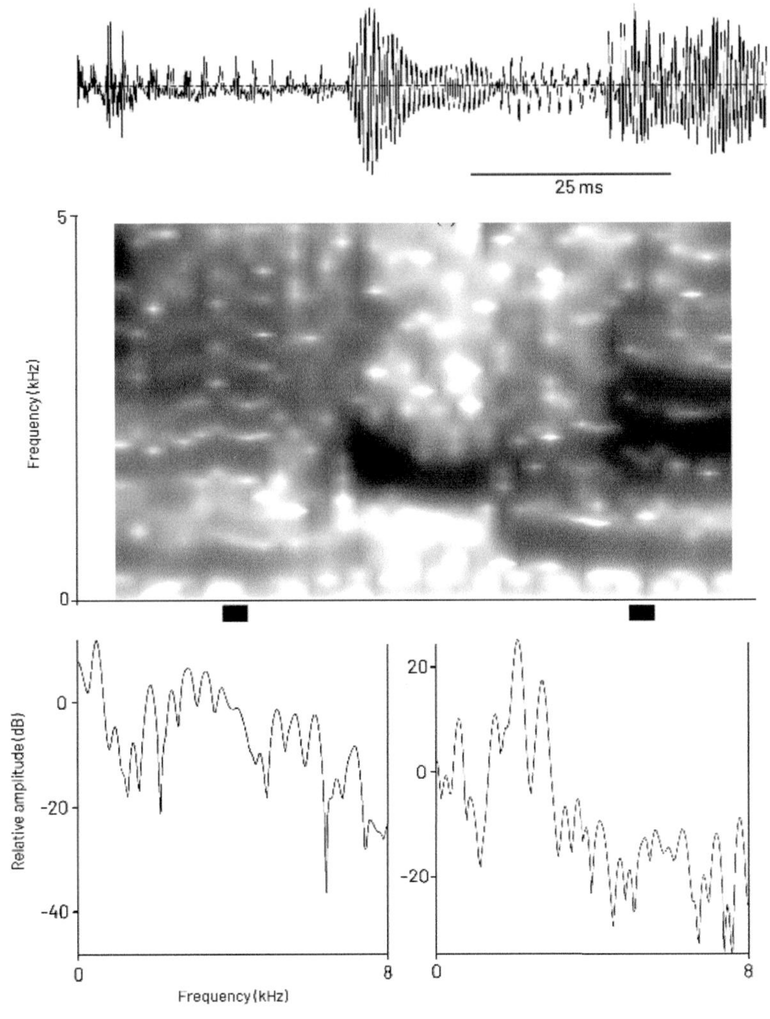

Figure 5.6 The short pulse-tone sequence (illustrated in Fig. 5.5) generates a different energy distribution across the upper harmonic spectrum than the interaction of sounds generated by the two independent sources. The short sound segment is illustrated as oscillogram (top panel) and below as spectrogram. The black horizontal bars identify the segments depicted as power spectra (left bar, pulse tone, lower left panel; right bar, interaction, bottom right panel).

A major difference between pulse tone and modal sounds in the vocal repertoires of the investigated avian species appears to be sound amplitude. Although we do not have comparative data on absolute sound intensity in the field, the relative amplitude differences in recordings where both mechanisms contribute song syllables illustrate this point. Both in European starlings and in zebra finches, pulse tone song syllables are much softer than the modal song sequences. Furthermore, the pulse rate and, therefore, the fundamental frequency are dependent on the driving air sac pressure. This relationship is nicely illustrated in the song of the white-tipped plantcutter (*Phytotoma rutila*), where pulse rate increases dramatically at the onset of each song (Uribarri et al. 2020). Therefore, the frequency reflects effort, which may be a perceptually salient feature for receivers.

A song is a signal used in male–male interactions and in attracting and courting females. It is, therefore, thought to be linked to sexual selection, and song features are thought to reveal male quality (Andersson 1994; Podos and Sung 2020). To what degree individual syllables provide information about the sender is poorly understood, but some correlative evidence suggests this possibility (Rehsteiner *et al.* 1988). Although zebra finch song is predominantly composed of pulse tone syllables, these are only a small component of song in the European starling. Accordingly, their role in communicating the sender's qualities may differ. Zebra finches are likely to indicate vocal prowess in the harmonic stack syllables, and song amplitude has been identified as an important signal to females (Holveck and Riebel 2007; Riebel 2009). In the case of the European starling, it is possible that the few pulse tone syllables and the ability to generate them with relatively high amplitude may constitute a specific signal that is indicative of a male's fitness. This possibility, however, needs to be thoroughly tested.

Pulse tone vocalizations are rich in upper harmonic content, and this feature may provide a specific substrate for modulation by the upper vocal tract to encode subtle differences in meaning. Of course, as is evident in human speech, the use of upper harmonic energy and its dynamic filtering is not restricted to sounds generated with the pulse tone mechanism. However, as suggested above for small birds, this may be the main mechanism for generating harmonically rich, low-frequency sounds that then can be modulated subtly by upper vocal tract filtering. The sounds of many corvids and the zebra finch provide prominent examples that support this notion, and behavioral and neural evidence show the perceptual importance of these subtle modifications (Elie and Theunissen 2015a,b). These examples, from humans to songbirds, lead to the speculation that the evolution of social behavior has favored the use of rich harmonic sounds for communication.

References

Allen, S. E. and Suthers, R. A. (1994). "Lateralization and motor stereotypy of song production in the brown-headed cowbird." *Journal of Neurobiology*, 25, 1154–1166.

Amador, A. and G. B. Mindlin (2008). "Beyond harmonic sounds in a simple model for birdsong production." *Chaos*, **18**, 043123.

Andersson, M. (1994). *Sexual Selection*. Princeton University Press.

Birkholz, P., B. J. Kröger, and C. Neuscheafer-Rube (2011). "Synthesis of breathy, normal, and pressed phonation using a two-mass model with a triangular glottis." *Proceedings of the Interspeech 2011* (Florence, Italy, 2011), pp. 2681–2684.

Blomgren, M., Y. Chen, M. L. Ng, and H. R. Gilbert (1998). "Acoustic, aerodynamic, physiologic, and perceptual properties of modal and vocal fry registers." *J. Acoust. Soc. Am.*, **103**, 2649–2658.

Brainard, M. S. and A. J. Doupe (2002). "What songbirds teach us about learning." *Nature*, **417**, 351–358.

Cataldo, E., F. R. Leta, J. Lucero, and L. Nicolato (2006). "Synthesis of voiced sounds using low-dimensional models of the vocal cords and time-varying subglottal pressure." *Mech. Res. Commun.*, **33**, 250–260. Retrieved from doi:10.1016/j.mechrescom.2005.05.007.

Catchpole, C. K. and P. J. B. Slater (2010). *Bird Song: Biological Themes and Variation*, 2nd edition. Cambridge University Press.

Echternach, M., F. Burk, M. Köberlein, C. T. Herbst, M. Döllinger, M. Burdumy, and B. Richter (2017). "Oscillatory characteristics of the vocal folds across the tenor passaggio." *J. Voice*, **31**, 381.e5–381.e14.

Elemans, C. P., J. H. Rasmussen, C. Herbst, D. N. Düring, S. A. Zollinger, H. Brumm, K. Srivastava, N. Svane, M. Ding, O. N. Larsen, S. J. Sober, and J. G. Švec (2015). "Universal mechanisms of sound production and control in birds and mammals." *Nat. Commun.*, **6**, 8978.

Elie, J. E. and F. E. Theunissen (2015a). "Meaning in the avian auditory cortex: Neural representation of communication calls." *Eur. J. Neurosci.*, **41**, 546–567. Retrieved from doi:10.1111/ejn.12812.

Elie, J. E. and F. E. Theunissen (2015b). "Zebra finches identify individuals using vocal signatures unique to each call type." *Nat. Commun.*, **9**, 4026. Retrieved from DOI: 10.1038/s41467-018-06394-9.

Faunes, M., J. Botelho, and J. M. Wild (2017). "Innervation of the syrinx of the zebra finch (*Taeniopygia guttata*)." *J. Comp. Neurol.*, **525**, 2847–2860. Retrieved from DOI: 10.1002/cne.24236.

Goller, F. (2016). "Sound production and modification in birds: Mechanisms, methodology and open questions," in: *Comparative Bioacoustics: An Overview*. Bentham Science Publishers, pp. 165–230.

Goller, F. (2022). "Vocal athletics: From birdsong production mechanisms to sexy songs." *Anim. Behav.*, **184**, 173–184.

Goller, F. and B. G. Cooper (2004). "Peripheral motor dynamics of song production in the zebra finch." *Ann. N.Y. Acad. Sci.*, **1016**, 130–152.

Henrich, N. (2006). "Mirroring the voice from Garcia to the present day: Some insights into singing voice registers." *Logoped. Phoniatr. Vocol.*, **31**, 3–14. Retrieved from DOI: 10.1080/14015430500344844.

Hirano, M. (1988). "Vocal mechanisms in singing: Laryngological and phoniatric aspects." *J. Voice*, **2**, 51–69.

Holveck, M-J. and K. Riebel (2007). "Preferred songs predict preferred males: Consistency and repeatability of zebra finch females across three test contexts." *Anim. Behav.*, **74**, 297–309.

Jarvis, E. D. (2013). "Evolution of brain pathways for vocal learning in birds and humans," in: Bolhuis, J. J. and Everaert, M. (Eds.), *Birdsong, Speech, and Language*. MIT Press, pp. 63–107.

Jensen, K. K., B. G. Cooper, O. N. Larsen, and F. Goller (2007). "Songbirds use pulse tone register in two voices to generate low-frequency sound." *Proc. Royal Soc. B*, **274**, 2703–2710.

Kingsley E. P., C. M. Eliason, T. Riede, Z. Li, T. W. Hiscock, M. Farnsworth, S. L. Thomson, F. Goller, C. J. Tabin, and J. A. Clarke (2018). "Identity and novelty in the avian syrinx." *Proc. Natl. Acad. Sci. USA*, **115**, 10209–10217.

Mack, A. L. and J. Jones (2003). "Low-frequency vocalizations by cassowaries (*Casuarius* spp.)." *The Auk*, **120**, 1062–1068.

Mindlin, G. B. (2017). "Nonlinear dynamics in the study of birdsong." *Chaos*, **27**, 092101.

Nowicki, S. and W. A. Searcy (2014). "The evolution of vocal learning." *Curr. Opin. Neurobiol.*, **28**, 48–53.

Podos, J. and H.-C. Sung (2020). "Vocal performance in songbirds: From mechanisms to evolution," in: Sakata, J. T. *et al.* (Eds.), *The Neuroethology of Birdsong, Springer Handbook of Auditory Research 71.*

Prince, B., T. Riede, and F. Goller (2011). "Sexual dimorphism and bilateral asymmetry of syrinx and vocal tract in the European starling (*Sturnus vulgaris*)." *J. Morphol.*, **272**, 1527–1536.

Rehsteiner, U., H. Geisser, and H. U. Reyer (1988). "Singing and mating success in water pipits: One specific song element makes all the difference." *Anim. Behav.*, **55**, 1471–1481.

Riebel, K. (2009). "Song and female mate choice in zebra finches: A review." *Adv. Study Behav.*, **40**, 197–238. DOI: 10.1016/S0065-3454(09)40006-8.

Riede, T. and F. Goller (2014). "Morphological basis for the evolution of acoustic diversity in oscine songbirds." *Proc. Royal Soc. B*, **281**, 20132306.

Riede, T. and Zuberbühler, K. (2003), "Pulse register phonation in Diana monkey alarm calls," *J. Acoust. Soc. Am.*, **113**, 2919–2926.

Roubeau, B., N. Henrich, and M. Castellengo (2009). "Laryngeal vibratory mechanisms: The notion of vocal register revisited." *J. Voice*, **23**, 425–438. doi:10.1016/j.jvoice.2007.10.014.

Shapiro, A. D. (2006). "Preliminary evidence for signature vocalizations among free-ranging narwhals (*Monodon monoceros*)." *J. Acoust. Soc. Am.*, **120**, 1695–1705.

Sitt, J. D., A. Amador, F. Goller, and G. B. Mindlin (2008). "Dynamical origin of spectrally rich vocalizations in birdsong." *Phys. Rev. E*, **78**, 011905.

Suthers, R. A. and S. A. Zollinger (2008). "From brain to song: The vocal organ and vocal tract," in: Zeigler, H. P. and Marler, P. (Eds.), *Neuroscience of Birdsong*. Cambridge University Press, pp. 78–98.

Ŝvec, J. G. and H. K. Schutte (1996). "Videokymography: High-speed line scanning of vocal fold vibration." *J. Voice*, **10**, 201–205.

Titze, I. R. (1988). "A framework for the study of vocal registers." *J. Voice*, **2**, 183–194.

Titze, I. R. (2006). "The myoelastic aerodynamic theory of phonation." *The National Center for Voice and Speech.*

Titze, I. R. (2011). "Vocal fold mass is not a useful quantity for describing F0 in vocalization." *Journal of Speech, Language, and Hearing Research*, **54**(2), 520–522.

Titze, I. R. and B. H. Story (2002). "Rules for controlling low-dimensional vocal fold models with muscle activation." *J. Acoust. Soc. Am.*, **112**, 1064–1076. DOI: 10.1121/1.1496080.

Uribarri, G., M. J Rodríguez-Cajarville, P. L. Tubaro, F. Goller, and G. B. Mindlin (2020). "Unusual avian vocal mechanism facilitates encoding of body size." *Phys. Rev. Lett.*, **124**, 098101.

Vilkman, E., P. Alku, and A.-M. Laukkanen (1995). "Vocal-fold collision mass as a differentiator between registers in the low-pitch range." *J. Voice*, **9**, 66–73.

PART II

Narratives

Chapter Six

A case study of "vocal fry" in Suffolk

Intersections and contradictions in the sociolinguistic salience and dialectological prevalence of pulse phonation

Madeleine Masters and Christopher Strelluf

Introduction

The chapters in this book reflect the interest that nonmodal, pulse phonation—particularly under the label "vocal fry"—has drawn in recent years. Oft-cited accounts from popular press bemoan that "American women can't stop speaking in vocal fry" (Rhodan 2014); that "'Vocal fry' has joined more traditional young-women voice mannerisms such as run-ons, breathiness and the dreaded question marks in sentences (known by linguists as uptalk) to undermine these women's authority in newly distinctive ways"

Creak: Theories and practices of pulse phonation
Edited by Francesco Venturi

ISBN 978-981-5129-29-8 (Hardcover), 978-1-003-61606-1 (eBook)
www.jennystanford.com

(Wolf 2015); and that "women in the UK are changing the way they speak by adopting 'vocal fry' to sound more like A list celebs such as Katy Perry and Britney Spears" (Claire 2016).

To be fair, the volume of complaints over vocal fry may, in part, be the effect of an echo chamber. Stories in popular press about vocal fry often cite other stories in popular press as evidence of the "epidemic" of vocal fry (to borrow *Marie Claire*'s heading) and often derive credibility from ambiguous references to "a recent study," occasionally linking either to Wolk *et al.* (2012) or Anderson *et al.* (2014) (for a review of media coverage, see Lawson 2016: 174–176). Dallaston and Docherty (2020: 14; citing Lawson 2016) summarize the circularity that "researchers' focus on creaky voice among young American women has fueled public discourse on the topic" and researchers have grounded their studies of creaky voice among young American women in that same public discourse.

Regardless of the sources of attention, pulse phonation nevertheless sits at a fascinating intersection between phonetics and language ideologies. It is at once a potentially new feature of innovative Englishes and a traditional feature of conservative varieties. It is derided as a new force manipulating voice quality and an entirely phonetically natural feature of speech. It is ideologically associated with young women but was regularly used by the late Queen Elizabeth II (Lindsey 2015). It is linked to American Englishes and the spread of Americanisms but is also a feature of regional varieties (e.g., Wells 1982: 91–93). And, like many sociophonetic and other linguistic features, pulse phonation carries no linguistic meaning in English but is apparently rich in social meaning in English-language speech communities.

This chapter explores these intersections and contradictions in the context of a small study of language variation and change in a regional variety spoken in Suffolk, England. Creaky phonation has been noted as a feature of vernacular East Anglian accents (e.g., Norfolk in Trudgill 1974: 186). As such, pulse phonation qualities like creak or vocal fry may hold social meaning as traditional, local dialect features; potentially under threat from the spread of super-regional features of southeast England (see Beal 2010 for a comprehensive review of discourses around the loss of local dialects in England, and of the extent of and resistance to dialect leveling). At the same time, pulse phonation may be spreading as a feature of super-regional Englishes, making vocal fry a simultaneous carrier of social meaning as an external dialect feature that has made incursions into the local area. As such, pulse phonation presents a complex case in Suffolk of language variation and change within a dialect, where the direction of change is unclear and where changes in the social meanings of a dialect feature might be just as important as changes in the feature itself.

We approach these complexities through a small, traditional, sociolinguistic study of Suffolk English, which allows for examination of productions of creaky voice as a change in apparent time and according to the macro-level social category of biological sex. We also attempt to operationalize social meanings through a novel manipulation of the sociolinguistic interview itself—where some interviewees are exposed to a reading passage that invokes negative ideologies about vocal fry—and we examine the extent to which productions of pulse phonation vary following this intervention.

We find indications that pulse phonation does indeed represent a space of linguistic innovation in Suffolk, and one where social meanings interact with the implementation of creaky voice differentially according to the speaker's age and sex. In doing so, we not only reveal pulse phonation to be a fascinating space of linguistic and sociolinguistic change in Suffolk, but also increase attention to vocal fry and other phonatory features as sociolinguistic and dialectological variables. In turn, we can model new ways of building the traditional sociolinguistic interview to explore socially meaningful features like vocal fry.

Background

Defining vocal fry, creaky voice, and pulse phonation subtypes

As chapters by Abdelli-Beruh, Ishi, and others detail extensively, pulse phonation types differ from modal voice qualities in being produced via relatively slow and irregular open–close cycles in the vocal folds. These physiological characteristics manifest themselves in a range of acoustic features, including, most obviously, a drop in f0 as the acoustic correlate of pitch and reduced periodicity in waveforms. Careful articulatory and acoustic analyses delineate a range of pulse phonation subtypes (see Ishi, this collection). For instance, Keating *et al.* (2015) describe acoustic characteristics of six subtypes of creaky phonation, distinguishing vocal fry from prototypical creaky voice, multiply pulsed voice, aperiodic voice, non-constricted creak, and tense voice. They characterize vocal fry as acoustically distinct from commonly occurring prototypical creak on the basis of vocal fry's low noise, and as articulatorily distinct from non-constricted voice, which often occurs at the ends of utterances on the basis of the glottal restriction that produces vocal fry.

Despite articulatory and acoustic evidence for fine-grained categorization of creaky phonatory qualities, Dallaston and Docherty (2020: 3) summarize that these different voice qualities are typically treated as synonymous in sociophonetic literature. Crucially, they point out that while pulse phonation subtypes may be articulatorily and acoustically distinct, whether they are "auditorily distinct warrants further investigation, but experimental work so far supports the treatment of creaky voice as a unitary auditory phenomenon" (2020: 3; see also Davidson 2019; Nix and Chappell, this collection). In this chapter, we work with these looser uses of "vocal fry" and "creaky voice" as interchangeable labels for a range of pulse phonation phenomena. We will generally follow labels used in other sources we cite. In our own data, we will use the labels "creaky voice" or "pulse phonation" for the actual speech productions that we examine and use "vocal fry" as a more abstract label for pulse phonations that are sociolinguistically linked to the sorts of ideologies we described in the introduction to this chapter. We acknowledge, however, that a more careful delineation of voice qualities might lead to different results than the ones we will report.

In languages that do not use pulse phonations for phonemic contrast, such as English, creaky voicings occur naturally for a range of reasons. Creak is typical in normal speech in final positions of utterance, as pulmonic pressure decreases and the vocal folds are relaxed (Slifka 2006; see also Keating *et al.* 2015, 2–3). Creaky voice is more common in the speech of some older people, as their vocal folds become tougher over time and flex less, leaving the vocal folds adducted (Gibson 2017). Creak can also occur as a result of certain disorders or develop due to the treatment of some illnesses, e.g., radiation therapy or surgery for oropharyngeal cancer (Raeke 2021).

But of course, in the context of the present collection, speakers can also produce pulse phonation as a feature of their voice setting (Laver 1968) by speaking at lower pitch than is dictated by physiology, or to index social meanings (Chao and Bursten 2021). These sociolinguistic implementations of creaky voice are historically typical of some British Englishes, including traditionally high-prestige Received Pronunciation (Lindsey 2015).

Social meanings

Gibson (2017: 62) notes that creaky voice has historically been associated with male voices (see also Henton and Bladon 1988) and may be used by men to index masculinity (see also Pittam 1987). But as the chapters in this book make clear, public commentary on vocal fry has associated this voice quality with young females. There is some limited evidence that pulse phonation is indeed occurring more frequently in the speech of younger females relative to other groups of speakers. In New Zealand, Hornibrook *et al.* (2018) quantified the frequency of vocal fry usages by women in recordings of the Canterbury Corpus. They classified two age groups: speakers born between 1972 and 1976, and those born between 1980 and 1987. They reported that the younger group used vocal fry on average 146.7 times in 10-min speech samples, while the average for speakers in the older group was 51.3. Even when one younger speaker who used "extreme vocal fry" was excluded from the dataset, the younger group still produced vocal fry at about twice the rate of the older cohort.

However, as Dallaston and Docherty (2020) critique, scholarly evidence for increases in the use of creaky voice in apparent time or among females is surprisingly thin. As we noted above, Dallaston and Docherty report that sociophonetic studies of creaky voice have largely followed (or driven) popular ideologies by examining creaky voice among young (often American and enrolled in university) women (cf. Wolk *et al.* 2012; Yuasa 2010;

Gibson 2017; Hornibrook *et al.* 2018), to the exclusion of male or non-binary speakers and with limited apparent-time depth. As such, creaky voice has been confirmed to be present in the speech of young women exactly because it has been looked for in the speech of young women, and little can be said about their use of creaky voice relative to anyone else. Moreover, Dallaston and Docherty (2020) warn of a lack of studies of creaky voice in spontaneous speech and of an absence of consistent measurement protocols across studies to assure validity.

But while the actual correlation between young women as a social category and increases in the production of creaky voice appears tenuous, the popular association between young women and vocal fry is firm. As Lawson (2016) reviews, the association of vocal fry with women has brought the condemnation and derision that often accompanies "feminine" linguistic features as a thinly veiled mechanism for condemning women and femininities (Lakoff 1973, for a seminal perspective; Hess 2013 for a general rejection of young female voices by older men as "annoying"; and Maestas 2019 for a recent perspective on another pitch-driven feature "uptalk").

Popular sources like those cited at the beginning of this chapter and throughout this volume offhandedly describe vocal fry as "downright annoying" and "not good for your voice and vocal fry causes long-term damage" before providing lists of women who are "the biggest vocal friers in Hollywood" (see, e.g., Matuso 2015, whose list of 10 Hollywood celebrities nicely reflects the credulity of such pieces by including Hillary Clinton).

A few studies suggest that such ideologies have been internalized by hearers. In their historical survey, Chao and Bursten (2021) report that extreme critical perceptions of vocal fry included people claiming the woman's usage of vocal fry was distracting to such an extent that they could not focus on the content of their utterances. Chao and Bursten argued that this occurred when women were seen to be speaking with an authority they were deemed not to possess. As pointed out by Lindsey (2015), the sexism encoded in such evaluations is especially obvious when they are compared with the desirability attributed to men in popular culture, such as Sean Connery or George Sanders who spoke with very low pitch and, consequently, often produced creaky voice. Women may also adopt creaky voice to index sexual desirability. Lindsey (2015) highlights Mae West as operationalizing vocal fry for sex appeal, and Shaw and Crocker (2015, cited in Jelle 2017) find that Scarlett Johansson uses more vocal fry in seductive and flirtatious roles than in less sexualized roles (she also uses vocal fry more as an American character than a British one). These examples suggest that women are licensed to use creaky voice to index sexual desirability and availability to men but are, otherwise, sanctioned for using creaky voice as they may be asserting authority that they are not allowed to claim. Moreover, as Hejna and Eaton (this collection) reveal, empirical investigation rejects the claim that women use more pulse phonation than men in fictional characterizations, further revealing the disparity between women's perceived versus actual use of vocal fry.

On a global scale, then, the coverage of vocal fry in media and popular culture has led to the ideology that creaky voice is harmful not only to women's vocal folds but also to society. This ideology works to reinforce sexist patriarchal structures that aim to control women in part by sharply critiquing their voices. These ideologies further advance by (and reinforce) narratives of American English spreading through popular culture and adoption by young people and, in doing so, threatening local identities and the integrity of the English language. Research has, to an extent, confirmed some factual points underlying these ideologies—i.e., young American women do indeed use pulse phonation in some speech contexts—but has potentially done so by being designed in fact to confirm these ideologies by only studying the speech of young American women.

Studying pulse phonation in Suffolk English

Ultimately, the abstracted global conceptualizations of the spread and meta-level critique of vocal fry matter most in the ways that they play out in individual speech and local speech communities. In other words, if pulse phonation is an innovation in Englishes, then it will appear as a quantitative change in dialects and idiolects of English. And if sociolinguistic evaluations of vocal fry harm affected speakers, then those consequences will be experienced and navigated by real speakers within speech communities.

Accordingly, we explore the linguistic and sociolinguistic practices of speakers within a speech community as sites where vocal fry is negotiated linguistically and sociolinguistically. We turn to Suffolk, England to study both pulse phonation and the ideologies that surround it in a real locality. In doing so, we respond to calls by Dallaston and Docherty (2020) and by our review above for increased empirical examination of creaky voice outside the United States, as a change in time, beyond just females, and in relatively spontaneous speech.

The county of Suffolk, England is in East Anglia along with Norfolk, Essex, and Cambridgeshire. Despite the region's proximity to London and other areas of southeast England, Wells (1982: 337) highlights that the English of East Anglia "sounds very different from that of London and the home counties." At the same time, Trudgill's (1974, 2021) extensive work in Norwich specifically and throughout the region more generally has shown the region to be a dynamic dialectological space where speakers navigate among marked local variants that trace their origins to the transition from Middle to Modern English and new super-regional innovations, and do so across complex negotiations of macro-level social categories such as sex, socioeconomic class, and region, as well as local identity practices (see also works collected in Fisiak and Trudgill (2001) and Britain's (2005) work in the nearby Fens).

Among the best-known features of East Anglian English is resistance to the merger of the Middle English vowels /a:/ and /ɛi/ to /e/ and the merger of vowels /ɔ:/ and /ɔu/ to /o/. East Anglia joins a very small set of other local traditional dialects in keeping these phonemes distinct, especially in the case of /o/, making minimal pairs such as *moan~mown, sole~soul,* and *nose~knows*—produced, respectively, as [ʊu] and [ʌu]. Wells (1982: 338–340) also characterizes East Anglian English as distinct from other varieties for reassignment of some /o/ lexical types to /ʊ/; variable merger of /iə/ and /ɛə/; centralization of /ɛ/ to [ɜ] or [ʌ]; and extensive deletion of /j/ in words such as *pure, beauty,* and *few.* Butcher (2020: 50) notes a long-attested East Anglian phonotactic constraint, which requires [ə] in unstressed syllables where other varieties allow [ɪ] (e.g., [ɛksət] for *exit;* see also Potter 2018). Recent work by Trudgill (2021) describes a "simultaneous oral and glottal closure" that occurs in intervocalic and word final /p, k/—e.g., *Baker* is produced as [bæikʔə̰]—as well as East Anglian smoothing, in which triphthongs ending in /ə/ are transformed into monophthongs—e.g., *fire* is produced as [fa:] rather than [faɪə]. Beyond this list of segmental features, Wells says "it is a prosodic feature that furnishes one of the most typical characteristics setting East Anglian accents apart from those of elsewhere" (1982: 341). Trudgill labels this characteristic "extreme stress timedness" (2021: 60). Wells (1982: 341) describes this feature as an "extreme lengthening of stressed long vowels, and the compensatory reduction or elimination of certain unstressed vowels."

Descriptions of regional dialects in England and indeed in a broad range of languages and varieties across the history of dialectology (Chambers and Trudgill 2002; Kretzschmar 2018) have focused largely on lexical and segmental variation, rather than suprasegmental features such as voicing modalities. As the works cited in this section make clear, research in East Anglian Englishes has generally mirrored this focus on segments. But these observations about suprasegmental variation between East Anglia and other varieties of English highlight the value of new work to look beyond segments to features such as creaky voice—and indeed to develop methods to support sociophonetic, sociolinguistic, and dialectological examination of pulse phonation.

These features obviously attest to the uniqueness of traditional East Anglian dialects. But these traditional features are also under pressure from supra-regional features and other linguistic innovations. Importantly, competition between local and non-local linguistic features, particularly those associated with London, is metalinguistically salient within Suffolk. Trudgill (2021: 196) attests to the perceived threat posed by London English to local accents in Norwich, citing familiar criticism directed at young people by older Norwich residents for "talking Cockney." Much of this criticism focuses on the sociolinguistically marked replacement of the interdental consonants /ð/ and /θ/ with [v] and [f]. Trudgill (2021: 96) reports, "complainers invariably make the same point: that young people tend to say *fing* and *muvver* rather than *thing* and *mother*."

This discussion hints at the complex landscape of linguistic and sociolinguistic change in East Anglia. It is a region with well-established and locally salient traditional dialect features. While many of these features are segmental, pulse phonation is potentially among them. These traditional features appear to be under pressure from supra-regional leveling and other sources of innovation. There are indications of metalinguistic awareness of some of these changes and in particular awareness of changes that are sociolinguistically associated with innovations from the Englishes of young people in London.

As such, we see potential for pulse phonation and vocal fry to add to this complexity. If pulse phonation is becoming more common in East Anglian Englishes, it might be an additional instance of a supra-regional (or perhaps global) variant replacing traditional characteristics of local dialects. Popular ideologies around vocal fry might make listeners especially sensitive to the voicing modality and potentially critical of it, in parallel with the criticism of pronunciations of <th> in Norwich. On the other hand, if creaky voicing is actually part of local East Anglian dialects, then pulse phonation could exist in the speech community as a conservative feature rather than an innovation. Or pulse phonation could be retreated from on the basis of its stigmatized or perceived status as a non-local variant, which would ironically lead to a retreat from the local feature. Many possibilities exist, and all of them raise interesting possibilities for the ways that phonation, locality, sociophonetic variation, and sociolinguistic ideologies might intersect in voicing modalities.

Methods

This project asks the following questions:

1. To what extent do people in Suffolk use creaky voice?
2. Does conscious awareness of vocal fry by a speaker affect their use of creaky voice?

To answer these questions, the first author (Madeleine Masters) conducted sociolinguistic interviews with eight Suffolk residents. Participants were recruited to form an apparent time sample of four older participants born between 1955 and 1962 and four younger participants born between 1993 and 2000. Participants were balanced between the two groups by biological sex (with all participants also identifying with the gender they were assigned at birth), with two females and two males in each group. All participants were born in Suffolk and lived there throughout childhood and adolescence, when most of their phonetic and phonological acquisition would have taken place (Sankoff and Blondeau 2007).

Participants were interviewed following a sociolinguistic interview script designed by Masters. Consistent with the model that has dominated sociolinguistic methods since Labov's (1966/2006) foundational work, the interview was designed to capture participants' speech across different "contextual styles" by presenting tasks that are likely to vary the level to which participants are consciously aware of their own speech (Becker 2017). It comprised five sections, which moved through semi-structured open-ended questions, evaluations of personal engagement with Suffolk dialect phrases, a reading passage—including some unstructured discussion of the content of the passage—a word list, and judgments about the phonemic statuses of 18 minimal pairs.

While our sample is relatively small, the study design does allow us to address several of the deficiencies identified in Dallaston and Docherty's (2020) review of sociophonetic literature. We explore a meaningful apparent-time interval to examine pulse phonation as a change in progress. We sample males and females to inform pulse phonation's status as a gendered sound change. We research a regional variety of English outside of North America. And we capture data from a range of speech styles, including relatively spontaneous and unmonitored speech in the semi-structured, open-question section of the interview.

In order to specifically probe ideologies connected with vocal fry and examine interactions between participants' speech and the activation of these ideologies, the first author introduced an additional innovation to the disciplinary-standard model of the sociolinguistic interview. The reading passage that half of the participants—balanced across age group and biological sex—engaged with was presented as a critical story about vocal fry from popular media. The passage was created by Masters by combining extracts from two articles—"Vocal fry: What it is and how to get rid of it" (Van Edwards 2021) and "Is vocal fry ruining my voice?" (Johns Hopkins Medicine 2022)—to mimic the style of popular news coverage of vocal fry (e.g., Rhodan 2014; Matuso 2015; Wolf 2015; Marie Claire 2016). Specifically, the passage indexed the stigmatized associations between vocal fry and American women that are typical of these pieces. The other half of participants read a different news-style passage that focused on Storm Arwen, which had passed through the United Kingdom about one month earlier. Masters compiled this passage from excerpts of Lee (2021). Both passages are included as an appendix to this chapter.

This innovative approach to the traditional reading passage plays into the sociolinguistic interview's intended function of manipulating attention to linguistic variables and attention to speech, as a relatively novel approach to examining the interactions between a sociophonetic variable and the ideologies it carries. In this chapter, participants' use of pulse phonation in the word list that followed the reading passages will be quantitatively compared to identify whether the activation of ideologies about vocal fry affects the production of creaky voice. Additionally, the task of reading passage was followed with brief unstructured discussion of the content of passages. Masters asked questions like: "What do you think of the article? Do you agree with what is said?" for vocal-fry-activated participants and simply "what do you think of the article?" for non-activated participants. This naturally yielded qualitative data that offer insights into the meanings that the users attach to vocal fry, while also potentially further heightening the attention of participants to the sociophonetic variable of pulse phonation.

Table 6.1 categorizes each participant in the study according to their macro-level social group and whether they read the vocal-fry-activated passage or weather passage.

Interviews were recorded in Audacity using a lavalier microphone attached directly to a laptop via USB. Interviews were orthographically transcribed in ELAN (Sloetjes and Wittenberg 2008; ELAN 2022) and then automatically aligned at the segment level with FAVE (Rosenfelder *et al.* 2014). Recordings were analyzed in Praat (Boersma and Weenink 2022).

Table 6.1 Participants by group and reading passage

	Older (60–67 years)	Younger (21–28 years)
Female	Vocal fry passage: Participant A Weather passage: Participant B	Vocal fry passage: Participant E Weather passage: Participant F
Male	Vocal fry passage: Participant C Weather passage: Participant D	Vocal fry passage: Participant G Weather passage: Participant H

To limit differences in the susceptibility of different vowels to pulse phonation and imbalances in the numbers of different vowels produced during responses to open questions, we limited analysis to allophones of /ɔ/ produced in primary stress position.

We focused analysis on relatively spontaneous, unmonitored, low-attention-to-speech conversation during the semi-structured, open-question portion of interviews (we often refer to this in the text below as "Section A" of the interview) and the elicited, monitored, and high-attention-to-speech reading of the word list (Section D). Because Section A solicited spontaneous responses to open questions, the amounts participants spoke varied widely. To analyze similar amounts of data from each participant, Masters extracted a 10-minute portion of the interviews from around the middle of Section A for each speaker and focused Section A analyses on these extracts.

Masters captured an average f0 for each such occurrence of /ɔ/ by selecting the entire vowel as marked by FAVE-Align and taking a single averaged measurement of pitch from Praat. Masters then auditorily coded each token as either modal or potentially creaky. She then visually examined segments coded as potentially creaky in Praat. If auditory creakiness was accompanied visually by irregular glottal pulsing in Praat's spectrogram and waveform, and if Praat's pitch tracker showed a drop in f0 during the segment, then the token was coded as creaky. If glottal pulsing was visually periodic and f0 remained stable, then the token was recoded to modal voice.

These procedures allow us to report creaky voicing as a proportion relative to modal voicing and to give overall acoustic characterizations of f0 for utterances of /ɔ/ coded as creaky or modal. As such, we can explore in general terms how often speakers use pulse phonation and, when they do use pulse phonation, how different it is from modal voicing. We examine these as group averages according to our two macro-level social categories of age and biological sex, as well as through our experimental intervention in reading passages.

Results

Table 6.2 shows the number of occurrences of creaky voicing and the overall proportion of pulse to modal phonation according to our three group-level predictor variables. It also reports the proportions of creaky and modal voice in the two interview sections where we took measurements. To facilitate comparison, we also list differences in creaky voice proportions between interview sections. The proportions reported for the predictor variables reflect the means of proportions returned by each speaker in that group. For example, the average proportion of creaky to modal voice for the older group was calculated by adding together the data for participants A, B, C, and D—who were all in the older variable group—and dividing by four. Finally, we report this information for each speaker. We will explore results within each predictor category in more detail through the rest of this section.

Looking at the overall proportions of creaky voice for all participants in both interview contexts, on average 21.6% of /ɔ/ vowels were produced with creak. The proportion of creaky vowels decreased when participants were reading the word list relative to their more casual speech.

In general, the use of pulse phonation was quite low in Section D. Here we are unable to disentangle stylistic effects of self-monitoring from pulmonary and articulatory effects of the interview tasks. It is unsurprising, for instance, that speakers would produce less pulse phonation in reading a word list than in casual speech, since words read in isolation can be produced with greater intensity and with less likelihood of premature glottal release. Nevertheless, in the discussion, we will identify between-contexts differences that might reflect sociolinguistic self-monitoring of vocal fry.

Table 6.2 Proportion of vocal fry to modal voice by social category and participant

		Section A (Semi-structured Open Questions)			Section D (Word List)			
	Overall Proportion	Occurrences of /ɔ/ (n)	Vocal Fry	Modal Voice	Occurrences of /ɔ/ (n)	Vocal Fry	Modal Voice	Change in v. f. from A to D
All participants	21.6%	485	26.2%	73.8%	142	16.9%	83.1%	Decrease (9.3)
Women	23.8%	265	23.5%	76.5%	70	24%	76.0%	Increase (0.5)
Men	19.4%	220	28.9%	71.1%	72	9.8%	90.2%	Decrease (19.1)
Older	16.2%	294	12.9%	87.1%	71	19.5%	80.5%	Increase (6.6)
Younger	26.9%	191	39.5%	60.5%	71	14.3%	85.7%	Decrease (25.2)
Vocal fry passage	20.9%	239	22.3%	77.7%	71	19.5%	80.5%	Decrease (2.8)
Weather passage	22.2%	246	30.1%	69.9%	71	14.3 %	85.7%	Decrease (15.8)
Participant A	42%	69	16 (23%)	53 (77%)	18	11 (61%)	7 (39%)	Increase (38)
Participant B	0%	74	0 (0%)	74 (100%)	17	0 (0%)	17 (100%)	No change

(Continued)

Table 6.2 *(Continued)*

		Section A (Semi-structured Open Questions)			Section D (Word List)			
	Overall Proportion	Occurrences of /ɔ/ (n)	Vocal Fry	Modal Voice	Occurrences of /ɔ/ (n)	Vocal Fry	Modal Voice	Change in v. f. from A to D
Participant C	13.5%	71	15 (21%)	56 (79%)	17	1 (6%)	16 (94%)	Decrease (15)
Participant D	9.3%	80	6 (7.5 %)	74 (92.5%)	19	2 (11%)	17 (89%)	Increase (3.5)
Participant E	17.5%	71	17 (24 %)	54 (76%)	18	2 (11%)	16 (89%)	Decrease (13)
Participant F	35.5%	51	24 (47 %)	27 (53%)	17	4 (24%)	13 (76%)	Decrease (23)
Participant G	10.5%	28	6 (21%)	22 (79%)	18	0 (0%)	18 (100%)	Decrease (21)
Participant H	44%	41	27 (66%)	14 (34%)	18	4 (22%)	14 (78%)	Decrease (44)

Overall, the participants' creaky voice f0 increased from Section A to Section D. What is, perhaps, more important is the relative difference between the creaky and modal voice of speakers. This difference decreased by 26.7 Hz between Section A, where the difference was 86.4 Hz, and Section D, where it was 59.7 Hz. This shows that as their attention to speech increased, the speakers made a greater effort to speak in a more consistent manner with their modal voice. Creaky voice and modal voice frequencies converge, showing an overall shift in participants making their vocal fry less creaky and more akin to their modal voice.

The group averages that we report in this section naturally gloss over substantial interspeaker variability. Given that we are working from a very small sample, such differences may affect group averages quite substantially. Of particular note, Participant B was a "knock out" (e.g., Tagliamonte 2006), not producing a single token coded as creaky in either Section A or Section D. We also note that, compared with other speakers in our older group, Participant A appears to be a heavy user of pulse phonation. Despite skewing that these two may enter our small dataset, we will include their measurements in group averages throughout this section. However, we will note their effects in discussing results and consider these potentially anomalous speakers and other matters of interspeaker variation as we try to make sense of our data below.

Age

Younger participants produced a greater proportion of vowels with creaky voice than did older participants. Across both interview sections, 26.9% of the younger group's vowels were creaky, compared with 16.2% of the older group's vowels. The older group's creaky voice proportions were naturally depressed by Participant B's avoidance of pulse phonation. But the proportions provided for individual participants in Table 6.2 make it clear that younger speakers (Participants E–H) produced greater proportions of creaky voice than older speakers (Participants A–D). The two groups were strikingly different in the ways that their use of pulse phonation changed between interview sections.

The younger group produced a high proportion of /ɔ/ with creaky voice during casual speech, which dropped substantially during the higher-attention-to-speech word list reading.

This reflects an obvious shift away from pulse phonation during more careful speech. By contrast, the older group produced little creaky voice during Section A and increased the proportion of creaky voice slightly during Section D.

While this increase was largely driven by Participant A's high proportion of creak in Section D, it at least shows interspeaker variation in the implementation of pulse phonation across stylistic contexts. We also acknowledge the physiological possibility that the conventional placement of a word list near the end of a sociolinguistic interview might also have influenced this result—the vocal folds of speakers might have been overtaxed, which could inhibit modal pulsing, and this could disproportionately affect older speakers.

Nevertheless, it is noteworthy that the proportions of younger and older speakers varied across stylistic contexts in opposite directions, with younger speakers decreasing their proportion of pulse phonation while older speakers increased theirs. To whatever (presently unknowable) extent these shifts were stylistic effects of attention to speech, this would raise the possibility of a generational difference in the stylistic evaluation of pulse phonation. The younger group's decrease in creak in more careful speech is consistent with the avoidance of a sociolinguistically stigmatized variant, while the older group's increase is consistent with adjustment toward a prestigious variant. To the extent that the data show stylistic variation, they suggest that our two groups have opposite sociolinguistic evaluations of pulse phonation.

In actual acoustic terms, when speakers produced creaky voice, they unsurprisingly did so with greater f0 during Section D speech, where (as noted above) the reading of isolated words would facilitate greater pulmonic pressure and glottal control. This means that when pulse phonation was produced during more careful speech, it was objectively more similar to modal voice than it was during speech in the semi-structured portion of the interview. The older group increased the average f0 of their pulse phonation during Section D more than the younger group did. Across all four speakers in the older group, the average f0 of tokens categorized as creaky increased from 55.9 Hz to 87.9 Hz, reducing the difference between creaky and modal f0 from 94.6 Hz to 67.6 Hz. The younger group's creaky voice f0 increased from 69.0 Hz in Section A to 80.3 Hz in Section D. This reduced the modal-creaky difference from 97.6 Hz to 80.1 Hz. However, while the younger group's creaky f0 in careful speech reduced the difference between speech contexts, the younger speakers did not increase their creaky f0 as much as the older group did. In other words, the pulse phonation of younger speakers during Section D was "creakier" than that of older speakers.

Although we are left with very speculative interpretations of this difference, we suggest more careful exploration of acoustic qualities could reveal more nuanced differences in voicing between the age groups, perhaps even along the lines of careful distinction among voicing modalities described in works like Keating *et al.* (2015). It may, for instance, be that younger speakers are truly producing vocal fry, while older speakers are producing non-constricted creak. We leave this speculative possibility for future research.

In general, though, our data suggest increased use of pulse phonation among younger speakers in Suffolk. This proportion shows up both in terms of pulse phonation as a proportion of voicing modality and in terms of the acoustic quality of creakiness.

Biological sex

Across both interview contexts, men and women produced pulse phonation at similar proportions. About one-fifth of men's vowels were creaky, compared with about one-quarter of women's vowels. Women's proportions are lowered by Participant B's avoidance of creaky voice.

Beneath this overall similarity, however, women and men responded differently across interview tasks. During the relatively low attention-to-speech Section A, men produced pulse phonation at similar proportions to women (technically, men's Section A proportions were slightly greater than women's—again a consequence of Participant B). However, in the higher attention-to-speech word list, men's creaky voice dropped to just 9.8% of vowels. Women's creaky voice proportion actually increased very slightly as their attention to speech was increased. In more casual speech, then, men exceeded women in using creaky voice, but then stylistically adjusted toward modal voice in more careful speech, while women retained consistent levels of production. Here, too, this overall proportion was affected by Participant A's relatively high use of pulse phonation in Section D.

The actual acoustic quality of pulse phonation also differed by biological sex. When males produced tokens with creaky voice, their collective overall f0 was 65.6 Hz. The average f0 of females in all tokens produced with creaky voice was 83.8 Hz. These compare with average f0 in the modal voice of 114.5 Hz for men and 202.1 Hz for women. So, for both groups, during the reading list, even tokens produced with pulse phonation were produced more like modal voice than were tokens produced during loosely structured interviews.

As we noted above, it is not surprising that f0 increased even in tokens produced with creaky voice in Section D, since speakers are likely to maintain greater pulmonic pressure while reading words in isolation in a list than they do when talking relatively normally. However, the difference between averaged f0 of creaky and modal voice for men was 62 Hz during Section A, which decreased to 35.7 Hz in Section D. While women also reduced the relative difference between creaky voice and modal voice from 124.6 Hz in casual speech to 111.9 Hz during the word list, between tasks this was an increase of only 12.7 Hz.

As such, men not only avoided pulse phonation at greater rates than women during careful speech, but when they did produce pulse phonation during careful speech, they produced it with an acoustic quality relatively more like modal voice than women did. In terms of both the categorical usage of creaky or modal voice, and in the acoustic characteristics of "how creaky" their pulse phonation was, men shifted away from creaky voice when reading a word list more than women. Women, by contrast, overall produced creaky voice relatively similarly—proportionally and qualitatively—across the two interview tasks.

Reading passage

The overall proportions of pulse phonation were similar between speakers who read the vocal-fry-activated and weather reading passages—both producing about one-fifth of /ɔ/ tokens with creaky voice. This is expected as long as there were not fundamental differences in the way speakers assigned to these two stimuli used creaky voice, since vocal fry would not be activated during Section A.

However, the two groups are, in actuality, more different than they appear from this overall view. Participant B, with a categorical modal voice, was assigned to the weather passage, which depressed their overall average, but because Participants F and H used relatively high proportions of pulse phonation, the weather passage group's Section A proportion of creaky voice was greater than the vocal-fry-activated group's.

In shifting to Section D, the weather group reduced their proportion of pulse phonation—as expected, given the articulatory characteristics of reading a word list and what would be expected if speakers were avoiding a stigmatized variant as attention to speech increased. Surprisingly, however, the vocal-fry-activated group maintained a proportion of 19.5% creaky voice in Section D. This was largely an effect of Participant A, who dramatically increased her proportion of creaky voice after being exposed to a reading passage that evoked critical ideologies about vocal fry.

The vocal-fry-activated group also looks stable in terms of f0 across contextual styles. In Section A, the activated group produced creaky voice with an average f0 of 69.7 Hz, which increased by a minimal 1.5 Hz during Section D. The weather passage group behaved more like other social groups we have explored above regarding this measure. In Section A, their average creaky f0 was 109.1 Hz less than modal f0. Creaky f0 increased during Section D, reducing the modal-creaky difference to 65.4 Hz.

Generally, then, invoking negative ideologies surrounding vocal fry had little effect on the implementation of pulse phonation, which was generally suppressed in the higher-attention-to-speech task of reading a word list. However, we find the surprising response of Participant A, who increased her implementation of pulse phonation following engagement with a passage critical of vocal fry. Accordingly, at the group level, her elevated levels of creak create the impression that the vocal-fry-activated group responded to the critical passage by sustaining pulse phonation as a proportion of voicing modality from more casual speech and by sustaining the actual objective acoustic quality of their creaky productions.

Interestingly, in metalinguistic commentary, Participant A invoked some concerns about the loss of local dialects due to the influence of London English in East Anglia. However, she generally adopted an agnostic acceptance of these innovations.

> Participant A: I'm not sure it really irritates me but people can't say ['wɔːtə]...[wɔːʔə] you know (laughing). ... I think when people drop... the consonants like that it can be a bit irri- but not really irritating, d'you know what I mean, you <u>notice</u> it.
>
> Interviewer: What do you think of the Suffolk, Norfolk and Essex dialect?
>
> Participant A: Um I much prefer the Suffolk one, and I love the Norfolk one ... but Essex, it doesn't bear any relationship to the other two which I find quite odd and I think it's very heavily influenced by the London, and you know...and that's I think maybe partly due to because so many people are moving out of London now into this direction and it's—it does worry me that eventually it's going to um, you know, impact on you know, the- the local accent side of things it could do because everywhere you go now you do hear the Essex accent and I think it's just that so many people have moved in this direction, which is fine, it's just something I've noticed.

While she does not invoke vocal fry or other labels connected with pulse phonation explicitly, Participant A's neutrality toward the counter-urbanization of London into Suffolk—even if she worries about its effects on the local dialects of Suffolk and Norfolk—may find a productive parallel in her own phonatory choices, as she adopts the international innovation of pulse phonation in careful speech.

Discussion

The small study of creaky voice in Suffolk that we have presented here offers several insights on pulse phonation and vocal fry and suggests directions for future exploration. In this section, we summarize our key findings and offer some speculative interpretations of and explanations for them.

First, our findings suggest that young people in Suffolk are using creaky voice as a greater proportion of vocalic productions in unmonitored speech than older people, consistent with statements in popular media that vocal fry is spreading. We see the further possibility of an apparent-time shift in the "meaning" of creaky voice, with younger and older speakers shifting in opposite directions with increased self-monitoring—though this may also be an effect of one apparently particularly innovative speaker, Participant A. If, however, the findings were replicated in larger samples or in other speech communities, it could indicate that, in addition to creaky voice spreading among younger speakers of Englishes, ideologies associated with vocal fry might also be spreading and replacing previously existing ideologies associated with acoustically and perceptually similar pulse phonation subtypes. In other words, if pulse phonation is indeed moving into speech communities where prestige variants like Received Pronunciation or local accents included creaky voice, negative sociolinguistic evaluations of vocal fry might also be moving into speech communities and displacing the previous associations with prestige.

Second, in contrast to the popular associations between vocal fry and young women, our findings indicate that men use creaky voice as a greater proportion of voicing than women (see also Hejna and Eaton, this collection). Again, with the caveat that Participants A and B may skew results, we also note the possibility that men may avoid pulse phonation more than women do when attention to speech is increased. This raises the possibility that men are responding to the associations in popular discourse between vocal fry and women and shifting away from sociolinguistic indexicalities of femininities. This would point toward fascinating complexities in the sociolinguistic associations between speakers as they produce pulse phonation and their negotiation of gendering. As described above, women face linguistic penalties for producing pulse phonation because of the sociolinguistic ideologies connected with vocal fry—regardless of the empirical fact that they may produce less pulse phonation than men. However, men may also face penalties for producing pulse phonation, perhaps specifically by being associated with femininities indexed in popular discourse around vocal fry. We suggest that this points toward fascinating possibilities for future research into the sociolinguistic meanings of vocal fry.

Indeed, the metalinguistic comments of younger participants provide evidence that they have adopted the negative ideologies that surround vocal fry in popular discourse. Both younger vocal-fry-activated participants imitated vocal fry in a stereotypical manner during or after the vocal fry reading passage. While the young male (G) produced an exaggerated, drawn out, extremely low creaky voice, the female (E) mimicked it in a valley girl style accent.

> Participant G (accentuated extreme vocal fry represented in italics): (reading the passage) I want to explain what vocal fry is and how to stop it happening to you, *ho::w yo::u* (laughing) is that going to ruin the experiment
>
> Participant E (American accent represented in italics): Is it like when the voice goes, kind of like *tails off* I'm trying to think of an example, *tails off at the end and it sounds like this yeah,* I thought that's what it was (laughing) and I yeah, I suppose perhaps it is slightly irritating, maybe I don't know?

Later in the interview, these two participants comment on the extreme nature of the reading passage and claim to be unphased by the others' use of vocal fry:

> Participant G: People are people aren't they? People are going to speak the way they wanna speak and you can't and shouldn't try and change them, and this article suggests you should.
>
> Participant E: I don't think it's something that's ever particularly annoyed me but yeah … the person who wrote this clearly, it's clearly something that's a real pet peeve for them

Despite expressing relaxed attitudes toward vocal fry, these younger speakers demonstrated the adoption of negative attitudes surrounding vocal fry, internalizing these popular associations. Both speakers had similar levels of pulse phonation during Section A (21% and 24%), and while participant E reduced her proportion of creak to 11% of vowels in Section D, participant G eliminated his creak entirely showing the impact of popular associations on their own phonation, especially on the young male. Comments by Participant C, a man from the older group, show that the negative evaluations of younger speakers are not shared by all members of the speech community. The older male, who read the vocal fry reading passage, overtly disagreed with the content of the passage, defending vocal fry as something that occurs naturally due to age or disability:

> Participant C: I mean thas obviously you're studying but I think it's a load of rubbish (laughing) to be perfectly honest with you (laughing) um people speak and they speak naturally and sometimes people's voices are a bit croaky.

However, despite his pluralistic attitude toward vocal fry, Participant C avoided using creaky voice in his word list reading. Moreover, he shifted away from creak in this task after joking before he started to read, "alright croaky voice coming up" (sic) while attempting to clear his throat. This hints at the complex relationships between attitudes and productions that underlie the sociolinguistics of vocal fry.

Third, in contradiction with our speculations that interspeaker variation in pulse phonation points toward speakers navigating the stigmatization of vocal fry in popular ideologies, stigmatization was explicitly activated for speakers and that activation did not lead them to avoid pulse phonation. Participant A, in particular, increased her implementation of creaky voice after encountering the negative ideologies that are typical in popular discourse. For us, her pluralistic comments toward linguistic features recall Bourhis and Giles's (1977) classic study of adult learners of Welsh who were studying the language for cultural connections, who spoke with more heavily Welsh-accented English after they heard a recording of a speaker with a Southern England accent denigrating Welsh. We think it possible that participants who read the vocal-fry-activated passage might have resisted the passage's negative evaluations of others. In this case, their maintenance of conversational creaky voicing levels, both in terms of proportions of pulse phonation to modal voicing and of f0, may have represented active rejection of negative evaluation of speakers who use vocal fry and active sociophonetic alignment with them.

The maintenance of creaky voice by at least one vocal-fry-activated speaker surprised us, but it offers interesting possibilities for future exploration of ways that speakers and speech communities might navigate the changes in voicing modalities and sociolinguistic ideologies connected with them. At the least, it reminds us of the complex interactions between sociophonetic variation, sociolinguistic ideologies, and notions like stigma.

No doubt even in this small sample there is much more information to glean from navigations of pulse phonation and sociolinguistic ideologies of individual speakers. Results presented in this chapter point toward fascinating possibilities for the ways that speakers may operationalize and interact with vocal fry.

Conclusion

This chapter has presented a small study of creaky voice in Suffolk, England. Despite its limitations in terms of the number of speakers sampled and the amount of acoustic data measured for each speaker, we believe we have revealed some fascinating insights on pulse phonation as an innovation in the speech community, both as a phonetic feature and as a sociolinguistic one. We have in a very modest way answered the call for more research on pulse phonation as a change in progress, a feature of men's speech, and an innovation in Englishes outside the United States. We have contributed to findings in each of these areas. Moreover, in Suffolk we have identified surprising and complex interactions between the activation of popular ideologies associated with vocal fry and production of creaky voice; indeed further interactions have been noted between these responses to the activation of ideologies and speaker categories of sex and age. We hope these findings might encourage new exploration of pulse phonation as a sociophonetic feature.

Additionally, we hope this project has modeled new possibilities for scholars interested in vocal fry and other aspects of variation beyond traditional segments. By examining pulse phonation in Suffolk, we have framed creaky voice as an incoming feature of a local dialect—an approach typically taken with phonetic segments. Our findings indicate that pulse phonation is an incoming innovation of Suffolk English, as are ideologies associated with vocal fry. We encourage other researchers to actively study creaky voicing and the sociolinguistic meanings of vocal fry as real features. And we have reached these observations through a novel manipulation of the sociolinguistic interview itself, which capitalizes upon the genre's goal of strategically adjusting the self-monitoring of the speaker's own linguistic productions. We are excited by the possibilities afforded by the modifications of core sociolinguistic tools and methods to continue to map the fascinating landscapes of sociolinguistic and sociophonetic variation.

Acknowledgments

This chapter is a revision of Madeline Masters' undergraduate dissertation at the University of Warwick, which was supervised by Christopher Strelluf. Masters designed the project, conducted all fieldwork, completed all analyses, and drafted the first write-up. Christopher substantially revised the paper for publication. The authors mutually endorse the findings and analyses presented in this revision.

Appendix A. Vocal fry reading passage

What do the Kardashians, Zooey Deschanel and Katy Perry all have in common? They all are well-known for their use of vocal fry, a creaky voice tone. But women aren't the only ones using vocal fry—men use it too.

Have you ever heard someone who has vocal fry?

It's when the edge of their voice cracks, creaks and fries—it's quite unpleasant (many would say down right annoying) to listen to.

I want to explain what vocal fry is and how to stop it from happening to you!

How you say something can be just as important as what you say. Vocal fry, or a glottal fry as it's sometimes called, has become a cultural epidemic that undermines our vocal power.

What is Vocal Fry?

Vocal fry happens when someone's voice sounds raspy or creaky. It's called "fry" because it sounds like bacon sizzling in a frying pan. Vocal fry occurs when there is not enough breath being pushed through the vocal cords. When we breathe, our vocal cords separate. Then when we speak, those cords rub together and the vibration creates sound. If you speak without enough breath, your vocal cords cannot rub together and they create a creaky, hollow sound. In vocal fry, it's as if you are hearing someone's vocal cords rattling next to each other.

Here's the problem: Vocal fry is not just annoying for others to listen to, it's also undermining you and your message. It is incredibly hard for people to take you seriously, listen to you and believe you when you speak with vocal fry.

Appendix B. Weather reading passage

Some of those who lost power due to Storm Arwen will have to wait at least two more days to be reconnected electricity companies have said.

About 30,000 households have spent five nights without power so far since gusts of nearly 100mph hit parts of the UK.

The energy companies said they had reconnected about a million homes.

He said he wanted to reassure people "who are exhausted, who are worried and who are angry" that energy companies and the government are working hard to restore power.

Welfare centres and hot food are being provided in some affected places, with energy companies working with emergency services, local councils and the British Red Cross.

But Dr Lily Fulton-Humble, who was at home near Alnwick in Northumberland with a seven-week-old baby and a sick toddler, said her family was "losing the stamina" to endure further nights without power.

"It's pretty cold - and when you're feeding a baby every two hours it's even colder", she said.

She said communication from Northern Powergrid during the outage had been "pretty appalling".

References

Anderson, R. C., Klofstad, C. A., Mayew, W. J., and Venkatachalam, M. (2014). "Vocal fry may undermine the success of young women in the labor market." *PLOS One*, **9**(5): e97506.

Beal, J. (2010). *An Introduction to Regional Englishes: Dialect Variation in England*. Edinburgh: Edinburgh University Press.

Becker, K. (2017). "The sociolinguistic interview." In C. Mallinson, B. Childs, and G. Van Herk (Eds.), *Data Collection in Sociolinguistics: Methods and Applications* (2nd edn.), pp. 99–107. New York: Routledge.

Boersma, P. and Weenink, D. (2022). *Praat: doing phonetics by computer* [Computer program]. Version 6.2.09, Retrieved 16 February 2022 from http://www.praat.org.

Bourhis, R. Y. and Giles, H. (1977). "The language of intergroup distinctiveness." In H. Giles (Ed.), *Language, Ethnicity, and Intergroup Relations*, pp. 119–135. London: Academic Press.

Britain, D. (2005). "Innovation diffusion: 'Estuary English' and local dialect differentiation: The survival of Fenland Englishes," *Linguistics*, **43**(5): 995–1022.

Butcher, K.-A. (2020). "Dialect maintenance in East Anglia: Singin' the same old tune," *English Today*, **36**(3): 48–58. Doi:10.1017/S0266078420000231.

Chambers, J. K. and Trudgill, P. (2002). *Dialectology* (2nd edn.). Cambridge: Cambridge University Press.

Chao, M., and Bursten, J. R. S. (2021). "Girl talk: Understanding negative reactions to female vocal fry." *Hypatia*, **36**(1): 42–59.

Dallaston, K. and Docherty, G. (2020). "The quantitative prevalence of creaky voice (vocal fry) in varieties of English: A systematic review of the literature." *PLoS ONE*, **15**(3): e0229960.

Davidson, L. (2019). "Perceptual coherence of creaky voice qualities." In S. Calhoun, P. Escudero, M. Tabain, and P. Warren (Eds.), *Proceedings of the 19th International Congress of Phonetic Sciences*, Canberra, Australia: Australasian Speech Science and Technology Association Inc: 147–151.

ELAN (Version 6.2) [Computer software]. (2022). Nijmegen: Max Planck Institute for Psycholinguistics, The Language Archive. Retrieved from https://archive.mpi.nl/tla/elan.

Fisiak, J. and Trudgill, P. (Eds.). (2001). *East Anglian English*. Woodbridge: D.S Brewer.

Gibson, T. A. (2017). "The role of lexical stress on the use of vocal fry in young adult female speakers." *Journal of Voice*, **31**(1): 62–66.

Henton, C. G. and Bladon, A. (1988). "Creak as a sociophonetic marker." In L. M. Hyman and C. N. Li (Eds.), *Language, Speech and Mind: Studies in Honour of Victoria A. Fromkin*. London: Routledge.

Hess, A. (2013). "Why old men find young women's voices so annoying." *Slate Magazine*. https://slate.com/human-interest/2013/01/vocal-fry-and-valley-girls-why-old-men-find-young-women-s-voices-so-annoying.html.

Hornibrook, J., Ormond, T., and Maclagan, M. (2018). "Creaky voice or extreme vocal fry in young women." *NZ Medical Journal*, 36–40.

Jelle, J. (2017). "A study in Scarlett: Creaky voice and romantic intention in Spike Jonze's Her." *Leviathan: Interdisciplinary Journal in English*, **1**: 35–44.

Johns Hopkins Medicine. (2022). "Is vocal fry ruining my voice?" *Johns Hopkins Medicine*, March 1, 2022. Retrieved from https://www.hopkinsmedicine.org/health/conditions-and-diseases/is-vocal-fry-ruining-my-voice.

Keating, P., Garellek, M., and Kreiman, J. (2015). "Acoustic properties of different kinds of creaky voice." *18th International Congress of Phonetic Sciences*, Glasgow, Scotland.

Kretzschmar Jr, W. A. (2018). "Linguistic atlases." In C. Boberg, J. Nerbonne, and D. Watt (Eds.), *The Handbook of Dialectology*, pp. 57–72. Oxford: Wiley.

Labov, W. (1966/2006). *The Social Stratification of English in New York City*. Cambridge: Cambridge University Press. doi:10.1017/CBO9780511618208.

Lakoff, R. (1973). "Language and woman's place." *Language in Society*, **2**(1): 45–80.

Laver, J. (1968). "Voice quality and indexical information." *British Journal of Disorders of Communication*, **3**: 43–54.

Lawson, R. (2016). "A different drum: Social media and the communication of sociolinguistic research." In R. Lawson and D. Sayers (Eds), *Sociolinguistic Research: Application and Impact*, pp. 171–192. Abingdon, UK: Routledge.

Lee, B. J. (2021). "Storm Arwen power cuts: 30,000 still waiting to be reconnected." *BBC News*, December 1, 2021. Retrieved from https://www.bbc.co.uk/news/uk-59396135.

Lindsey, G. (2015). "Britney, Pitney and vocal fry." *English Speech Services*, July 25, 2015. Retrieved from www.englishspeechservices.com/blog/britney-pitney-and-vocal-fry/.

Maestas, R. (2019). "In defense of upspeak: Reclaiming 'feminine' communication styles at work." *Public Humanities | Brown University*, March 14, 2019. Brown.Edu. Retrieved from https://www.brown.edu/academics/public-humanities/blog/defense-upspeak-reclaiming-"feminine"-communication-styles-work.

Marie Claire. (2016). "Why women are changing the way they speak to sound like Kim Kardashian." *Marie Claire*, May 3, 2016. Accessed May 22, 2022. Retrieved from https://www.marieclaire.co.uk/news/vocal-fry-women-are-changing-the-way-they-speak-to-sound-like-kim-kardashian-9551.

Matuso, A. (2015). "10 most irritating female celebrity voices." *The Richest*, March 15, 2015. Retrieved from https://www.therichest.com/entertainment/10-most-irritating-celebrity-vocal-fry-offenders.

Pittam, J. (1987). "Listeners evaluations of voice quality in Australian English-speakers." *Language and Speech*, **30**: 99–113.

Potter, R. (2018). "A variationist multilocality study of unstressed vowels and verbal -s marking in the peripheral dialect of East Suffolk." Unpublished PhD Thesis. Colchester, UK: University of Essex.

Raeke, M. (2021). "Managing voice and speech side effects after oropharyngeal cancer treatment." *MD Anderson Cancer Center*, May 6, 2021. Retrieved from https://www.mdanderson.org/cancerwise/managing-voice-and-speech-side-effects-after-oropharyngeal-cancer-treatment.h00-159460845.html.

Rhodan, M. (2014). "3 speech habits that are worse than vocal fry in job interviews." *Time*, June 4, 2014. Retrieved from https://time.com/2820087/3-speech-habits-that-are-worse-than-vocal-fry-in-job-interviews.

Rosenfelder, I., Fruehwald, J., Evanini, K., Seyfarth, S., Gorman, K., Prichard, H., and Yuan, J. (2014). *FAVE (Forced Alignment and Vowel Extraction) Suite Version 1.1.3*. Software.

Sankoff, G. and Blondeau, H. (2007). "Language change across the lifespan: /r/ in Montreal French." *Language*, **83**(3): 560–588.

Shaw, F. and Crocker, V. (2015). "Creaky voice as a stylistic feature of young American female speech: An intraspeaker variation study of Scarlett Johansson." *Lifespans and Styles*, **1**, 21–27.

Slifka, J. (2006). "Some physiological correlates to regular and irregular phonation at the end of an utterance." *Journal of Voice*, **20**: 171–186.

Sloetjes, H. and Wittenburg, P. (2008). "Annotation by category: ELAN and ISO DCR." *Proceedings of the 6th International Conference on Language Resources and Evaluation (LREC 2008).*

Tagliamonte, S. A. (2006). *Analysing Sociolinguistic Variation*. Cambridge: Cambridge University Press.

Trudgill, P. (1974). *The Social Differentiation of English in Norwich*. Cambridge: Cambridge University Press.

Trudgill, P. (2021). *East Anglian English (Dialects of English [DOE] Book 21) (1st edn.)*. De Gruyter Mouton.

Van Edwards, V. (2021). "Vocal fry: What it is and how to get rid of it." *Science of People*, October 25, 2021. Retrieved from https://www.scienceofpeople.com/vocal-fry.

Wells, J. C. (1982). *Accents of English 1: An Introduction*. Cambridge: Cambridge University Press.

Wolf, N. (2015). "Young women, give up the vocal fry and reclaim your strong female voice." *The Guardian*, July 24, 2015. Retrieved from https://www.theguardian.com/commentisfree/2015/jul/24/vocal-fry-strong-female-voice.

Wolk, L., Abdelli-Beruh, N. B., and Slavin, D. (2012). "Habitual use of vocal fry in young adult female speakers." *Journal of Voice*, **26**(3): e111–e116.

Yuasa, I. P. (2010). "Creaky voice: A new feminine voice quality for young urban-oriented upwardly mobile American women?" *American Speech*, **85**(2): 315–337.

Chapter Seven

Creak and embodying gender

A case study of media interviews with an actor before and after coming out as transgender

Robert James Hellyer

Introduction

When we change—change our identity, change our views, change our relationships, change the way we present ourselves in the eyes of others—how far does the way we speak change? In sociolinguistics, the speaker or speech community stands at the center of investigation for such questions. Research on language use and shifts in language change among communities of place, class, gender, or religion, such as in early seminal studies in English, traditionally deal with these facets of an individual speaker's identity as fixed, static, and sometimes even predetermined and unchangeable (Labov 1966; Trudgill 1974; L. Milroy 1980; J. Milroy 1992). Viewing the complexity of societies and cultures

Creak: Theories and practices of pulse phonation
Edited by Francesco Venturi

ISBN 978-981-5129-29-8 (Hardcover), 978-1-003-61606-1 (eBook)
www.jennystanford.com

through synchronic investigations of sociolinguistic change may be perfectly reasonable for some enquiries. However, if language is considered so closely tied with aspects of a speaker's identity, or their relationship with certain social groups—e.g., class, sex, or gender assigned at birth—we must also consider what happens when these speakers change their social groups, move through classes, or openly identify as a gender they were not assigned at birth.

In this chapter on creak and gender identity, I draw greatly on research by Lal Zimman, including his comprehensive investigation in 2012 of a group of transmasculine speakers who adjusted their vocal presentation as their social presentation developed across 2 years. Among this group, the most masculine presenting were not necessarily the ones who used more features typically associated with masculinity, such as lowered vocal pitch. Instead, it seems that constructing gender involved far more than just the use of gendered linguistic features, but a complex process of embodiment, drawing on fluid sets of styles, both linguistic and otherwise. I aim to build on this foundation, while looking beyond the moment of transitioning, addressing speakers before they made their trans identity public knowledge, marking a distinct shift in the identity they negotiate with that public audience.

The following sections begin with a review of linguistic, anthropological, and sociological literature concerning language variation. I am foremost concerned with how voice quality and creak have been investigated in terms of speakers' sex and gender. The third section ("Gender and style") then introduces a theoretical basis for analyzing gender and a speaker's constructed style, and in the fourth section ("The present study"), I propose that taking coming-out events as a significant social juncture in the lives of some queer people supports interpretations of how both identity and these sets of styles shift at significant personal events. The fifth section ("Results") then outlines the rationale and procedures of the study, in which data were collected from television and online interviews with the young transgender actor Josie Totah, spanning 7 years.

This study considers how coming out as a life event and subsequent changes in an individual's gender presentation across time influence their use of creak as part of a gendered system, but also whether this change may be linear and sustained—given the range of situational factors that constrain variation.

Voice quality and creak

In popular language and gender discourses, perhaps few linguistic features are more divisive than the voice quality popularly known as vocal fry, more accurately termed creaky voice, creak phonation, creak, glottalization, or pulse phonation. Headlines of creaky voice in the media range from "Women should avoid using 'vocal fry' in the workplace" (Mo 2016) to "These vocal techniques used by rappers & Kardashians might make you sound sexier" (Gillies 2018), "This is why ridiculing vocal fry is sexist" (Feinn 2016), "What is 'vocal fry,' and why doesn't anyone care when men talk like that?" (Dunn 2015), and "Young women, give up the vocal fry and reclaim your strong female voice" (Wolf 2015). Here we see that vocal fry and creaky voicing are not only a particularly salient feature in some varieties of English, but the voice quality is clearly strongly associated with women, and often negatively so.

Insofar as linguistic literature is concerned, creaky voice is typically characterized by high glottal tension as the arytenoid cartilages of the larynx are pressed inward, adducting and compressing the posterior portions of the vocal folds, while enough of the anterior length of the vocal cords open and vibrate, allowing for voicing (Ladefoged 1971: 14–15). In English, Luthern and Clopper (2011) found that creaky voice is used regularly in phrase-final syllables and marginally in phrase-initial syllables, regardless of speaking style. This aligns with the expectation that lower airflow through the larynx is more likely to cause irregular phonation for some speakers, such as at the very beginning of an utterance or at the end of an utterance, as the speaker depletes their stock of air. The vocal folds then vibrate at an irregular rate as low as 20–70 Hz (Henton and Bladon 1988).

In one of the earliest large-scale sociolinguistic studies of voice quality, Henton and Bladon (1988) investigated the use of breathy voice and creaky voice in the speech of 80 informants (40 male and 40 female) in two dialects of British English: Received Pronunciation and Modified Northern. The authors found that, in both dialects, men produced significantly more creak than women. An additional difference between the level of creak used by the Modified Northern men and Received Pronunciation men was attributed to the latter being "less masculine" and the use of creak in Modified Northern English as a "robust marker of male speech" that is "hyper-masculine" (1988: 21). It is worth noting, however, that the variation in the norms of masculinity between the two communities does not necessarily result in a causal link between creaky voice and masculinity.

Contrastingly, Wolk, Abdelli-Beruh, and Slavin (2012) investigated the use of creaky voice in American English, recording sustained vowel production and sentence-reading tasks for women aged 18–25. They found that approximately two-thirds of the women in this sample used creaky voice without any underlying vocal pathology or abnormal laryngeal physiology. Indeed then, this is a linguistic feature subject to sociolinguistic variability, rather than a mere physiological condition. Furthermore, in a sample of male and female speakers of American English or Japanese in same-gender pairings, Yuasa (2010) notes that American women produce significantly more creak than either the American men or the Japanese women. In a following matched-guise perception study, listeners from both northern California and eastern Iowa perceive female creaky voice as "hesitant, nonaggressive, and informal but also educated, urban-oriented, and upwardly mobile" (Yuasa 2010: 315). In line with Henton and Bladon, however, Yuasa also contributes women's use of creaky voice in American English to masculinity and authoritativeness, even citing her own observations of speeches by Hilary Clinton as evidence for speakers using creak to "project a male-like authoritative image" (2010: 318). This, of course, remains questionable in the absence of further investigation into the correlation between the culturally situated norms of masculinity and use of creaky voice.

Podesva (2018) reports on an ongoing project in California, in which participants used higher rates of creak during less enjoyable conversations, regardless of gender. Podesva also remarks that young female speakers creak more than other participants. By computerized visual tracking of stillness versus body movement as an indicator of comfort or discomfort of speakers sat in pairs, the findings of the experiment suggest that the young female speakers prefer using creak for enacting a negative or disengaged stance. Although creak is not a feature used exclusively by women in the Californian study, Podesva asserts that it does appear interactionally useful in expressing affect, not just gender (2018).

As seen in this section, research on creak phonation spans a wide range of fields, highlighting gender-stratified usage in different English-speaking contexts, as well as being utilized to enact an affective stance by some younger women. But how do linguistic features such as creak become so associated with particular identities, stances, or affects? We turn to this question in the following section.

Gender and style

Many early seminal works on language and gender focused on broad and ultimately essentialist, sex-based categorizations of both the speaker and their speech (e.g., Lakoff 1975; Spender 1980; Tannen 1990). Theories of language would remain tied to views of gender as biologically determined, binary, and static, well into the 1990s (Angouri 2021, for an overview). Eckert (1989) argues instead that, "the correlations of sex with linguistic variables are only a reflection of the effects on linguistic behavior of gender—the complex social construction of sex—and it is in this construction that one must seek explanations for such correlations" (1989: 245). This section discusses how language is used as part of a speaker's construction of gender and style, but also how different contexts allow for the same speaker to draw on the same linguistic variables to achieve different communicative effects.

Gendering language

Gender is not a tangible object. It has no static genesis. Instead, gender is continually constructed through "psychological, cultural and social means" (West and Zimmerman 1987: 125). Butler reiterates, "gender is always a doing, though not a doing by a subject who might be said to preexist the deed" (Butler 1990: 33). In this sense, gender is "produced by the regulation of attributes along culturally established lines of coherence" (1990: 32–33).

This is also facilitated through language. Certain features of speech become indexed for gender, where we see "relationships between linguistic forms and social meanings" (Jaffe 2016). Variables themselves do not index social groups, group categories, or membership directly, but rather are indirectly associated with qualities and stances involved in the construction of categories (Eckert 2008).

As for gender identity, a single linguistic feature alone cannot index something as complex. Instead, this is achieved through bricolage (Hebdige 1991) wherein speakers combine—agentively or otherwise—potential but separate linguistic resources into something that creates a more complex meaning beyond the sum of its components (Eckert 2004, 2008). The results of such actions are not static after completion: Whenever speakers make social-semiotic moves, they are "reinterpreting variables and combining and recombining them in a continual process of bricolage" (Eckert 2012: 94).

Such a theoretical trajectory aids in guiding variationist research away from static, one-to-one correlations between variables and gender, instead directing it toward how variables are embodied and situated contextually.

Style and constructing gendered voice

When observing and analyzing just one specific moment, the concept of bricolage may be useful in explaining how an individual draws on different resources to create more complex meaning around their identity. This is, of course, only part of the system of production and perception of identity through language (for a discussion of the perception of gender and creak, see Pearce in this collection). However, understanding why a particular speaker may call upon certain resources in particular settings is a matter of linguistic style.

Podesva, Roberts, and Campbell-Kibler (2001: 179) assert that a "style may be viewed as a collage of co-occurring linguistic features which, while unfixed and variable, work together to constitute meaning in coherent and socially intelligible ways. Style simultaneously gives linguistic substance to a given identity and allows the identity to be socially meaningful." From this perspective, identity and style are constructed together. The matter of which linguistic resources a speaker draws upon is then motivated by the speaker's attempts to convey identity in the moment (see also Nix and Chappell's discussion of creak and musical style in Chapter 2).

Using style to allow "identity to be socially meaningful," a speaker is able to negotiate how they might be perceived in a particular setting. This is useful for discussing how people go about negotiating gender in different spaces, as the following examples highlight.

Zimman emphasises how the voice "acts as a critical mediator of gender perceptions, making it hugely important to transgender people who are pursuing a transition from one gender role to another" (2017: 340). In an analysis of transmasculine speakers' metalinguistic commentaries concerning their own voice, Zimman (2016) noted that some transmasculine speakers undergoing testosterone treatment were reluctant or ideologically opposed to agentively constructing masculinized speech styles. Instead, these speakers rely on, or rather trust in, the effects of testosterone treatment to masculinize both their voices and appearance. However, this is far from the whole picture. The transmasculine speakers in Zimman's 2017 study also vary by the fundamental frequencies of their voices, as well as how high or low pitched the frequency of their /s/ pronunciation is, although not all of these speakers are drawing or able to draw on the same symbolic resources. How much other people categorize a speaker's physiological appearance or the sound of their voice as physiologically male impacts how they engage in feminine activities without being perceived as female (2017: 261).

Indeed, individual context plays a significant role in trans or non-cis-heteronormative gender presentation and linguistic intra-speaker variation. Gratton's (2016) study of two nonbinary trans speakers in queer spaces observed an even distribution of variation in word final (ING) between the two speakers. Comparatively, regarding the same speakers in non-queer-friendly spaces, where the risk of misgendering was higher, Gratton found they used either higher or lower rates of gendered variation of (ING) to distance themselves from being perceived as the gender they were assigned at birth. Gratton remarks that "the situations that call for more conscious identity work are those in which they are around people with whom they felt uncomfortable" (2016: 55). The relationship between comfort and conversation, or affect, style, and stance shall be returned to later. These findings highlight the significance of both a speaker's feelings of being perceived in a gendered way and also how gendered linguistic variation is highly situational.

As these examples have shown, trans voices require us to reconsider preconceptions about gendered voices, with Zimman calling for "a reframing of the gendered voice as a fluid set of multidimensional styles rather than a static property determined by speaker sex" (2017: 341). Clearly, the use of creak phonation is not a static property predetermined by speaker sex, though how people perceive the relationship between such a salient feature and gender may impact a speaker's use of a voice quality as part of their set of styles in constructing gendered voice. We could possibly investigate this with cisgender speakers; however, taking a trans celebrity—particularly someone used to being perceived by the public as one gender at a younger age, then shifts some years later in the perception of her audience—offers us an opportunity to investigate how a speaker constructs a gendered voice.

The following section considers the moment of these shifts through the case of a celebrity *coming out* as trans—as a social juncture around which to investigate identity and linguistic change.

Coming out

Coming out, a widely understood turn of phrase, comes from the idiom coming out of the closet "and refers to lesbian, gay, and bisexual people's disclosure of their sexual orientation, and to transgender people's disclosure of their (trans)gender identity" (Sauntson 2015: 197). In linguistics, these life events are typically analyzed through personal narratives of individuals' own coming-out experiences (Sedgwich 1990; Rust 1993; Liang 1997; Bacon 1998; Sauntson 2007; Zimman 2009; Levon 2015; DiDomenico 2015). For some, it seems that coming-out events are "not a linear, goal-oriented, developmental process," but rather coming out is a "process of describing one's social location within a changing social context" (Rust 1993: 50).

As social junctures, coming-out events require the speaker and the listener to renegotiate presentation and perception of the gendered self. It is at this intersection where a speaker's established sets of styles, formed of both the linguistic and non-linguistic resources at their disposal, are reassessed and restructured, drawing on and constituting the speaker's negotiated identity, which is momentarily new. The present study considers how significant an effect coming out may have on the voice, both at the point of restructuring of gendered styles and beyond.[1]

The present study

Josie Totah

California-born actor and producer Josie Totah starred in a number of national US television series, beginning her career at 11 years old, while still presenting as male. Assigned male at birth, Totah came out as a trans woman in August 2018 and has since starred in five television and film productions and several online interviews and shows. Using methods outlined below, this study analyzes creak in vowel tokens in seven interviews from 2015 to 2021, before and after coming out as trans.

All interviews used were available on YouTube (Table 7.1). The interviews were recorded with Audacity (Audacity Team 2008), with a sample rate of 44.1 kHz and 32-bit sample .wav format. Each file was then manually orthographically transcribed using Transcriber (Barras *et al.* 2001), before isolating each speaker and TextGrid using ELAN (ELAN 2021). Tokens with significant background noise or overlapping speech were excluded from analysis and removed from the final TextGrids using Praat (Boersma and Weenink 2018).

Table 7.1 Recording schedule for selected interviews with Josie Totah

Interview	Interlocutor	Year
Celebified	Danny Hoit	2015
Build	Ricky Camilleri	2018
The Wendy Williams Show	Wendy Williams	2018
(Totah comes out as trans in August 2018)		
It Gets Better Project	N/A	2019
GLAAD	Anthony Ramos	2020
CornbreadSays	Taylor Henderson	2020
ScreenRant	Tatiana Hullender	2021

Methods of measuring creak

As the chapters in this book reveal, a range of methods can be utilised for measuring creak. Across studies such as this, a systematic review by Dallaston and Docherty (2020) identifies three main measurement methods (auditory, auditory–visual, and automated acoustic methods) and five different formulae. Measured at the sentence level, Oliviera *et al.* (2016) and Abdelli-Beruh, Wolk, and Slavin (2014) perceptually and auditorily coded their dataset for the amount of creaky voice identified per minute, observing creak being used both sentence-finally and sentence-medially. For a further historical review of creak in speech-language pathology, see also Abdelli-Beruh in Chapter 1. In the analysis of California speakers, Podesva (2018) reports how each syllable was manually coded for creak auditorily, then automatically processed at every 10 ms for differences between a range of measures, including spectral tilt and CPPS. The results were then calculated as a percentage of the intonational phrase that contained vowels identified as creaky. For further measures of creak, see Ishi in Chapter 4.

For small-scale studies of individual variation across unbalanced sets of prerecorded, spontaneous speech, the present study analyzes Josie Totah's variation at a phoneme-by-phoneme level, further accommodating comparison across recordings and utterances of different lengths. This also allows for the analysis of individual vowel tokens, where voicing will always occur and, therefore, creak is always possible (Gorden and Ladefoged 2001: 386).

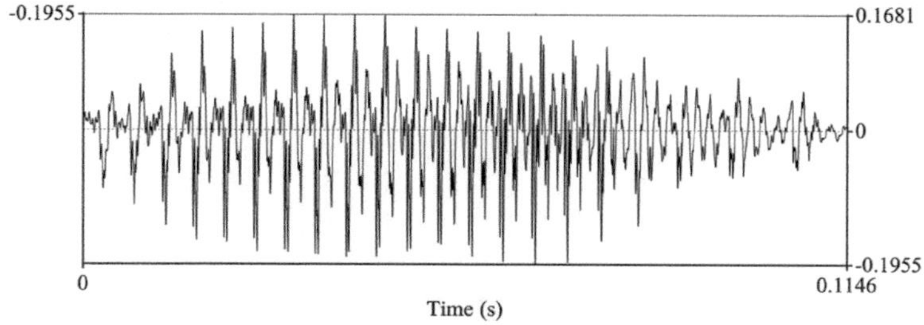

Figure 7.1 Waveform of medial vowel /ɪ/ in modal voice quality (Totah corpus).

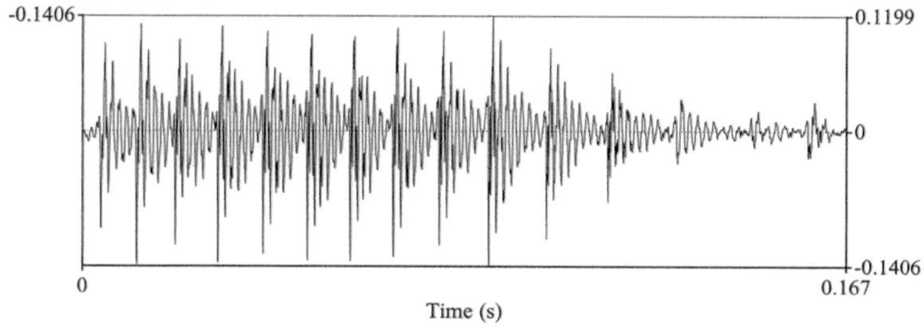

Figure 7.2 Waveform of medial vowel /ɪ/ in creaky voice quality (Totah corpus).

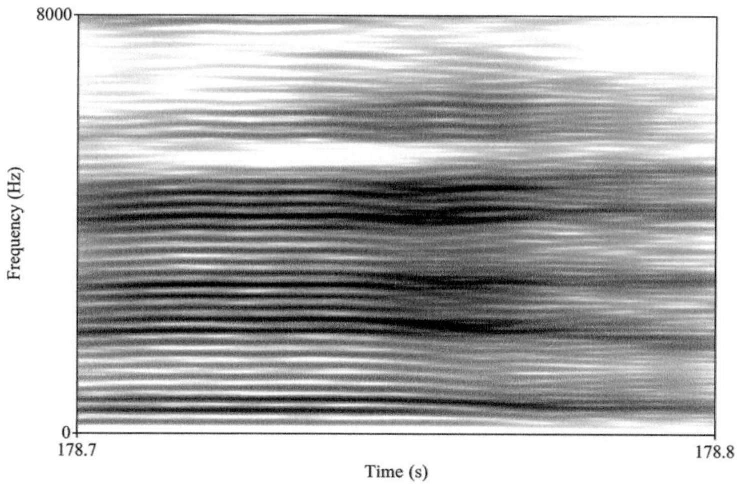

Figure 7.3 Spectrogram of medial vowel /ɪ/ in modal voice quality (Totah corpus).

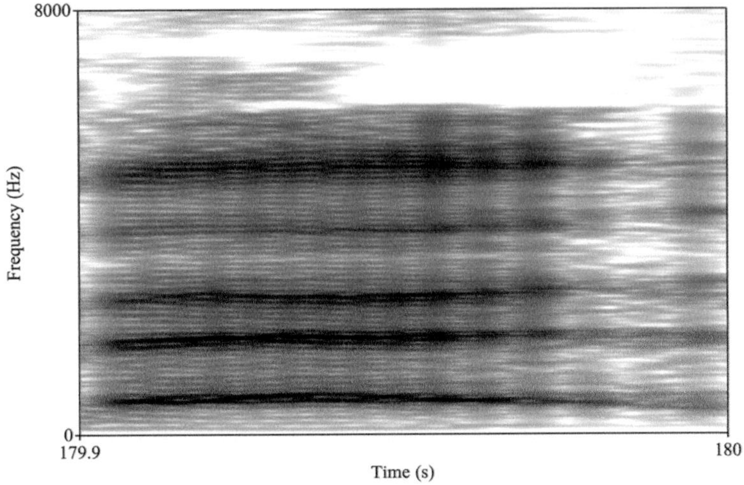

Figure 7.4 Spectrogram of medial vowel /ɪ/ in creaky voice quality (Totah corpus).

Automated measures of difference in spectral tilt, fundamental frequency, and formants F1 and F2 were taken across 10 equally spaced intervals within each vowel to identify creaky voice. An initial sample of four minutes was used to compare tokens against auditory and audiovisual spectrographic manual processing. The irregular waveform produced (Figs. 7.1 and 7.2) and the dark and light pulses throughout a spectrogram of the sound (Figs. 7.3 and 7.4) visually indicate the presence of creak in the token, whereas listening to the vowel in its sentence context provides more accurate auditory classification (Abdelli-Beruh *et al.* 2014). The two manually validated samples were then compared to the vowels identified through the automated measures before processing the remaining data.

Coding and analysis

Using the methods described above, each vowel token was coded as ±creak for all speakers, including Totah and interviewers. To measure for interaction with contextual factors, additional independent variables were also coded for each token (Table 7.2). Sentence position and affect were coded for each token, following Podesva's (2018) finding that phrase position, gender, age, and affect were significant factors in the use of creak for English speakers in California. In an analysis of one speaker's discourse content and discourse management, Zimman (2018) also finds higher rates of creak during moments where the speaker expresses a negative or disengaged affectual stance. In the present study, affect was coded for using a manual, categorical classification rather than computer vision analysis as reported by Podesva (2018), and a thematic analysis was then performed to code for discourse content and discourse management, in order to observe whether an interaction with creak observable at the discourse level in the data might emerge also.

Table 7.2 Independent variables included in the regression analysis

Variable	Description
Sentence position	Initial, medial, or final
Out as trans	Coded as YES or NO, relative to the interview before or after coming out as trans
Affect	Data were coded through visual analysis, whereby COMFORTABLE indicates minimal body movement, smiling and UNCOMFORTABLE indicates increased body movement when not smiling, frowning, or downward averted gaze.
Topic	Themes identified were grouped into four primary discourses of ACTING, TRANS IDENTITY, DISCRIMINATION (including preventing bullying, racisms, and anti-bullying charity work), and IDENTITY OTHER (including themes of family, friends, hobbies).
Interlocutor gender	Female, male, or no interlocutor

The data were then subjected to post-hoc statistical analysis, with each factor added to a generalized linear regression model for the binary variable of ±CREAK, using the glmer function of the lme4 package in R (Bates *et al.* 2015; R Core Team 2021). The "drop1" function of the same package was then used to remove factors and their interaction terms that were not significant when modeling for the interaction such as affect and topic. Post-hoc within-factor contrasts were computed using estimated marginal means (Lenth 2022). The results of the final model will be described in the following section.

Results

The compiled data arising from the Josie Totah interviews (hereafter Totah corpus) totaled 7956 vowel tokens processed using F0 and H1–H2 automatic voice quality detection methods and manually checked using both auditory perceptual and visual spectrographic assessment. Of these tokens, 939 (11.80%) were identified as containing creaky voice. Figure 7.5 shows a relatively consistent proportion of creaky voice used in five of the seven interviews, although the fourth interview by far has a highly increased number of syllables produced with creaky voice. This interview contained between two to three times the frequency of creaky voice usage in comparison to all other interviews. However, the following interviews clearly indicate that this is not a sustained change, as the rate of creak returns within the range of Totah's rates of creak before coming out.

As two separate samples—*before* and *after* coming out—this distinction alone is insufficient to claim that Totah being out as trans correlates with a wholesale increase in the use of creak. Figure 7.6 displays the mean average proportion of creaky voice across the corpus, subset into the two periods that encompass data before and after Totah publicly came out as trans.

There seems to be little difference in the proportion of creak used before and after coming out. This is reflected in the regression model too (Fig. 7.6), where the model estimate reflects, if anything, a nonsignificant decrease ($p < 0.277$). If this is the case, what other factors could begin to explain the variation in creak from interview to interview?

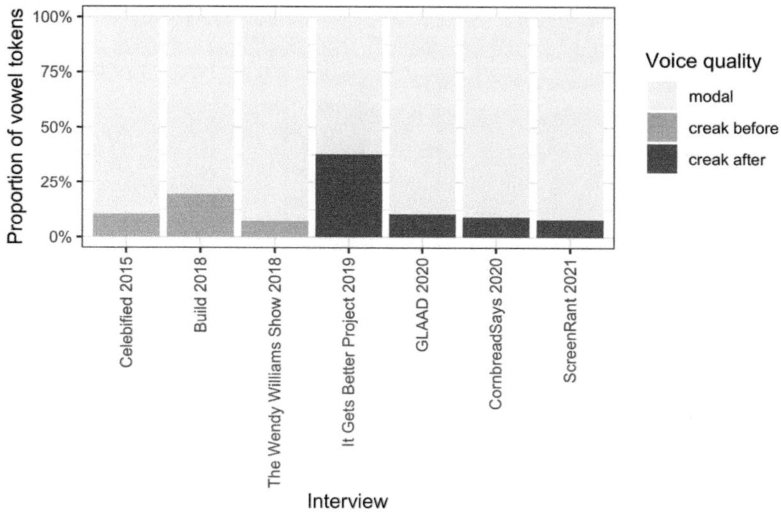

Figure 7.5 Proportion of vowel tokens produced with creaky voice by interview.

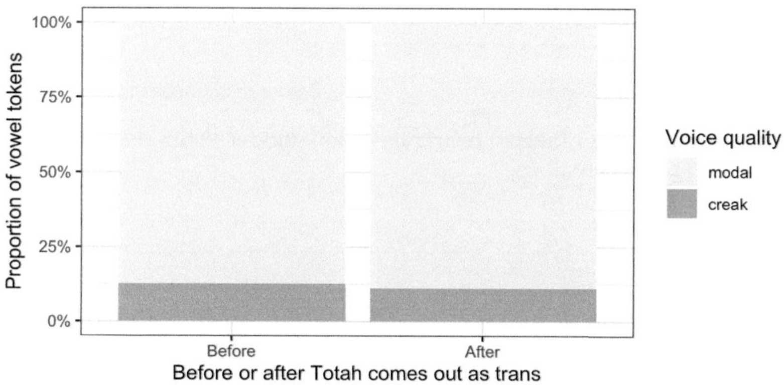

Figure 7.6 Proportion of syllables produced with creaky voice before and after coming out as transgender.

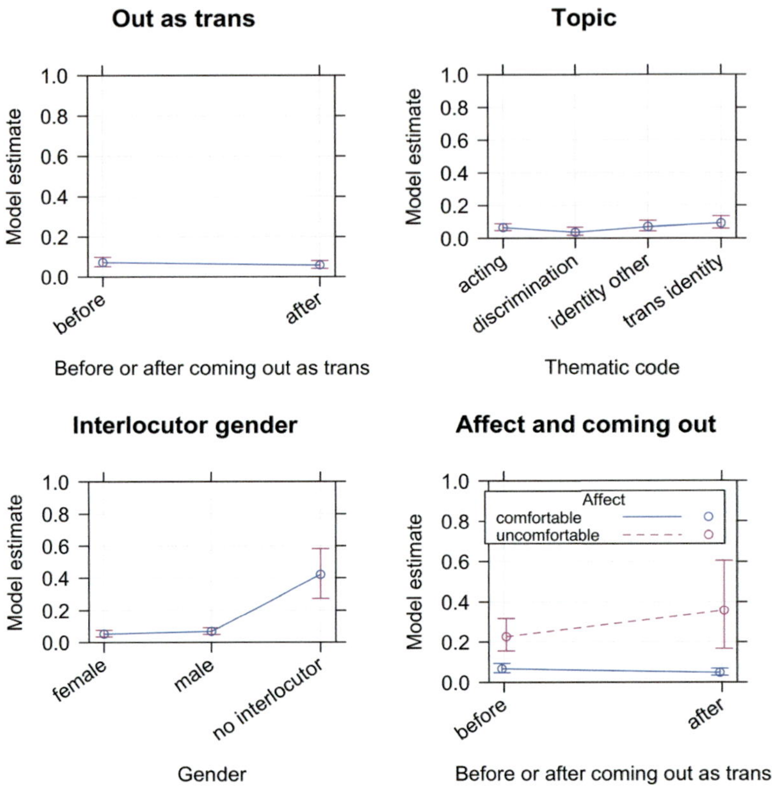

Figure 7.7 Generalized linear regression model estimates for factor effect on rate of creak use.

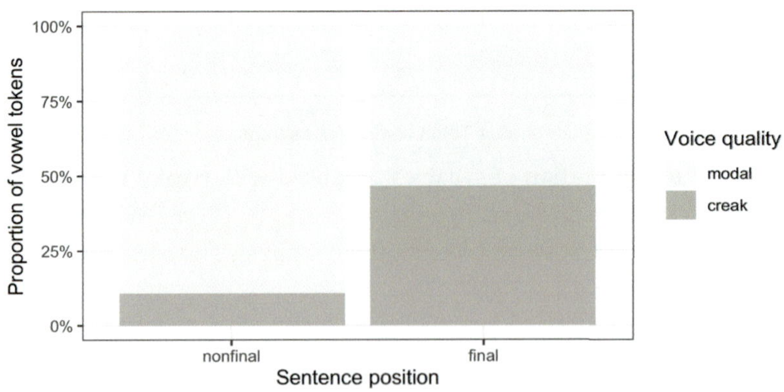

Figure 7.8 Proportion of vowel tokens produced with creaky voice in final or nonfinal position in the utterance.

Most vowel tokens produced with creak occurred in syllables in sentence-final position, as to be expected in naturalistic English speech (Henton and Bladon 1988). However, sentence-initial and sentence-medial syllables were still found to contain vowels produced with creak (Fig. 7.8). They indeed may well still be used stylistically but constrained by other factors. The following section turns to the social context coded throughout the data—namely, theme, affect, and the interlocutor's identity.

Social factors: theme

The thematic analysis revealed a narrow range of discourses within the interviews. These were broadly collapsed into four primary themes. The theme of ACTING included discussions of experiences of acting and discussions of characters. DISCRIMINATION included conversations around bullying and other forms of identity-based social injustice.

The category IDENTITY OTHER was used to collapse together discussions around her friends, family, hobbies, and interests, while TRANS IDENTITY included topics such as Totah's own trans identity, broader trans representation in television and the media, as well as supporting trans friends.

As can be seen in Fig. 7.9, these four categories do not appear to vary greatly in the use of creak from topic to topic. Within-factor

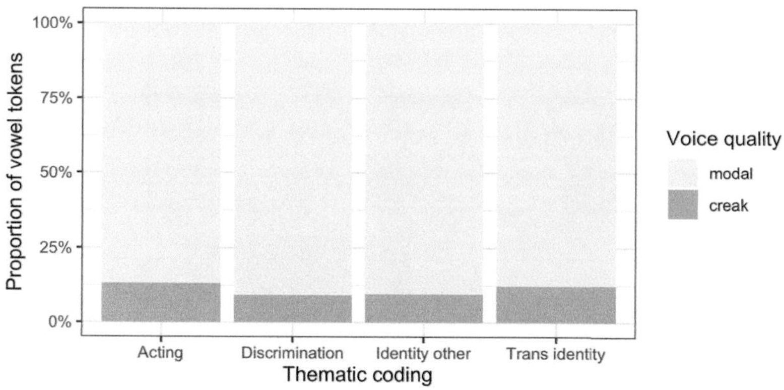

Figure 7.9 Proportion of vowel tokens produced with creaky voice in final or nonfinal position in the utterance.

comparison does highlight a statistically significant difference between DISCRIMINATION and TRANS IDENTITY ($p = 0.0284$), suggesting that in fact there may be some variation in the rate of use of creak when discussing different topics. However, it is worth bearing in mind that discussions around trans identity only occurred after Totah had come out publicly as trans herself. A comparison between rates of creak in different themes shows no significant change ($p = 0.16$).

What is noticeable here at least is that there is a marked shift in the discourses occurring within these interviews with Josie Totah and the sorts of interviews Totah then engages with. This presents not only issues for comparability across the dataset, but also a point of reflection regarding the degree to which the contexts of Totah's interviews change in of themselves after publicly coming out as trans.

Social factors: affect

Figure 7.10 also supports the findings presented by Podesva (2018), suggesting a speaker's affect may correlate with their use of creaky voice. Of the samples coded uncomfortable, 28.3% of these vowels were produced with creak, as opposed to 10.1% produced in comfortable tokens. The linear regression model found that this change is statistically significant ($p < 0.001$). Figure 7.7 also shows

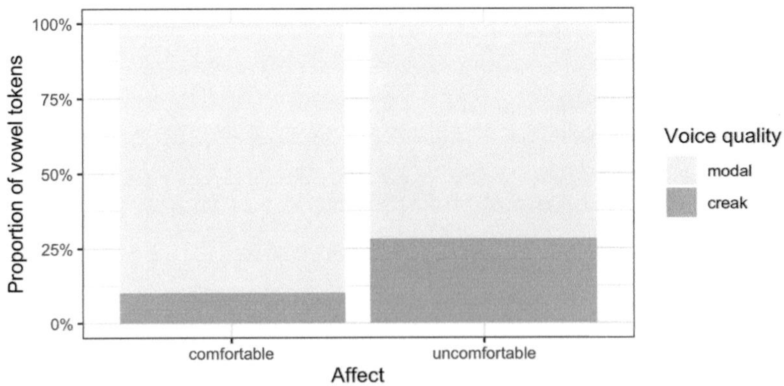

Figure 7.10 Proportion of vowel tokens produced with creaky voice during syllables coded for affect.

how when accommodating for the interaction between affect and whether or not Totah is out as trans, a low count of UNCOMFORTABLE tokens once again prohibits a significant result, although it does approach significance ($p = 0.09$).

Social factors: interlocutor

As already identified in the results so far, creaky voice did appear to be statistically significant in certain interviews, as opposed to others. By taking factors such as the interlocutor's gender presentation, sexuality, race, or own rates of creaky voice correlated with Totah's voice quality, the linear regression model suggested that in fact none of these factors appear to correlate in any variation in rates of creak. A clear distinction, however, occurs when we include all instances of tokens where no direct interlocutor is involved. Figure 7.7 compares the effect of interlocutor gender in comparison to when no direct interlocutor is included, displaying a highly significant correlation between the rates of use with a specific interviewer as opposed to in the context where Totah is not directing her speech toward a specific interlocutor at all ($p < 0.0001$). Again, we return to the issue mentioned earlier in this section. Interview four, the only interview without a direct interlocutor, has by far the highest rates of creak. The following section takes a look at this particular interview, which occurs just after Totah comes out as trans.

If positions within the utterance, theme, affect, and the interlocutor's identity do not vary significantly when comparing either side of this social juncture, why is it that this particular interview contains such high rates of creak? To answer this, we turn to the complexity of gender construction as part of a process of bricolage and perhaps here consider how such intricacies may not rise above the level of significance for individual factors alone across a larger sample.

Seen, visible, loved, accepted

Interview four, recorded seven months after Totah came out as trans, differs from the other interviews in a number of ways. Most other samples of Totah's speech were taken from interviews with Totah and an interviewing interlocutor, whereas this recording features Totah talking to the camera, with the public audience as an imagined listener, rather than one physically present. Totah is then able to construct, through her own discourse, who the imagined listener is. In terms of *audience design* (Bell 1984), the audience take the form of an *addressee* rather than *auditor*, one who is directly addressed rather than merely listening in on a conversation where Totah is positioned against another, very real addressee.

During the interview, Totah discusses her experiences with family, school, and coming to terms with her gender identity and society. Toward the end of the interview, Totah again talks directly to the addressee, but this time identifying the listener as a trans individual listener (Extract 7.1).

Extract 7.1 Extract from interview four with Totah for the Trans Day of Visibility (2019)

Totah	Set aside this day to be like, come on. Call your trans brother or sister or sibling. Hit them up and be like, what's up? Like, I love you. No matter where you identify on the LGBTQ+ spectrum, I want you to know that on this day, trans day of visibility, you are seen, you are visible, you are loved and you are accepted. There are people waiting to love you, I promise you even if you can't see it now. There are people lining up around the corner. You just have to fight the fight. You will get there, and it will get better. Hi, I'm Josie Totah, and I'm here to tell you that it gets better.

Lines 1–3 of this interview appear to suggest Totah is directing her speech to a listener who perhaps is not transgender, advising them to reach out to trans people they know and offer support. From lines 3–10, the remaining utterances of the interview, Totah then directs her speech toward a slightly different imagined audience—the trans individual. Though described as a person that identifies anywhere "on the LGBTQ+ spectrum," the framing of

the section as a message from Totah on the Trans Day of Visibility, Totah speaks as though from experience, "you will get there, and it will get better," suggesting that indeed, this addressee has something in common with Totah's experiences. Totah speaks these lines with a calm confidence, almost conveying a persona of a knowledgeable, supportive other.

By assuming the great variation in results correlates to the divergent nature of the interview, compared to others in the dataset, this may occlude a secondary and more compelling argument that additional data from around this timepoint may help to provide. Irvine's (2018) data from recordings and qualitative interviews with transmasculine speakers suggested that, while there was evidence of style shifting with /t/ glottalling, speakers who had only recently come out as trans and only just begun transitioning were those seen to style shift the most when in the presence of a cisgender female interviewer than another transgender male interviewer. Again, this highlights the importance of investigating data prior to individuals coming out or transitioning. Here, there may be an early period where speakers, much like the nonbinary speakers in Gratton's (2016) study, draw on phonetic features that index gender, to either disambiguate or foreground their gender identity. By using these gendered variables, the trans men in Irvine's (2018) study both distance themselves from their cisgender female interlocutor, while also asserting their masculine identity. Conversely, in the presence of another transmasculine speaker, divergence toward a masculine target is not necessitated.

In terms of indexing gender through a process of bricolage, Totah and the participants in the studies of Irvine, Gratton, and Zimman draw on much more than just linguistic resources. With an awareness of the other's perceptions and preconceptions about physiological appearances, the trans individuals in these studies can draw on the voice "as a critical mediator of gender perceptions" (Zimman 2017: 340), recontextualizing the embodiment of gender by bringing different features that are part of their "set of multidimensional styles" to the foreground (2017: 341). Indeed, Totah can be heard using creak phonation long before coming out, but at this point in interview four, Totah still has short hair, though by the following interview she presents below-shoulder length hair. While this is, of course, no indicator of gender identity, it does form part of the broader semiotic bricolage involved in her presentation at this point. By the following interview (GLAAD interview in 2020), the rates of creaky voice have again dropped much lower, suggesting a somewhat inverse relationship between resources that visually index and embody gender after coming out and the use of gendered linguistic variables. Just like trans and nonbinary speakers referenced throughout this chapter, it is the feeling of being "seen, visible, loved, accepted" that Totah describes here that motivates a conclusion that Totah's use of creak is indeed to make dimensions of her style "visible," drawing here on creak to achieve a gendered visibility.

Closing remarks

This chapter sets out to investigate whether an individual's use of creak, as a saliently gendered feature, could be influenced by a turning point in their gender presentation. The first interview after Josie Totah came out as trans provides an interesting example of how events in an individual's lifespan may challenge a speaker to renegotiate their identity and, in turn, draw from their available meaning-making resources in new ways. As a *social juncture*, coming-out events provide an insightful starting point for investigating repositionings and reorganizations of linguistic styles. For Josie Totah at least, this does not appear to result in a maintained change in the use of creak phonation. However, further research on speakers at these turning points of identity construction may uncover underlying characteristics involved in the mechanisms of negotiating reconfigurations of style sets. This chapter calls for investigations not just into the micro-level navigation of identity, but in the larger life-altering changes that lead to people wanting to express their identity in a way that is then seen, visible, loved, and accepted.

Endnote

1. For the present chapter, conceptualization of coming-out events will be within a framing that acknowledges that the Western white hegemonic conceptualizations of "coming out" are not the sole constitutive realization of such speech acts.

References

Abdelli-Beruh, N. B., Wolk, L., and Slavin, D. (2014). "Prevalence of vocal fry in young adult male American English speakers." *Journal of Voice*, **28**(2), 185–190.

Angouri, J. (2021). "Introduction: Language, gender, and sexuality: Sketching out the field." In *The Routledge Handbook of Language, Gender, and Sexuality*. London: Routledge.

Audacity Team. (2008). *Audacity* [computer program]. Version 2.4.2. Retrieved from http://audacityteam.org.

Bacon, J. (1998). "Getting the story straight: Coming out narratives and the possibility of a cultural rhetoric." *World Englishes*, **17**(2), 249–258.

Barras, C., Geoffrois, E., Wu, Z., and Liberman, M. (2001). "Transcriber: Development and use of a tool for assisting speech corpora production." *Speech Communication*, **33**(1–2), 5–22.

Bates, D., Mächler, M., Bolker, B., and Walker, S. (2015). "Fitting linear mixed-effects models using lme4." *Journal of Statistical Software*, **67**(1).

Bell, A. (1984). "Language style as audience design." *Language in Society*, **13**(2), 145–204.

Boersma, P. and Weenkink, D. (2018). *Praat* [computer program]. Version 6.0.37. Retrieved from http://praat.org.

Butler, J. P. (1990). *Gender Trouble*. London: Routledge.

Dallaston, K. and Docherty, G. (2020). "The quantitative prevalence of creaky voice (vocal fry) in varieties of English: A systematic review of the literature." *PLOS ONE*, **15**(3), e0229960.

DiDomenico, S. M. (2015). "Putting a face on a community: Genre, identity, and institutional regulation in the telling (and retelling) of oral coming-out narratives." *Language in Society*, **44**(5), 607–628.

Dunn, T. (2015). "What is 'vocal fry,' and why doesn't anyone care when men talk like that?" *Upworthy*. Retrieved March 16, 2021 from www.upworthy.com/what-is-vocal-fry-and-why-doesntanyone-care-when-men-talk-like-that.

Eckert, P. (1989). *Jocks and Burnouts: Social Categories and Identity in the High School*. New York: Teachers College Press.

Eckert, P. (2004). "The meaning of style." In *Proceedings of the Eleventh Annual Symposium about Language and Society*, **47**, 41–43. Austin, TX.

Eckert, P. (2008). "Variation and the indexical field." *Journal of Sociolinguistics*, **12**(4), 453–476.

Eckert, P. (2012). "Three waves of variation study: The emergence of meaning in the study of sociolinguistic variation." *Annual Review of Anthropology*, **41**(1), 87–100.

ELAN. (2021). *ELAN*. Max Planck Institute for Psycholinguistics, The Language Archive, Nijmegen. [Computer program]. Version 6.2. Retrieved from https://archive.mpi.nl/tla/elan.

Feinne, L. (2016). "Ridiculing vocal fry is sexist, and this awesome video explains why." *Bustle*. Retrieved August 15, 2021 from www.bustle.com/articles/157606-ridiculing-vocal-fry-is-sexist-and-this-awesome-video-explains-why.

Gillies, C. (2018). "Vocal fry & upspeak: Why does Kim Kardashian speak like that?" *Highsnobiety*. Retrieved June 15, 2021 from www.highsnobiety.com/p/vocal-fry-upspeakcelebrities-kardashian.

Gordon, M. and Ladefoged, P. (2001). "Phonation types: A cross-linguistic overview." *Journal of Voice*, **29**(4), 383–406.

Gratton, C. (2016). "Resisting the gender binary: The use of (ING) in the construction of non-binary transgender identities." *University of Pennsylvania Working Papers in Linguistics*, **22**(2), Article 7.

Hebdige, D. (1991). *Subculture: The Meaning of Style*. London: Routledge.

Henton, C. and Bladon, A. (1988). "Creak as a sociophonetic marker". In V. Fromkin, L. M. Hyman, and C. N. Li (Eds.), *Language, Speech, and Mind: Studies in Honour of Victoria A. Fromkin* (pp. 3–29). London: Routledge.

Irvine, I. (2018). "'I Think Tha[ʔ] Gender Is a Bullshi[ʔ] Thing': A Study of Scotland's Transgender Male Community and Where They 'Fit', If at All, into the Gender Paradox." MA Dissertation, University of Glasgow.

Jaffe, A. (2016). "Indexicality, stance and fields in sociolinguistics." In N. Coupland (Ed.), *Sociolinguistics: Theoretical Debates* (pp. 86–112). Cambridge: Cambridge University Press.

Labov, W. (1966). *The Social Stratification of English in New York City*. Washington, DC: Center for Applied Linguistics.

Ladefoged, P. (1971). *Preliminaries to Linguistic Phonetics*. Chicago: University of Chicago Press.

Lakoff, R. T. (1975). *Language and Woman's Place*. New York, NY: Harper & Row.

Lenth, R. (2022). *Emmeans: Estimated Marginal Means, aka Least-Squares Means*. Vienna, Austria. Retrieved from https://cran.r-project.org/package=emmeans.

Levon, E. (2015). "'The ideal gay man': Narrating masculinity and national identity in Israel." In T. M. Milani (Ed.), *Language and Masculinities: Performances, Intersections, Dislocations* (pp. 133–155). London: Routledge.

Liang, A. C. (1997). "The creation of coherence in coming-out stories." In A. Livia and K. Hall (Eds.), *Queerly Phrased: Language, Gender, and Sexuality* (pp. 287–309). New York, NY: Oxford University Press.

Luthern, E. and Clopper, C. G. (2011). "Variation in glottalization at prosodic boundaries in clear and plain lab speech." *Proceedings of the 18th International Congress of Phonetic Sciences* (p. 5). Glasgow: University of Glasgow.

Milroy, J. (1992). *Linguistic Variation and Change: On the Historical Sociolinguistics of English*. Oxford, UK, Cambridge, USA: B. Blackwell.

Milroy, L. (1980). *Language and Social Networks*. London, England: Blackwell.

Mo, K. (2016). "Women should avoid using 'vocal fry' in the workplace." *The Arizona State Press*. Retrieved August 15, 2021, from https://www.statepress.com/article/2016/10/spopinion-women-shouldstop-using-vocal-fry-in-the-workplace.

Oliveira, G., Davidson, A., Holczer, R., Kaplan, S., and Paretzky, A. (2016). "A comparison of the use of glottal fry in the spontaneous speech of young and middle-aged American women." *Journal of Voice: Official Journal of the Voice Foundation*, **30**(6), 684–687.

Podesva, R. J. (2018). "The affective roots of gender patterns in the use of creaky voice." Presented at the *Experimental and Theoretical Advances in Prosody 4*, October 13, University of Massachusetts Amherst.

Podesva, R. J., Roberts, S. J., and Campbell-Kibler, K. (2001). "Sharing resources and indexing meanings in the production of gay styles." In K. Campbell-Kibler, R. J. Podesva, S. J. Roberts, and A. Wong (Eds.), *Language and Sexuality: Contesting Meaning in Theory and Practice* (pp. 175–189). Stanford: Center for the Study of Language and Information.

R Core Team. (2021). *R: A Language and Environment for Statistical Computing*. Vienna, Austria: R Foundation for Statistical Computing. Retrieved from https://www.R-project.org.

Rust, P. C. R. (1993). "'Coming out' in the age of social constructivism: Sexual identity formation among lesbian and bisexual women." *Gender & Society*, **7**(1), 50–77.

Sauntson, H. (2007). "Education, culture and the construction of sexual identity: An appraisal analysis of lesbian coming out narratives." In H. Sauntson and Sakis Kyratzis (Eds.) *Language, Sexualities and Desires* (pp. 140–64). London: Palgrave Macmillan UK.

Sauntson, H. (2015). "Coming out." In *The International Encyclopedia of Human Sexuality* (pp. 197–290). Oxford, UK: John Wiley & Sons, Ltd.

Sedgwick, E. K. (1990). *Epistemology of the Closet*. London, England: Prentice-Hall.

Spender, D. (1980). *Man Made Language* (4th edn.). London, England: Rivers Oram Press/Pandora List.

Tannen, D. (1990). *You Just Don't Understand: Women and Men in Conversation*. New York, NY: William Morrow & Company.

Trudgill, P. (1974). "Linguistic change and diffusion: Description and explanation in sociolinguistic dialect geography." *Language in Society*, **3**(2), 215–246.

West, C. and Zimmerman, D. H. (1987). "Doing gender." *Gender & Society*, **1**(2), 125–151.

Wolf, N. (2015). "Young women, give up the vocal fry and reclaim your strong female voice." *The Guardian*. Retrieved June 16, 2021, from https://www.theguardian.com/commentisfree/2015/jul/24/vocal-fry-strong-femalevoice.

Wolk, L., Abdelli-Beruh, N. B., and Slavin, D. (2012). "Habitual use of vocal fry in young adult female speakers." *Journal of Voice*, **26**(3), e111–116.

Yuasa, I. P. (2010). "Creaky voice: A new feminine voice quality for young urban-oriented upwardly mobile American women?" *American Speech*, **85**(3), 315–337.

Zimman, L. (2009). "The other kind of coming out: Transgender people and the coming out narrative genre," *Gender & Society*, **3**(1), 53–80.

Zimman, L. (2016). "Sociolinguistic agency and the gendered voice: Metalinguistic negotiations of vocal masculinization among female-to-male transgender speakers." In A. M. Babel (Ed.), *Awareness and Control in Sociolinguistic Research* (pp. 253–277). Cambridge: Cambridge University Press.

Zimman, L. (2017). "Gender as stylistic bricolage: Transmasculine voices and the relationship between fundamental frequency and /s/." *Language in Society*, **46**(3), 339–370.

Zimman, L. (2018). "Affective stance and voice quality in a pervasively creaky speaker: Stance objects as a tool for investigating indexical meaning." Presented at the *Experimental and Theoretical Advances in Prosody 4*, University of Massachusetts Amherst.

Chapter Eight

"It's chiefly your eyes I think, and that throb you get in your voice"

The place of creaky voice in the soundscape of attractive female voices in twentieth and twenty-first century American cinematography

Míša Hejná and Mark Eaton

> The deep, dark secret about the femme fatale is that most of her fatale was located in her voice. (…) There's just something about a scratchy, deep, knowing voice that used to make a guy chuck a sweetheart like Jen Aniston to the curb and set off into the hot, dusty, dirty, Darfur night with Angelina.
>
> —Hunt 2009

Creak: Theories and practices of pulse phonation
Edited by Francesco Venturi

ISBN 978-981-5129-29-8 (Hardcover), 978-1-003-61606-1 (eBook)
www.jennystanford.com

Introduction

Academic literature on the multidirectional interconnections between social life and popular culture produced by scholars in sociology, anthropology, history, media studies, and other fields is voluminous (e.g., Nachbar and Lause 1992; Guins and Cruz 2005; ErkenBrack et al. 2012; Prasad 2018). However, one area that has received relatively scant attention from researchers—and this includes linguistics scholars—is the place and roles of the human voice in popular culture. The central aim of this chapter is to direct some focus onto selected features of the soundscapes of attractive characters played by American actresses and to indicate how they have evolved over time. More specifically still, the objects of our analysis are "femme fatale" character types who occupy central protagonist roles in American cinema from the 1950s to the 2010s. Through this focus we were able to closely study how personas are portrayed through vocal variation over time, zooming in on pitch. Our research questions are, therefore, as follows:

1. What vocal features are used to construct the femme fatale persona in American cinematography?
2. Has creaky voice been used in the construction of the femme fatale in this area? If so, how and what functions does it have exactly?
3. Do we see an increase in creaky voice across time, in accordance with the assumption that creaky voice has been on the rise in the speech of female speakers of American English?

In the remainder of this chapter, we first present a historical and contemporary background on the femme fatale trope to provide readers with necessary context and justify its choice as our object of analysis. This is followed by an analysis of the place and roles of creaky voice in the soundscapes of femme fatale character types. Next a methodology is presented, beginning with an outline of the characters and films selected for analyses. The chapter then proceeds to analyze how phonation variation was operationalized. We conclude that creaky voice has become more prevalent in the femme fatale persona over time; however, this increase in occurrence does not seem to be inherently linked to the construction of this trope. Instead, creak is used to signal emotional turmoil and/or as a strategy to do negative stance-taking work. Rather than creak, higher breathiness combined with low pitch are the constant vocal building blocks of the trope.

Background and theoretical perspectives

Why femme fatales?

There are many reasons why the femme fatale trope is an exciting object for academic research, including its long existence as an influential cultural frame used to express and (re-)enforce gender norms and identities. Notably, it did not suddenly appear with the invention of moving pictures. The trope has long been mobilized as a vehicle communicating the values, morals, norms, and anxieties of dominant groups in Western societies and cultures, and in particular, obsessions and fears surrounding powerful women [for example, as discussed in works on the biblical Eve (Edwards 2010), Cleopatra VII (Chiari 2019), and Elizabeth I (Richards 1997)]. What connects many cultural representations of these and other powerful women of the past is that they are portrayed as threatening to the established order.

The sense of dread and impending destruction associated with "dangerous" women in history is also present in femme fatale characters of the 20th and 21st century cinema (film noir or otherwise), with Farrimond (2017) concluding that the femme fatale represents "one of the most enduring character types in Hollywood cinema." The extant research into the trope falls into three broad and inter-related categories, presenting femme fatale characters as: (1) feminist celebrations of female power, independence, and uninhibited sexuality (Cowie 1993); (2) misunderstood victims as opposed to "bad girls" or highly motivated villains (Grossman 2009); and (3) the most repeated and enduring model, deeply misogynistic expressions of male desires and anxieties (Kaplan 1978; Cameron 1993; Copjec 1993; Boozer 1999/2000). More recent research has focused on the ways that the boundaries separating these perspectives can easily become blurred (See Grossman 2009: 4, in ref. to Gledhill 1978: 122).

According to Julie Grossman (2020), "[i]ntelligent, witty, able to role-play and perform, deceptive, enraged, frustrated, mercenary, seductive, overtly sexual, fearless and tough as nails, physically self-confident with a striking appearance..." are the qualities that come to mind when we imagine the femme fatale. Richard Dyer further alludes to the complex dualities inherent in the character type. For him, the femme fatale is a woman who while "exceedingly 'feminine' and flirtatious" is also "too ambitious and intelligent for the docile role society has decreed she play" (Jilliand 2020). These personal traits have been identified by scholars working on vocal attractiveness, voice, and emotions to be closely linked with variation found in the production of the human voice (more details are provided below). We would, therefore, expect a striking trope such as the femme fatale to be constructed through a set of specific vocal features. Furthermore, considering the extent to which the trope has been used to reflect gender politics in various societies, we would also expect the vocal characteristics used to portray the femme fatale to reflect broader societal debates about and changes taking place in the areas of gender roles and identity.

The femme fatale trope became more prominent in the post-1945 period as an expression of male anxieties about the Second World War's impact on gender roles, which had become more fluid and ambiguous. While men returned from the war, the femme fatale functioned as a device for the "demonization" of independent women "at a time when there was a concerted effort to persuade women to surrender the jobs that they had taken on during the war" (Cook 1978; Jancovich 2011: 100). Specifically, the trope served as a warning about the risks associated at the time with unchecked female sexuality, infidelity, materialism, and other perceived threats to the social order believed to have risen because of the war experience.

Fears about the changing status and roles of women were clearly reflected in the traits and behaviors of the post-war femme fatale, including being dangerous, aggressive, competent; cold and unattainable; exceptionally intelligent; unapologetically independent; averse toward marriage and motherhood; ruthlessly ambitious, deceptive, seductive; sexually promiscuous; often engaged in unhealthy habits such as smoking and drinking (see Bronfen 2004; Bell 2011; Andreetti 2019; Smith 2019). Although this list is not exhaustive, these characteristics all contribute to what Bell (2011) describes as the femme fatale's "mystery," "unknowability," and "undefinability" (632). For Grossman (2020), the mysterious nature of the femme fatale character type is compounded by the fact that it combines both feminine and masculine qualities. In order to "find meaning outside of oppressive social roles" (2020: 3), femme fatales adopt some of the qualities of their male counterparts, which transform them into "foul-mouthed" and "tough-talking" (2020: 12) "bad girls" (2020: 2–3). And indeed if, as Grossman (2020) asserts, the femme fatale offers an interesting object of analysis because it represents "insurgencies against conventional gender categories that are insidious" (2020: 4), is it possible that their voices also include unique, perhaps ambiguously gendered, features?

Gender-related ambiguity can be achieved in a range of ways. Focusing on vocal variation, the most obvious vocal characteristic to target is that of a speaker's pitch or f0,[1] with stereotypically feminine and attractive voices in female speakers associated with higher f0 and stereotypically masculine and attractive voices in male speakers with lower f0 (e.g., Pisanski and Feinberg 2018). It might well be the case that femme fatales find themselves between the average f0 values reported for male and female populations of their respective societies. Such gender-related vocal ambiguity might also extend to other variables known to index sex and/or gender and attractiveness, such as pitch dynamism (e.g., Anolli and Ciceri 2002), breathiness (e.g., Babel et al. 2014), vowel formant dispersion (e.g., Babel et al. 2014), and others (e.g., Anolli and Ciceri 2002). Through the process of bricolage (Zimman 2017), a femme fatale voice might utilize some vocal aspects with values associated with male speakers and others which are typically associated with female speakers. In what follows below, we discuss to what extent creaky phonation might contribute to the portrayal of the femme fatale character.

Place and roles of creak in soundscapes of cinematic femme fatale

The literature on creaky voice and attractiveness is rather limited, at least when compared to the research available on other vocal features. Xu et al. (2016) report that creaky voice is not perceived as attractive by Japanese listeners. Xu et al. (2018) conducted a comparable study with Mandarin Chinese, in which they found that creaky phonation was perceived as less attractive in female voices than modal or breathy phonation, whereas it was evaluated in the same way as breathy and modal phonations in male voices. More specifically for American English, Greer and Winters (2015) report that creaky voice in male voices is evaluated as cooler and more attractive by American English listeners, while breathiness is the preferred voice quality in female voices, assessed as attractive. Thus, with limited research available, it would seem that creaky voice is unlikely to be associated with romantically and/or sexually attractive female voices. The study of Shaw and Crocker (2015) presents findings that are at first glance somewhat at odds with the findings reported by the studies mentioned above. The authors analyze Scarlett Johansson, an American actress whose voice is often commented on in terms of attractiveness, as evidenced for example by surprisingly frequent webpage titles such as "Scarlett Johansson makes a Bible verse sound really sexy." Interestingly for our purposes, the authors find that in American roles (as opposed to British roles), Johansson uses more creaky voice. Furthermore, the actress employs this phonation more frequently in overtly sexual and flirtatious scenes (2015: 26).

Would we expect the American femme fatale to use creaky voice? On the one hand, this phonation does not seem to be associated with attractiveness in American English female voices, or with Mandarin and Japanese female voices, and possibly with female voices across cultures. However, it is likely that the relevant studies focus on normative attractiveness since the questions typically asked do not go beyond evaluation of voice samples in terms of "attractiveness" on a scale from a–z, i.e., without necessarily specifying *what type* of attractiveness. Considering, however, how much the social functions of creak seem to be understudied (see, e.g., Dallaston and Docherty 2020), it is difficult to make any strong predictions about the role of creak in the construction of the femme fatale. Its use in overtly sexual situations would be in line with the femme fatale's ambiguous nature since the trope portrays female protagonists as rebels who transgress context-based gender norms in their performance of "unconventional and unprescribed roles" (Grossman 2020: 4). Often these roles include many characteristics usually associated with male power/agency and masculinity. As a result, the femme fatale may be understood as occupying a space between traditional conceptions of femininity and masculinity, complex characters who are, on the one hand, strikingly beautiful, seductive, and irresistible; while simultaneously intelligent, ambitious, determined, hard-boiled, destructive, and violent, on the other. This complex duality of what are often regarded as typically feminine versus masculine traits has been the focus of research, primarily in film studies and film criticism, and exclusively with the analytical focus on primarily nonverbal qualities of the femme fatale. If creaky voice is indeed associated with more masculine voices, we might expect the femme fatale to display creakier phonation than normatively feminine tropes.

Early research on creaky voice does associate this vocal profile primarily with conceptions of masculinity and authority among male speakers (Henton and Bladon 1988). It would, therefore, seem that creak might index masculinity in American English, at least in the 1980s, and that the femme fatales from this period in particular might be portrayed with more creak than normatively feminine characters. However, although no consensus has been reached on whether we are currently witnessing a rise in creak use among young American women (Dallaston and Docherty 2020), recent research has nevertheless pointed toward its increasing prevalence (Yuasa 2010; Wolk, Abdelli-Beruh, and Slavin 2012; Hornibrook et al. 2018). Recent research has also indicated that the use of creak among female speakers of American English may be contributing to what have traditionally been understood as masculine dimensions to female voices (namely the deeper pitch associated with creak). For instance, Mendoza-Denton (2011) has concluded that creak has been employed in the construction of gang girl personas in line with the "qualities, styles … and stances of toughness and of being hardcore," and that acting hardcore serves for "self-protection and self-preservation" functions (2011: 262).

All things considered, given creak's undeniable rise as a subject of scholarly, public, and media fascination, scrutiny and panic in recent years (see Dean n.d.; Steinmetz 2011; Anderson et al. 2014;

Wolf 2015; Blum 2016)—evidenced by numerous studies citing its increased prevalence among female speakers and its apparent gender-bending qualities—we determined that among the many female cinematographic archetypes, it is interesting to investigate creak's place and roles in the soundscapes of the equally complex, controversial, and ambiguously gendered femme fatale.

Methodology

This section first outlines the characters and the films selected for analyses and the rationale behind the chronological structure of the analysis. It then proceeds to discuss how phonatory variation was analyzed and, finally, what overall procedure was adopted to process the data.

Selection of films and chronological structure

The films containing femme fatale characters as central protagonists analyzed in this study were identified through reviews of the academic and popular literature dealing with the genre, as well as through targeted internet searches utilizing search terms such as "femme fatale," "femme fatale films," "femme fatale actresses," and "most popular femme fatales." Criteria for our selection included that the films and their femme fatale characters would be widely known to audiences, either achieving significant box office success at the time of their release, or otherwise reaching large audiences by subsequently achieving a cult status.

The analysis is divided into three chronological periods that correspond with the waves of feminism in post-1945 American society. The first period, while not technically a "wave" of feminist activity, comprises roughly the first two decades following the end of the Second World War and was characterized by many contradictions in terms of the place and roles of women in public life. While the proportion of women working outside of the home rose during the war years out of necessity and continued to do so post-1945, the period also witnessed a strong conservative backlash characterized by a desire to reassert the importance of female domestic life and clearly defined and rigid gender roles (Tyler May 1988). The second chronological period, coinciding with second-wave feminism, an era ushered in by, among other things, the publication of Betty Friedan's critical work *The Feminist Mystique* (1963) and the passage of the Civil Rights Act of 1964, saw an intensification of efforts to build on the achievements of earlier feminists, but expanded on the earlier period with a focus on achieving gender equality when it came to legal rights (in the family and beyond), sexuality (including reproductive rights), the professions and other areas of inequality ("Second Wave" n.d.). The third chronological period coincides with third-wave feminism, an era which began in the early 1990s and was triggered by, among other factors, the perceived official mistreatment of Anita Hill during the Supreme Court confirmation process of Justice Clarence Thomas, whom Hill had accused of sexual harassment. This wave of feminist activity continued to work to establish equality between the sexes; however, the feminist tent was expanded to include women of racial, ethnic, and other minority groups ("Third Wave" n.d.).

The films analyzed in this study provide evidence that as the rights and social roles of women evolved through different waves of feminism, these changes were reflected in the evolving motivations, behaviors, and identities of the femme fatale. Films produced in the first wave (mid-1940 to early 1960s) tended to be situated and closely identified with domestic spaces (e.g., Bacall as Vivian Rutledge in 1946's *The Big Sleep*; Monroe as Rose Loomis in 1953's *Niagara*). With the coming of the third wave in the early 1990s, the femme fatale has adopted more multidimensional characteristics and identities, with traditional domestic themes fading more and more into the background. For example, as women became an increasingly permanent fixture in the business world, we see the emergence of what could be described as the professional femme fatale—for instance, female characters in the 1990s films portrayed as highly motivated and independent-minded professionals who will go to great lengths to accomplish their goals (e.g., Kidman as Suzanne Stone in 1995's *To Die For*). The femme fatale characters of this period are also adorned with increasingly complex hybrid characteristics and identities, such as impressive physical and, in some cases, supernatural powers, as well as an unapologetic willingness to use them (e.g., Stone as Lori Quaid in 1990's *Total Recall*). The analyses were based on the films summarized in Table 8.1.

Table 8.1 Summary of films analyzed by period

Post-war	Second-wave feminism	Third-wave feminism
The Big Sleep (1946)	*The Graduate* (1967)	*Total Recall* (1990)
Niagara (1953)	*The Postman Always Rings Twice* (1981)	*Basic Instinct* (1992)
Gentlemen Prefer Blondes (1953)	*Body Heat* (1981)	*The Crush* (1993)
Vertigo (1958)	*Top Gun* (1986)	*The Last Seduction* (1994)
Splendor in the Grass (1961)		*Disclosure* (1994)
		To Die For (1995)
		Double Jeopardy (1999)
		Matchpoint (2005)
		Black Widow character in Marvel film scenes, starting with *Iron Man 2* (2010)
		Sin City: A Dame to Die For (2014)
		La La Land (2016)

Phonatory variation

In our analyses of phonatory variation, we focused on four aspects of voice quality: (1) presence of creaky voice, (2) mean pitch, (3) pitch range, and (4) presence of breathiness. The presence of creaky voice and breathiness and the overall pitch and pitch range were evaluated primarily through auditory analyses. In order to outline how the four aspects above were identified and quantified here, we first need to explain what modal phonation is as it functions as a reference phonation. Indeed, modal phonation is often described as "the neutral mode of [vocal fold] vibration" (Laver 1980[2009]: 95). It consists of regular vocal fold vibration with little air passing through. As shown in Fig. 8.1, this phonatory profile manifests itself acoustically through the presence of dark striations (Y-axis in the spectrogram), which are regularly spaced out from each other—see the dark repeated striations in two intervals of the signal in Fig. 8.1 (corresponding to /ɛ/ and /ɪ/ in "detective"). The striations are associated with higher energies, which can be seen through the amount of darkness visible on the vertical axis. The presence of such higher energies across the frequencies means that the voice is not particularly soft. The darker the striations and the higher the frequencies at which these are visible on the Y-axis of a spectrogram, the tenser the phonatory setting.

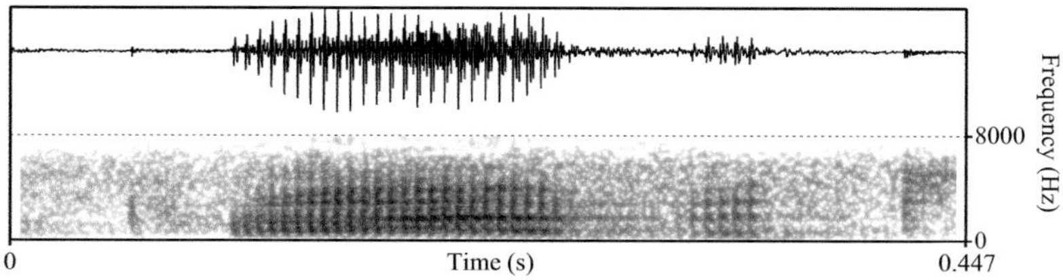

Figure 8.1 Acoustics of modal voice in "detective" (Bacall, *The Big Sleep* 1946). Two final syllables shown.

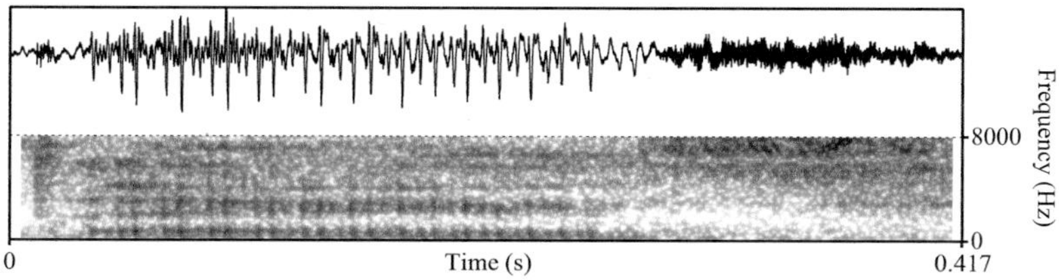

Figure 8.2 Acoustics of creak in "days" (Emma Stone, *La La Land* 2016).

Creaky voice is defined here as aperiodic vocal fold vibration, or an abrupt drop in pitch with respect to the preceding pitch levels. Included are also cases of diplophonia, which involves a period doubling (more below). This is in line with some other studies (e.g., Keating et al. 2015). An example of creak is shown in Fig. 8.2. The acoustic signal here shows the word "days" (Emma Stone in 2016's La La Land). The soundwave is more jagged than in Fig. 8.1, and the vocal fold pulses are not necessarily evenly timed with respect to each other. Figure 8.2 shows what could be analyzed as diplophonic creak, as we can see a mixture of two repeated vibratory patterns. Furthermore, the overall pitch is noticeably lower than is the case when Stone does not produce creak. The identification of creak is best observed when Fig. 8.2 is contrasted with Figs. 8.1 and 8.3.

Breathy voice involves vocal fold abduction, i.e., vocal folds spreading apart. This means that more air escapes through the larynx than is the case with modal phonation, and this is why breathiness gives rise to the perception of softness (e.g., Esling et al. 2019: 41–44). It is associated with periodic vocal fold vibration and a lower amplitude, and more glottal friction present. In the example shown in Fig. 8.3 ("thanks"), the two speech sounds that can show what happens with the speaker's voice are /a/ and /n/. The amount of friction (grayness with higher intensity across frequencies on the Y-axis of the spectrogram) is fairly high, and we can no longer see the dark striations that represent the vocal fold vibration, with the exception of the very bottom of the Y-axis of the spectrogram, where the so-called voicing bar can be seen. The vocal folds can later become so abducted that the voiced breathy phonation turns into voiceless breath, manifested in the absence of the voicing bar.

Creaky voice and breathiness can co-occur in whispery creak (e.g., Esling et al. 2019: 66, 127) or breathy/lax creak (e.g., Keating et al. 2015). In such types of creaks, the vocal fold vibration can be irregular while more air travels through the larynx. Figure 8.4 shows the acoustic representation of "the name Maretto" (Kidman as Suzanne, *To Die For* 1995). The signal contains a considerable degree of breathiness, visible through the dark energies spread across the frequencies on the Y-axis. We can also see at least 12 vocal fold pulses at the very bottom of the spectrogram. These are unevenly spaced from one another, resulting in aperiodic creak. The creak shown in Fig. 8.2 is also an example of lax creak, albeit one with a smaller degree of breathiness. Figure 8.4 also shows a type of creak that is not diplophonic.

Turning to pitch, this can vary in a range of ways regarding average pitch, but speakers also show variation in pitch minima and maxima, i.e., the lowest and highest pitch realizations present in their speech. The differences in these minima and maxima can then also result in pitch range differences. We conducted auditory analyses of pitch and supplemented these with quantitative analyses of the fundamental frequency (f0): the acoustic correlate of pitch.

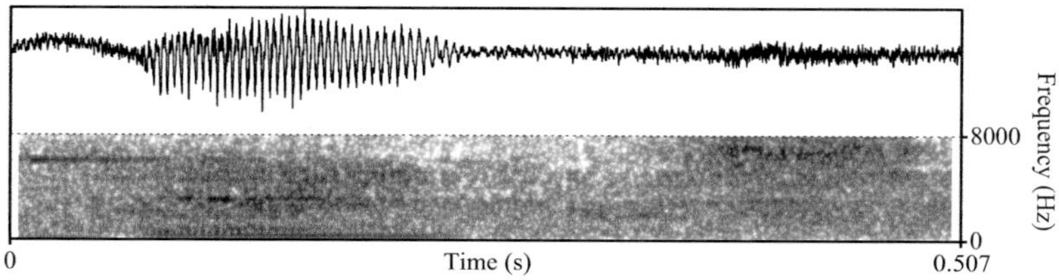

Figure 8.3 Acoustics of breathiness in "thanks" (Monroe, *Niagara* 1953).

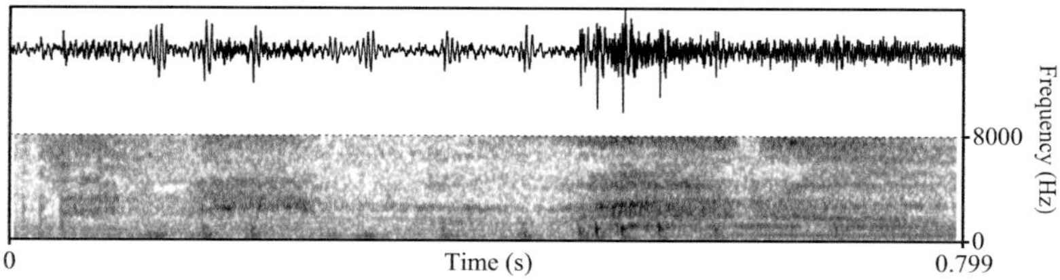

Figure 8.4 Acoustics of lax creak in "the name Maretto" (Kidman, *To Die For* 1995).

Procedure

The selected film clips were extracted from online sources via AudioHijack (Rogue Amoeba 2022). All wav recordings were scaled for intensity, using 70 dB, in Praat (Boersma and Weenink 2022). All vowels uttered by the femme fatale characters were manually segmented, as well as those of some of the other female and male characters present in the clips in question. The non-femme fatale characters and the male characters were annotated to serve as reference points when analyzing the vocal variation found in the femme fatale characters. The vast majority of clips contain background music, which made it considerably challenging to gather speech suitable for acoustic analyses. This being the case, the different vowels were coded for two contexts: speech with and without background music. For those characters who produced speech with and without music, the differences shown between the acoustic measures did not differ relationally. For instance, the differences between character types remain the same whether or not quieter background music is included. Hence, in the quantitative analyses, we report results based on speech produced with as well as without background music. Tables 8.2–8.4 summarize the number of vowel tokens available for the characters by the character and film for each period.[2]

Table 8.2 Number of vowel tokens by character and film, post-war films

	The Big Sleep	*Niagara*	*Gentlemen Prefer Blondes*	*Vertigo*	*Splendor in the Grass*
Femme fatale	308	34	123	109	—
Non-femme fatale	—	31	22	—	57 (Deanie) 117 (mother)
Male	46	10	107	106	—

Table 8.3 Number of vowel tokens by character and film, second-wave feminism

	The Graduate	*The Postman Always Rings Twice*	*Body Heat*	*Top Gun*
Femme fatale	78	23	152	—
Non-femme fatale	—	—	—	124
Male	178	9	193	34

Table 8.4 Number of vowel tokens by character and film, third-wave feminism

	Total Recall	*Basic Instinct*	*The Crush*	*The Last Seduction*	*Disclosure*	*To Die For*
Femme fatale	24	172	49	134	125	185
Non-femme fatale	—	—	—	—	—	—
Male	—	30	103	152	57	—

	Double Jeopardy	*Matchpoint*	*The Black Widow* (Marvel films)	*Sin City: A Dame to Kill For*	*La La Land*
Femme fatale	134	71	132	78	—
Non-femme fatale	—	—	5	—	61
Male	67	85	—	8	107

Analysis

We first present the general quantitatively established tendencies found across the three periods. We then proceed to discuss each period in more detail.

Quantitative differences across time

Figure 8.5 shows the number of sentences with creak by period for the femme fatale characters, non-femme fatale characters, and male characters, expressed as a percentage of the sentences uttered with creak. As we can see, the post-war femme fatales almost never creak in the vocal performances analyzed here. We see a rise in creak in films from the second-wave of feminism, and more creak still is found in films from the period of the third-wave of feminism. The most recent period is also associated with the highest variation, which could be due to two factors: (1) either the characters are more varied and this is reflected in their production of creak as well or (2) this is an outcome of the fact that more data were included in the analyses for this period.

Figure 8.6 shows the number of creak quantified in the same way as in Fig. 8.5, but this time we zoom in on the individual years in which the films were produced. The three periods are color coded for convenience. With the exception of *The Last Seduction* (1994, more below), all films from the third-wave period of feminism show the highest amounts of creak in the femme fatale characters.

Figure 8.7 shows f0 (in Hz) by character and period. The male characters produce the lowest f0 in all three periods. It would also appear to be the case that the femme fatales' f0 is comparable to that of other female characters in the post-war films and the most recent films, but higher in the films from the second wave. In the third-wave films, this is likely due to the high amount of periodic creak in *La La Land* (2016), which provides most of the non-femme-fatale data for the period.

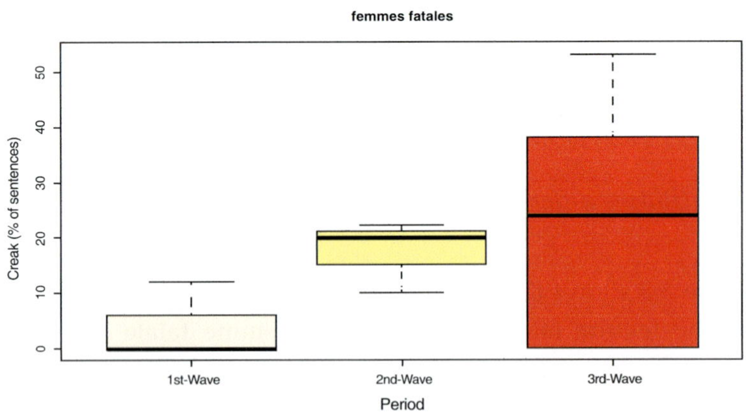

Figure 8.5 Number of sentences with creak (%) by period in the speech of the femme fatale characters.

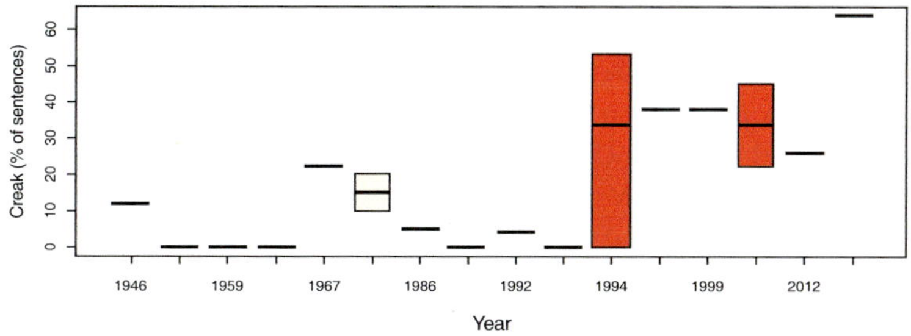

Figure 8.6 Number of sentences with creak (%) by year in the speech of the femme fatale characters.

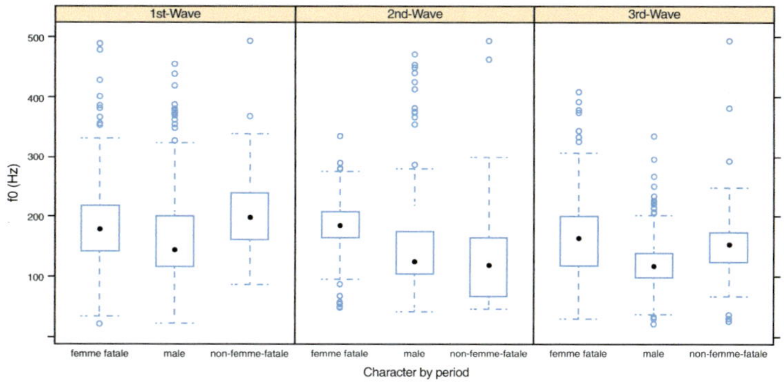

Figure 8.7 f0 (Hz) by period and character.

Post-war films

Starting with post-war films produced before the arrival of the second-wave of feminism (1965), we see an overall absence of creak. Of the five femme fatales analyzed here, only one produces creak in the scenes looked at, and this is so to a considerably limited extent. The overall phonatory profile of the characters is one with a fairly high amount of breathiness, although only three of the four femme fatales display this vocal characteristic. Interestingly, the one who does not also happens to be the only one who produces creak (Bacall, *The Big Sleep*). The individual characters generally show differences in their f0, with some using a fairly low f0 and others a fairly high f0, where "high" corresponds to the generally reported values for female speakers of American English (210–235 Hz; compare these to 121–143 Hz, reported for male speakers; Hillenbrand et al. 1995: 3103). Pitch dynamism is also employed differently by the characters. On the whole, however, the five characters analyzed show a similar phonation profile, especially when contrasted with the characters from the more recent films. Table 8.5 provides a summary of the quantitative analyses of the characters analyzed in the films in this period, including those of selected male characters and four female characters who are not femme fatales for comparison.

Table 8.5 Fundamental frequency (Hz and ST) and CPP (dB) by character and film, and by year (post-war films)

Character (Actress)	**Film (Year)**	**f0 Mean (Hz, ST)**	**f0 Range (Hz)**	**CPP (dB)**	**Creak (#sentences)**
Femme fatale (Bacall)	*The Big Sleep* (1946)	172; 20	455	20	4/34 (12%)
Male	*The Big Sleep* (1946)	175; 19	414	19	1/9 (11%)
Femme fatale (Monroe)	*Gentlemen Prefer Blondes* (1953)	227; 26	291	16 18	0/23 (0%)
	Niagara (1953)	237; 27	140		0/8 (0%)
Male	*Gentlemen Prefer Blondes* (1953)	219; 24	362	15	10/18 (56%)
	Niagara (1953)	257; 27	321	15	0/2 (0%)
Non-femme fatale	*Gentlemen Prefer Blondes* (1953)	226; 27	361	18	0/7 (14%)
	Niagara (1953)	246; 26	158	16	0/8 (0%)
Femme fatale (Novak)	*Vertigo* (1958)	173; 21	216	18	0/29 (0%)
Male	*Vertigo* (1958)	138; 15	383	19	2/29 (7%)

Character (Actress)	Film (Year)	f0 Mean (Hz, ST)	f0 Range (Hz)	CPP (dB)	Creak (#sentences)
Femme fatale (Wood)	*Splendor in the Grass* (1961)	236; 27	142	16	0/18
Non-femme fatale	*Splendor in the Grass* (1961)	175; 22	236	16	8/21 (38%)

Following a chronological order, we first start with *The Big Sleep* (1946), a complicated film noir crime drama starring Bacall as Vivian Routledge, whose father has hired detective Philip Marlowe—played by Humphrey Bogart—to investigate a case of potential blackmail involving Vivian's younger sister (Carmen) and his protégé, Sean Regan. In the analyzed scene, Vivian has a conversation with detective Marlowe to unearth the truth of his mission. She comes across as a very independent woman—her character wears long pants throughout, atypical for women in the late 1940s and perhaps used to symbolize her independent qualities—who is familiar with a certain amount of power—"People don't talk to me like that." On the whole, her pitch is fairly low—in fact, the lowest from the four femme fatales, and the only one which is equal to that of the male conversational partner. While her f0 range is the highest of the four femme fatale characters, it is again comparable to that of her male conversational partner. Her voice starts as modal in the scene analyzed here (see Sources for the links to the clips) and generally shows higher modality and tenseness in contexts associated with negative stance-taking, which is most of Vivian's speech here. Her voice does show utterance-final breathiness in other contexts: "You're a mess, aren't you?"; "broke dad's heart although he won't say much about it." To illustrate the links between negative stance-taking and tenser phonations, her voice is more modal when she mocks Marlowe ("or I'd have to be a comedian") and when she feels she is not being taken seriously ("I'm quite serious, Mr Marlowe"). When Marlowe starts listening and taking her more seriously, her voice gains a slight degree of breathiness again, although only utterance-finally.

Vivian is the only femme fatale analyzed here in the post-war era who produces creak. She produces five instances: (1) "Sean Regan"; (2) "Broke dad's heart although he won't say much about it."; (3) "Just drove off one afternoon without saying a word."; (4) "They found his car parked in some private garage."; (5) "I'm sure I don't care what you say, Mr Marlow." The second example can be explained through the character being emotionally affected by the negative impact Sean's action had on her father. At the same time, the presence of creak in this example may function as a way to signal a negative stance, whereby the character disapproves of Sean's action to drive off. Considering that she is not happy with Regan and not particularly positively disposed toward him, the presence of creak in the very first example may similarly express a negative stance. Example 4 is interesting because here Vivian is striving to establish her independence more explicitly, as signaled by her claim about her lack of interest in what Marlowe says. We interpret this as a case of negative stance-taking again. Finally, example 4 seems to contain creak, which functions primarily as a way to signal the end of the utterance.

In the classic film noir thriller *Niagara* (1953), Rose Loomis (played by Monroe) schemes with her lover Patrick to murder her much older husband (George) in a tunnel underneath the Falls. In the scene analyzed here, we find Rose dressed in a revealing hot pink dress, lying while supporting herself with her arms, while two new acquaintances sit beside her. This femme fatale is, therefore, presented rather differently from the one found in *The Big Sleep*, in the use of more stereotypically feminine visual features to construct the persona. George is depressed and jealous, and when Rose decides to join an impromptu party outside of their cabin and requests a song, he storms out of the cabin, angrily breaks the record, and hastily storms back to the cabin. Rose does not produce any creak in this scene. Her production displays a high degree of breathiness, which diminishes when she makes a non-positive remark about her husband: "Depends. You get right down to it if you give him a rocking chair and a corny old tune like in the Gloaming." Here, the tenser phonation is used to signal a negative stance, since gloaming refers to "gathering darkness" ("gloaming" n.d.), which does not reflect positively on her husband's tastes. Rose shows the highest f0 in the four femme fatales analyzed, and one of the highest f0 ranges. This meshes well with the fact that she is also visually portrayed as the most stereotypically feminine of the four adult femme fatales. The same description applies to the character of Lorelei Lee in *Gentlemen Prefer Blondes* (1953), played again by Monroe. Monroe is generally the breathiest of the actresses playing the femme fatales analyzed in this period, with frequent occurrences of completely voiceless signal (e.g., "Coupons?").

In *Vertigo* (1958), Gavin Elster hires his old friend, John "Scottie" Ferguson (played by Stewart), to follow a woman masquerading as his wife Madeleine (played by Novak) because she is acting strangely and he claims may be suicidal. Scottie eventually discovers, however, that he is just a pawn in Gavin's elaborate scheme to murder his wife, with the aid of Judy Barton (also played by Novak), who masquerades as Gavin's wife. In the analyzed scene, Madeleine does not produce a single utterance with creak (which is rather surprising considering that she had been pulled out of a bay in which she had been found drowning). Her f0 is one of the lowest observed for the four femme fatales analyzed here. Unlike in the case of Vivian in *The Big Sleep*, however, although her f0 is the lowest of the four femme fatale characters, it does not match that of the male conversational partner (173 Hz versus 138 Hz). Overall, the conversation of these two is fairly intimate, following Scottie saving Madeleine's life (=Judy's) by pulling her out of a bay, having changed her clothes, and dried her up while she was unconscious. This intimacy might explain the low f0 values. Madeleine's overall phonation contains a high degree of breathiness, even becoming voiceless on occasion ("I fell into the bay"; "No, I…"). Interestingly, her voice becomes the least breathy when she expresses criticism of Scottie: "You're terribly direct in your questions." Thus, a less breathy phonation signals a negative stance in this character's vocal portrayal, similar to that of Vivian in *The Big Sleep* and Rose in *Niagara*.

Finally, we look at *Splendor in the Grass* (1961), a tragic drama about high school star-crossed lovers Wilma Dean "Deanie" Loomis (played by Wood) and Bud Stamper (played by Beatty). Deanie is not a femme fatale, but she does not fit the normative happy feminine character either, considering she attempts to commit suicide and in the scene analyzed here questions her mother about whether and why it is wrong to enjoy sex as a woman. In this scene, Deanie has this conversation with her mother before going to bed, just after her boyfriend has dropped her off from a late date at the disapproval of her mother. Deanie produces no creak. Her f0 is the highest of all the female characters analyzed for this period. While it might be tempting to interpret the relatively high f0 as an indicator that this character is not a femme fatale, at the same time the high f0 is not surprising considering that she is a teenager. Her voice is also fairly breathy, although not as breathy as that of the character played by Novak or of the characters played by Monroe. In this case, the presence of breathiness could also be explained by the young age of the character (see, e.g., Stathopoulos et al. 2011: 1016).

Regarding the other three female characters who are not femme fatales, first we have Deanie's mother in *Splendor in the Grass* (1961) and second we have Polly Cutler (played by Jean Peters) in *Niagara* (1953) and third Dorothy Shaw (played by Jane Russell) in *Gentlemen Prefer Blondes* (1953). It is not surprising that the mother's f0 is one of the lowest due to her portrayed age (e.g., Hejná and Jespersen 2022). The voices of other females are comparable to that of Rose and Lorelei (played by Monroe), and these two are also much closer in age than is the case in *Splendor in the Grass*. The mother character produces very consistent creak when she explains how her husband never initiated physical contact with her until after marriage.

Second wave of feminism

Proceeding to the second period of films, we analyzed selected clips from films dating from 1965 until 1990. Here we observe a rise of creak in comparison to the post-war femme fatales.[3] These femme fatales produce between one and five sentences with creak (5–22%). The quantitative profiles of the characters are shown in Table 8.6. The functions of these creaks can be explained primarily to signal one's negative stance or due to emotional turmoil. Similar to those in the post-war era, femme fatales from the second wave of feminism employ breathy phonation and a lower pitch, although just how low the pitch is seems to greatly depend on the specific character, the male conversational partner's pitch, and the overall setting—e.g., how intimate the situation is. The f0 of the second wave of feminist femme fatales is on the whole lower than in the post-war period, and the f0 range of the characters analyzed here is also more limited than those of the post-war period—with some exceptions discussed below. This points to the use of less stereotypically normative feminine features of speech.

Table 8.6 Fundamental frequency (Hz and ST) and CPP (dB) by character, film, and year (second-wave feminism)

Character (Actress)	Film (Year)	f0 Mean (Hz, ST)	f0 Range (Hz)	CPP (dB)	Creak (#sentences)
Femme fatale (Bancroft)	*The Graduate* (1967)	184; 22	292	16	5/23 (22%)
Male	*The Graduate* (1967)	143; 18	273	18	0/30 (0%)
Femme fatale (Lange)	*The Postman Always Rings Twice* (1981)	154; 17	227	15	1/5 (20%)
Male	*The Postman Always Rings Twice* (1981)	107; 12	155	15	–
Non-femme fatale (McGillis)	*Top Gun* (1986)	129; 14	445	18	1/19 (5%)
Male	*Top Gun* (1986)	190; 18	410	17	0/9 (0%)
Femme fatale (Turner)	*Body Heat* (1981)	189; 23	277	18	3/31 (10%)
Male	*Body Heat* (1981)	143; 18	423	17	1/36 (3%)

In *The Graduate* (1967), we find recent college graduate Benjamin Braddock (played by Hoffman) being seduced by Mrs. Robinson (played by Bancroft), whose daughter he eventually falls for. In the scene analyzed here, Benjamin finds himself alone in Mrs. Robinson's house. Benjamin had driven Mrs. Robinson home after his graduation party—held at his parent's house—and she insists on him coming into the house—on the pretext of being afraid of the dark and of being alone—and staying with her until her husband returns. She repeatedly insists that he have a drink and despite Benjamin's repeated protestations succeeds by thrusting a glass in his hand. When it becomes clear that Mr. Robinson is not coming home any time soon, Benjamin accuses Mrs. Robinson of trying to seduce him. During the less explicit moments of his accusation, Mrs. Robinson engages in a bidirectional conversation with him to—at least superficially—better understand what it is that he is objecting to. Her phonation is fairly breathy right from the start, accompanied by a relatively low pitch.

Judging from her previous behavior it seems to be the case that she is quite aware of Benjamin's objections toward her efforts at seduction, and the more explicit her requests for "clarification" are the breathier her phonation. The fact that she is aware is also fairly explicitly signaled by the camera closeup of Benjamin being seen through the angle of her bent leg. She produces lax creak when portraying herself as lacking knowledge, i.e., when opposing Benjamin's indirect and later direct accusations, through the following responses: (1) "Well I don't know" while positioning her legs in a way which is more seductive; (2) "Well."; (3) "No"; (4) "I hadn't thought of it." When Benjamin starts vehemently apologizing for his accusations, she directs him to "sit down" with a raspy or hoarse phonation. This phonation displays higher degrees of jitter and shimmer, which means that the voice still exhibits periodic vocal fold vibrations, but the amplitude of the speaker's voice can oscillate somewhat more irregularly, and the glottal pulses may be slightly less evenly spaced out from each other. Mrs. Robinson employs the same raspy setting when reassuring him that his actions are not as negative as he thinks: "We'll forget it right now." This is immediately followed by an increased breathiness of the vocal performance: "Finish your drink." When she redirects the conversation to her daughter's portrait to change the subject, her phonation also becomes less breathy and more modal. Mrs. Robinson, therefore, produces creak primarily to signal a negative stance. Overall, her f0 is the highest of the femme fatales analyzed here, which might be explained through its closer proximity to the post-war films, which generally display higher f0 in their femme fatales. On the other hand, this might be a vocal attractiveness strategy employed to mitigate possible vocal effects of ageing.

In *The Postman Always Rings Twice* (1981), Cora Smith (played by Lange) has an affair and falls in love with a drifter named Frank Chambers (played by Nicholson). Tired and frustrated with her loveless marriage to an older man as well as her career, Cora and Frank conspire and then succeed in murdering her husband so that they can start a new life together. In the scene analyzed, she finds herself in a conversation with Frank and indirectly raises the idea of murdering her husband—her husband's voice can be heard in the background calling for her. Cora's phonation is extremely breathy—explaining why her f0 is the lowest of the female characters—and even voiceless most of the time. She produces lax creak, i.e., creak overlaid with breathiness, when she clearly expresses her fatigue with the moral norms of her society: "I'm tired of what's right and wrong." In this example, lax creak, therefore, functions as a negative stance-taking device as she—here explicitly—proclaims that she dis-aligns from the norms of her community. This is consistent with the function of lax creak found in *The Graduate* (see above).

Moving to the neo-noir erotic thriller *Body Heat* (1981), Matty Walker (played by Turner) assumes a dual identity and dupes Ned Racine (played by Hurt) into helping her murder the man Ned is led to believe is her husband. Matty's pitch is neither particularly low nor high, and her phonation is relatively modal as she fends Ned off by telling him that she is "a married woman," "meaning [she has] a husband." When she states: "You're not that smart, are you?", her phonation becomes tenser as she utters "I like that in a man." Thus, while she proclaims not to be interested in his company due to her marital status, she also makes it obvious that she is, in fact, rather interested. Her phonation becomes breathier as she changes the tone of the conversation to flirt: "Tell me, does chat like this work with most women?" When she concedes to him buying her something to drink—and thus accepting his company—her phonation becomes breathier, while still producing two of the three utterances that are exhibited in lax creak: (1) "You wanna buy me some<u>thing</u>?"; (2) "I'll take one of <u>those</u>." This lax creak is extremely short in duration, and its function is not entirely clear to us. Similarly, the third case of creak is found in the word "town," as she announces that her husband is out of town. Here creak is very momentary, being shorter than the duration of the diphthong and can be explained by the fact that she is putting a cigarette in her mouth while talking, which is also reflected in a heavier coarticulation of the vowels and consonants. In the rest of the scene, her phonation oscillates between more modal and more breathy, with the former being used in directives and/or when discussing topics which might be seen as getting in the way of the two characters getting involved—i.e., "get married." On the other hand, breathiness is used when signaling likeability and interest in mutual engagement, as in when Matty asks Ned, having spilt some wine on her white dress, to "get [her] a paper towel or something to dip it in some cool water." When he suggests he "will even wipe it off for [her]," she responds with "You don't wanna lick it?", with "lick it" being produced in an extremely breathy phonatory setting, reaching voicelessness.

Finally, we analyze two scenes from a non-femme fatale character in *Top Gun* (1986). Here, we find Charlotte "Charlie" Blackwood (played by McGillis), who is an astrophysicist and civilian "Top Gun" instructor who eventually falls in love with Pete "Maverick" Mitchell (played by Cruise), an elite naval pilot. Charlie represents a female character who is not a femme fatale because despite her intelligence and professional credentials, she plays a rather passive and submissive role, eventually succumbing to Maverick's charms. She is rather interesting, however, because her job puts her in a somewhat atypical situation for a woman of her (or indeed our) times. In the first scene analyzed, she reproaches Maverick, who purposely ignores her and claims not to hear her while starting his motorcycle. To his astonishment, she rather recklessly pursues him in her car to "finish [her] sentence"—with tense phonation. Her phonation changes, however, when she reveals that despite the negative report on his performance, she "[sees] a real genius in [his] flying" and did not voice this in the report for fear of others discovering that she is "falling for [him]."

In another scene, the two characters find themselves in a lift, flirting with each other before they become romantically involved. Charlie starts the conversation with "It looks like you need another shower," where lax creak occurs very briefly in the word "shower." Considering its brevity, the only interpretation we can suggest as to the function of this creak is that Charlie might be in some emotional turmoil, considering that this is the first time she sees Maverick after inviting him to her house for dinner. Overall, her phonation is fairly modal with some degree of breathiness. The amount of breathiness nevertheless increases when the nature of their personal relationship is touched upon: "I don't normally invite students to my house." This character produces the fewest creaks—only one instance—which is interesting considering that in her stereotypically masculine job as a civilian naval pilot instructor, she is expected to deliver unpleasant messages—primarily to men—and engage in conversations in which she is likely to express a negative stance. Her very low f0 might reflect this situational context. She produces the lowest f0 among the actresses analyzed here, and she is also the only character who finds herself in a very masculine environment, in which, what is more, she has a higher rank than many of her male students—one of whom she falls for. Rather than using creak to express a negative stance, this character, therefore, seems to be using less breathiness—tenser phonation—and lower f0.

To conclude, using tenser phonation and less breathiness to express a more negative stance is observed in all the femme fatales analyzed in this chapter, for all three periods.

Third-wave feminism

Looking at characters falling within the third wave of feminism (1990+), we find that, similar to the femme fatales of the previous two periods, their vocal performance is associated with low-pitched voices combined with breathiness. The characters presented here also present us with the highest amount of creaky phonation. Three out of the 10 femme fatales resemble those of the previous two periods in that they produce hardly any or no creak. However, the other femme fatales produce creak in 26–53% of the sentences analyzed. As in the previous periods, the use of creak signals a negative stance or an emotional turmoil (or a mixture of the two). Table 8.7 provides a quantitative overview. The f0 range shows the lowest values across the three periods, and many of the femme fatales show f0 not too distant from that of their male conversational partner.

Table 8.7 Fundamental frequency (Hz and ST) and CPP (dB) by character and film, and by year (third-wave feminism)

Character (Actress)	Film (Year)	f0 Mean (Hz, ST)	f0 Range (Hz)	CPP (dB)	Creak (#sentences)
Non-femme fatale (Emma Stone)	*La La Land* (2016)	147; 17	223	16	7/11 (64%)
Male	*La La Land* (2016)	128; 16	200	17	9/26 (35%)
Femme fatale (Fiorentino)	*The Last Seduction* (1994)	142; 18	282	17	0/32 (0%)
Male	*The Last Seduction* (1994)	133; 16	291	17	11/35 (31%)
Femme fatale (Green)	*Sin City: A Dame to Kill For* (2014)	169; 20	265	17	2/9 (22%)
Male	*Sin City: A Dame to Kill For* (2014)	168; 20	203	15	0/2 (0%)
Femme fatale (Johansson)	*Matchpoint* (2005)	213; 25	204	17	5/11 (45%)
	The Black Widow (2010)	126; 13	233	18	5/19 (26%)
Male	*Matchpoint* (2005)	116; 14	101	17	0/10 (0%)
	The Black Widow (2010)	Not measured (12 males)	Not measured (12 males)	Not measured (12 males)	2/23 (9%), both from 1 of the 12 males
Non-femme fatale	*The Black Widow* (2010)	245; 19	442	20	0/2 (0%)

Character (Actress)	Film (Year)	f0 Mean (Hz, ST)	f0 Range (Hz)	CPP (dB)	Creak (#sentences)
Femme fatale (Judd)	*Double Jeopardy* (1999)	136; 16	208	19	5/13 (38%)
Male	*Double Jeopardy* (1999)	104; 11	200	17	2/12 (17%)
Femme fatale (Kidman)	*To Die For* (1995)	162; 18	329	16	11/18 (38%)
Femme fatale (Moore)	*Disclosure* (1994)	188; 22	346	18	9/17 (53%)
Male	*Disclosure* (1994)	112; 13	204	20	1/11 (9%)
Femme fatale (Sharon Stone)	*Basic Instinct* (1992)	139; 15	376	17	1/26 (4%)
	Total Recall (1990)	240; 27	172	15	0/7 (0%)
Male	*Basic Instinct* (1992)	118; 14	130	19	2/5 (40%)
Femme fatale (Silverstone)	*The Crush* (1993)	197; 23	224	16	0/13
Male	*The Crush* (1993)	122; 15	162	16	6/30 (20%)

We start with *Total Recall* (1990), where we encounter a scene in which the character of Douglas (played by Schwarzenegger) discovers his wife Lori's (played by Sharon Stone) true secret agent persona. After Lori beats his partner unconscious and is about to slit her throat, the handcuffed Douglas manages to take hold of a gun and shoots the knife out of his wife's hand. He keeps his gun aimed at her while she explains her side. The voice we hear is one with a considerable degree of breathiness and a higher pitch, which is associated with a normatively feminine persona.

Importantly, this character is trying to persuade her husband that she is an innocent loving wife—while secretly reaching for the gun behind her back—which means that the vocal characteristics just described are used for manipulative purposes to lull her husband into a feeling of safety. As she becomes more confident, having reached the gun hidden behind her back, her voice becomes more tense, and we also observe a rise in the intensity of the voice and a higher degree of pitch dynamism. The character produces no creakiness in this scene, which is suggestive of the fact that creak may not be perceived as normatively feminine in American English.

Similarly, in the neo-noir erotic thriller *Basic Instinct* (1992), the character of Catherine (again played by Sharon Stone) is in a powerful yet vulnerable position. She is a famous crime novelist but also the prime suspect in a murder, mainly because the murder closely resembles one in her novel. In the analyzed scene, she clearly utilizes her sex appeal to get the character of Nick (played by Douglas) to confess to his accidental killing of tourists while under the influence of cocaine. The voice employed by the character is again considerably breathy, and it becomes breathier and lower in pitch when she more explicitly challenges Nick. In fact, her voice becomes so breathy that it becomes voiceless. The pitch range is significantly limited in the scene. The only instance of a creak observed in the scene is when the character nearly kisses Nick, daring him to acknowledge that he enjoyed killing the tourists and that the events contributed to his wife's suicide: "I think you liked it." The presence of creak is limited to this one utterance. At this moment, the two characters nearly kiss while Nick is holding Catherine's hand behind her back. Yet as Catherine switches out of the creaky phonation back to considerably breathy phonation—"That's why she killed herself"—Nick disengages from the embrace. The creak observed here is a rather extreme case of a lax creak.

The contrast of the two characters played by Sharon Stone is revealing: The first clip (*Total Recall*) shows a clearly performed persona by the femme fatale within the narrative of the film itself, whereas the second clip (*Basic Instinct*) reveals a character who shows the male conversational partner what seems to resemble her true self more closely. These two characters are played by the same actress, and yet Lori shows a fairly high f0 (240 Hz), clearly contrasting with that of Catherine's (139 Hz).

In *The Crush* (1993), we analyze Adrian (played by Silverstone), who first encounters Nick (played by Elwes) when he nearly hits her with his car while distracted as she rollerblades across the road. While Nick is considerably taken aback by the incident, Adrian gives him a direct cool and intense look as she rollerblades on. The scene we analyze features Adrian visiting Nick in his study. He is a 28-year-old writer, and she is the 14-year-old teenage daughter of his landlords. The body language of the two characters signals a scenario in which Adrian has come to seduce Nick, with whom she has become infatuated. However, Nick is pursuing a romantic relationship with a colleague and is not interested in Adrian. While Adrian is much younger than Nick, she seems extremely comfortable in the role of a seductress. Her seductee, on the other hand, finds himself feeling uncomfortable, especially with topics involving physical relationships.

Adrian's pitch is the second highest of the characters analyzed for this period—after Lori in *Total Recall*—which can be accounted for by the fact that she is only 14 years old, unlike all the other characters. Her phonation reveals a considerable degree of breathiness. This is, at least at first blush, trickier to interpret because children and teenagers, together with speakers in their later adulthood—65 years of age and higher—have been reported to show breathier phonation than the middle section of the population (e.g., Stathopoulos et al. 2011: 1016). However, the correlation between the degree of breathiness and the topic of the conversation and/or the body language of the scene is telling: The more explicit the seduction or a discussion of sexual and romantic subjects, the breathier Adrian's phonation. For instance, "Very attractive" and "Try me" are both produced with a breathier phonation than "So what're you doing?" or the utterances found in the introductory exchange about a specific legal case that Nick is writing about. Interestingly, the highest degree of breathiness is found where Adrian takes a chair to sit directly in front of Nick with her legs on its side and, almost patronizingly, asks him to clarify "You mean a lover?", to Nick's embarrassment. When Nick appeals to certain age-normative expectations by asking Adrian "Are you sure you're only fourteen?" and "Isn't it past your bedtime?", Adrian always answers in an increasingly breathier phonation, even reaching voicelessness, far from embarrassed and flustered by these questions. This character does not produce any creak during the scene analyzed.

In *The Last Seduction* (1994), we find Bridget (played by Fiorentino), a telemarketer who leaves her husband and hatches a plan to profit from the murder of cheating husbands she identifies through their personal credit records. In the analyzed scene, Bridget enters a bar that happens to be full of men. The only visible woman in the scene, she confidently marches in and orders a drink—being ignored by the bartender to her dismay. The bartender ignores her very explicitly and consistently, which infuriates her: "What does a girl have to suck around here to get a drink?" She is undoubtedly not the normative woman expected in this setting, or indeed in most settings. Her pitch is fairly low throughout the scene and in what follows as well, with the exception of her pitch rising when she shows a sudden interest in a young male character who pays for her drink. Despite this, her f0 is very close to that of her male conversational partner. Overall, her voice is produced with some breathiness, but this breathiness increases and decreases according to the situations in which she finds herself. First, she produces a higher degree of breathiness when she is exasperated: "Jesus Christ" (uttered when she realizes the bartender is set on ignoring her). Second, she uses increased breathiness to signal to the young man that she is not interested in engaging with him further: "fuck off," a mocking "please," "meet a nice little cowgirl, make nice little cow babies and leave me alone." Finally, breathiness increases with more explicitly sexual utterances, produced as she verifies his claim about his penis possessing equine qualities, less than two minutes after he bought her the drink and she instructed him to "fuck off." When she shows a sudden change in her interest in the male character by asking him to her table after he has made his claims about the qualities of his penis—"Let's see"—her pitch is higher and remains higher also when she tries to convince the man that she is in earnest: "I can be very nice when I try. Sit down." Creak is not produced in the scene beyond what could be attributed to a segmentally conditioned creak, i.e., glottal stop like "Manha<u>tt</u>an."

Moving on to *Disclosure* (1994), we analyze the character of Meredith (played by Moore), who presents as a successful businesswoman who falsely accuses her former lover Tom (played by Douglas) of sexual harassment to replace him in his position after the merger of their two companies. Interestingly, the first glimpse the viewer gets of Meredith in the scene analyzed here is that of her long smooth legs in chic stockings and stiletto shoes. Meredith's voice is not only low and breathy but also consistently creaky. More specifically, it is realized as lax creak. She becomes momentarily less breathy when she addresses the newly formed company's CEO Bob (played by Sutherland) but returns to the initial amount of breathiness and creakiness after Tom suggests that the two of them meet up to "catch up," to which Meredith replies "I'd love that." Interestingly, the sentences that contain most creak in terms of its prevalence within each sentence are those that most explicitly refer to the intimate relationship between Meredith and Tom in the past: (1) "Sanders and I go way back."; (2) "He broke my heart."; (3) "I'd love that." One last aspect of the vocal representation of this character is the fact that the voice is raspy, or hoarse, which means that even if prototypical creak is not present, shimmer and possibly jitter are higher in this case. This means that the amplitude of the soundwave oscillates a bit more irregularly (and the distance of the vocal pulses is ever so slightly irregularly spaced out from each other), which leads to the percept of raspiness.

In *To Die For* (1995), the ambitious Susanne (played by Kidman) delivers one of her numerous monologues, describing the nature of her relationships with her deceased husband, his family and hers, all while relating them to her career aspirations. This character presents us with a range of phonatory changes that accompany specific stances and aspects of this persona. Susanne's overall phonation is extremely breathy, so much so that some utterances are voiceless. In addition, she displays a high pitch dynamism (one of the highest in this period). The character's vocal profile does involve the presence of creak in some utterances, and creak is realized as lax in such cases—as is the case for all femme fatales analyzed in this chapter. The presence of creak can be explained by three factors in this scene. First, three utterances contain a rather brief final creak, which does not lend itself to any immediately obvious social function and seems to have a primarily, if not solely, prosodic and conversation-organizing role—signaling the end of a sentence or a turn. Second, creak is associated with emotional turmoil. On one occasion, Suzanne fakes crying, at which point her voice exhibits creak "whom I loved very very much" as well as other vocal changes associated with emotional turmoil. She also adopts lax creak when explicitly referring to a negative event in her life, which is bound to be emotionally disturbing: "particularly since my recent tragedy." However, as the viewer later discovers, the "tragedy" she has experienced refers to her husband's death, which she planned and is not disturbed about in the least. The viewer nevertheless does not have this knowledge at the time of the interview, and so we discover, as the film develops, that Susanne is extremely manipulative, and the emotional turmoil is very much an act to display a vulnerable and more normatively feminine persona, which is convenient for the character at the time.

Finally, Suzanne uses creak for the purpose of negative stance-taking. The first such example is encountered in "No, wait, I'm sorry," where she begins to distance herself from the identity implied by the surname she had taken from her husband. Her second "sorry" is also creaky. In addition, she produces one whole utterance with lax creak, which explicitly touches upon her feelings about her late husband's surname: "It's not like I have any negative feelings about the name Maretto." Although she proclaims otherwise, the viewer knows that she does have negative feelings about that surname, most immediately because she has expressed the need to proclaim Stone as her real surname. Interestingly, when discussing her late husband's family, she creaks at very specific points: "parents" and "sister Jane Maretto." Contrary to what she says (but not how she says it), Suzanne harbors negative feelings toward her late husband's family—illustrated, for example, in her husband's father who is convinced Suzanne is responsible for his son's death and uses his criminal connections to plot her murder.

In the scene analyzed for *Double Jeopardy* (1999), Nick Parsons (played by Greenwood) finds himself in a room with parole officer Travis Lehman (played by Lee Jones) and Libby Parson (played by Judd). To Nick's surprise, Libby is still alive. Years earlier, Nick staged his death so that Libby would be charged with his murder. His plan succeeded; however, after 6 years in prison, Libby is paroled and quickly evades the watching eye of Lehman and hits the road with a plan to exact revenge on her former husband who has started a new life as an influential hotel owner in New Orleans. In the analyzed scene, Libby finds herself in a position of power. Initially, her voice is fairly breathy and low in pitch—one of the lowest of the femme fatales of the three periods. However, the voice becomes more modal when Libby explains her plan to Nick, which clearly exhibits her higher moral ground in the situation. The pitch dynamism is the most limited in "No they won't." Thus, modal voice seems to be used to disalign herself further from Nick by signaling a power differential, which is reflected in the degree of pitch dynamism as well. Furthermore, the highest degree of breathiness is used to signal a caring relationship toward the character's son: "Where is my son?" At the same time, "Where is my son?" is produced with a fairly low pitch dynamism, which suggests that some degree of hostility is being communicated through this vocal performance. While this character is portrayed with a relatively higher number of sentences that contain creak, this creak is very limited in terms of duration, usually affecting only a single word. In this character, the presence of this durationally short creak does not seem to be correlated with any changes in the messages she conveys through changes in her persona. Instead, it seems that here creak functions solely as a prosodic marker signaling the end of the character's turn in the conversation.

In the psychological thriller *Matchpoint* (2005), we find Nola (played by Johansson), who attracts Chris at first sight. Nola is dating Chris's friend Tom, whose sister Chris is dating. Both Nola and Chris are, therefore, in a steady relationship, and both of them have also climbed up the social ladder by getting into these romantic relationships. When Nola and Chris first meet, Nola has defeated one of the male characters at a table tennis game, asking who is going to be her next "victim." Her sports prowess places her in a more powerful position than her male players. However, she is also presented as a very seductive female character, well aware of her charms. Her f0 is one of the highest and within the range of what could be considered normatively feminine voices.

Overall, Nola's general phonation could also be characterized by a degree of raspiness. Chris makes several moves on Nola, for instance by calling her "beautiful" right after pulling her closer to teach her a table tennis move. A tense flirtatious situation is interrupted by Tom. Nola playfully explains to Tom that Chris was trying to "have his way with [her]" and leaves the room by telling Chris that she "will be ready for [him] next time."

Nola's overall phonation shows a high degree of breathiness, accompanied by a higher pitch dynamism. When Chris manages against all expectations to score right at the first (and the last) go at the game, Nola asks "What did I walk into?" Her phonation is more modal here, i.e., more tense, and changes into a breathier and creakier phonation as the utterance takes place. The creak is realized as lax, which means that it is overlaid with breathiness. Nola is obviously on guard here, and this seems to be signaled by the creaky phonation as well. Creaky phonation happens simultaneously with breathy phonation in the more tense and physical conversation, occurring primarily toward the end of the sentences. In the conversation during which Tom is present, Nola does not produce any creak when addressing her fiancé, but when she tells Chris that she "will be ready for [him] next time," the phonation of the whole utterance is that of a lax creak. While it might be tempting to conclude that creak is used by the character for flirtatious purposes, the scene presents a rather tense encounter, and Nola is disaligning herself from Chris in the utterances where creak is most prominent. We, therefore, conclude that she uses creak for negative stance-taking purposes.

In numerous Marvel Cinematic Universe superhero films, Scarlett Johansson plays the role of Natasha, also known as the Black Widow. In contrast to *Matchpoint* (see above), her pitch is considerably lower, which corresponds to her portrayal as a very powerful female character (in contrast to Nola in *Matchpoint*, who ends up unwanted, pregnant, and ultimately dead). This is likely to signal a power difference between Natasha and Nola. Natasha has been trained as a professional "superpower" spy and is a powerful opponent. In the first scene, she analyses a specific situation the team of heroes find themselves in. Her pitch is fairly low, and her pitch dynamism is limited. In this scene, her voice is also relatively modal. In the second scene, we find Natasha in an intimate situation with Bruce Banner/Hulk (played by Ruffalo). The physical intimacy is reflected in her voice, which becomes extremely breathy and even lower in pitch. In the next scene, she is addressing her driver (Happy Hogen played by Favreau), who is frantically transporting her to the rescue. While he is distracted by her changing clothes in the backseat, she comments "Watch the road." Her voice is primarily modal here. In the next scene, Natasha finds herself with Steve hiding out at a friend's house. She thanks him for saving her life, and she asks him if the tables were turned whether he would have done the same for her (to which he responds in the affirmative). Her phonation is highly breathy, although not as breathy as in the intimate scene. Her pitch dynamism is a bit higher than in the previous scenes, and she produces lax creak in "for a guy that died for nothing." We then move on to a bar scene with Bruce/Hulk, which is accompanied by an increase in breathiness. This nonetheless does not reach the degree observed during the intimate scene with this character (above). Finally, we get Natasha sharing her story with Steve/Captain America about how she got shot through while trying to shield her engineer colleague. The situation is clearly emotional for her, and this is reflected in her voice, which shows a limited pitch dynamism, a more extreme amount of breathiness, and notably lax creak. Importantly, Tony Stark/Iron Man's (played by Downey Jr.) girlfriend Pepper Potts (played by Paltrow), who is not a femme fatale character, produces a much higher pitch dynamism partially because she allows the situation to shock her in a way that manifests itself through her use of falsetto. Her pitch is higher than Natasha's and also much less breathy. She also produces creak more consistently.

In the scene analyzed for *Sin City: A Dame to Kill For* (2014), the character of Ava Lord (played by Green) encounters Dwight McCarthy (played by Brolin). She reveals that she has manipulated him into killing an innocent man whom she claimed was mentally and physically torturing her. It is clear from her words and his immediate nonverbal response that he had clearly been manipulated sexually and otherwise: "Sex always made you stupid, ready to believe anything." Ava presents a very violent and calculating femme fatale, who uses sex to manipulate others, only to remove them when they become inconvenient. She aims a gun at Dwight, asking him to stay still long enough for her to "blow [his] brains out." The different aspects of this character are accompanied by specific vocal profiles and changes therein. Generally, Ava presents us with a relatively low pitch—in fact as low as her male conversational partner's—which nevertheless lowers further when she announces her true, morally dubious, motive: "You just made me a very rich woman." This lowering is also associated with a reduction in pitch dynamism. Interestingly, this character seems to use a higher pitch, a higher pitch dynamism, and more breathiness when she positions herself as a lover: (1) "I knew I could count on you."; (2) "Do me one last favor lover." The breathiness is mainly voiced but is not infrequently realized as either voiceless glottal or supraglottal, i.e., whispery friction—for instance in the case of "man," "count on you." In "Stay still long enough for me to blow your brains out," the pitch becomes radically lower in the "for me to blow your brains out" and lowers gradually as the message becomes more explicitly violent and hostile. This is accompanied by a brief interval of lax creak. The only other occasion on which creak is present is in "a very rich woman," where we find a lax creak again. This character uses creak in a limited way in this scene, but its perceptual effects, accompanied by those related to the other vocal phenomena, strike us as fairly salient: The presence of creak seems to enhance hostility and a cool calculating mind in this case. These are easily linked to negative stance-taking as this is an essential aspect of being hostile.

In comparison, the aspiring actress Amelia (played by Emma Stone) in *La La Land* (2016) presents us with a female character that is not a femme fatale. Her voice pitch is considerably low as well, and unlike the femme fatale characters, Amelia produces a considerably consistent creaky phonation, which is produced as *periodic* creak. This means that the presence of creak in Stone's performance affects the f0 and results in a lower pitch as well, i.e., the presence of periodic creak drives the low f0, unlike in the femme fatale characters, whose creak is not periodic—at least in the scenes analyzed here—and whose f0 is, therefore, lower in non-creaky environments. The scene analyzed here for the character of Amelia is telling in another respect as well: The two characters are saying goodbye, and the scene is very emotional on a deeper level. Amelia displays degrees of emotional turmoil throughout, and the presence of creak is not only higher than in the speech of the femme fatales analyzed, but also more varied. We, therefore, conclude that the primary function of creak related to persona work is that of a negative stance and/or to signal emotional turmoil.

Discussion

This chapter started by posing the overarching question of how the femme fatale persona has been portrayed in American cinematography through vocal variation across the years (RQ1). In particular, we ask whether creaky voice has been used in the construction of this trope and, if so, what functions it has exactly (RQ2). Finally, we also ask to what extent and in what ways creaky voice might be reflected or mirrored in popular culture, i.e., whether we see a rise in creaky voice in the femme fatale characters in the more recent films (RQ3).

Is creaky voice employed to construct the femme fatale trope in American cinematography?

Creaky voice is not employed to construct the femme fatale trope in the films and scenes analyzed here. Instead, what the femme fatales have in common is a low f0. Their f0 is lower than that typically reported in the literature for female speakers of American English (e.g., Esling et al. 2019: 46). This lower f0 is combined with a breathy voice, which can involve so much airflow that the outcome occasionally becomes voiceless breath. Notably, these two vocal characteristics are consistently present in the performance of the femme fatale persona across the three periods analyzed here, although one femme fatale is not particularly breathy—Vivian played by Bacall—and some femme fatales have a higher f0—those played by Monroe. Considering that breathiness is associated with intimacy, at least in English (Laver

1980[2009]: 135), this finding is consistent with the seductive nature of femme fatale characters.

Although the femme fatales analyzed here produce a lower f0, only one is portrayed with f0, which matches the f0 of the male protagonist/antagonist—Vivian played by Bacall in *The Big Sleep*. Thus, femme fatales may not be normatively feminine, but they do not speak the same way male characters do. While higher f0 is associated with higher femininity ratings in perceptual experiments, and with stronger perceptions of positive femininity traits such as tenderness and likeability (e.g., Krahé and Papakonstantinou 2020) and attractiveness (e.g., Xu et al. 2016), breathiness is similarly associated with stronger perceptions of femininity (e.g., van Borsel et al. 2009) and attractiveness (e.g., Babel et al. 2014; Greer and Winters 2015). Thus, on the one hand, the femme fatale vocally diverges from normative femininity through pitch variation, and on the other hand, exaggerating another normatively feminine vocal feature, that of breathiness, so much so that in various cases the amount of breathiness might be seen as hyperfeminine.

Even where the femme fatales produce creak, this creak is overlaid with breathiness, which strongly suggests that increased breathiness is one of the key components of the vocal construction of a femme fatale. While it is tempting to see the lower f0 of the femme fatale voice through the lens of agency and social constructivism, it is also possible that the f0 shows lower values due to the amount of breathiness, which is synergistic with larynx lowering and f0 lowering due to "the shortened length of vocal fold vibration and their relative separation and looseness" (Esling et al. 2019: 56). This articulatory link, nevertheless, does not preclude later indexical interpretations of the co-occurrence of lower f0 and higher breathiness.

What are the functions of creaky voice in the data analyzed here exactly?

As has just been discussed, creaky voice is not used to construct the femme fatale persona as such, at least not in the films and scenes analyzed here. The characters, nevertheless, use creak in interesting ways.

Apart from identifying some instances of creak, which occur at the end of a turn and/or an utterance, we have consistently identified examples that correspond either to situations with emotional turmoil—not only in the femme fatale characters but also other female characters (Amelia in *La La Land*)—or to expressions of a negative stance (Kiesling 2022 on stance-taking). This happens even in cases where emotional turmoil is part of a performance for other characters in the film, e.g., Susanne played by Kidman in *To Die For*. These findings are in line with the literature that shows that creaky voice can be used to signal vulnerability bound with emotional turmoil and negative stances. For instance, Esposito (2016) argues that Lady Gaga uses more creaky voice in interviews when discussing emotionally difficult topics in which she puts herself in a more vulnerable position by sharing very personal and traumatic experiences, thereby constructing a sincere pop star persona. Another case in point is Lacasse's (2010) study of the song "Breathe Me" by the singer Sia. Lacasse proposes that creaky voice is used by the performer to deliver an emotionally highly saturated delivery, which seems to revolve around a character who may have attempted to commit suicide.

Considering the very nature of the femme fatale as an opposer of various societal norms, it may not be surprising that even in the rather short scenes analyzed here, a relatively high number of negative stances were identified. Most of the creaks produced can be explained through negative stance-taking work, namely as a way to signal one's disalignment with and/or opposition to an object such as an opinion or indeed a person. Other researchers have similarly identified creak to have the potential to signal a negative stance in English (e.g., Grivicic and Nilep 2004).

As mentioned in the introduction, it is possible that creaky voice may signal masculinity. This, nevertheless, does not seem to be the case in the scenes analyzed here within the speech of the selected femme fatales. The only potential link to masculinity is an indirectly indexical one: If women are not expected to generally express disagreement, especially with men, or opinions of their own (see, e.g., Luke 1994 on the gender politics of speech and silence), it follows that under this stereotype any linguistic strategy employed for negative stance-taking would not meet normatively stereotypical feminine behavior and might, therefore, be perceived as masculine.

Finally, it has been suggested that creaky voice might be used to signal *overt* female heterosexual seductiveness and contemporariness in American English (Shaw and Crocker 2015: 26). Creak is not used to index this communicative intention in the scenes analyzed here. While our analysis did not explicitly contrast flirtatious and non-flirtatious contexts, or contexts with overt sexual intentions with those with implicit or absent ones, the scenes analyzed do provide an array of variably seductive scenarios, including those where "sexiness is a central facet of the character" (2015: 26). For instance, Bridget in *The Last Seduction*, which falls within the period of third-wave feminism, produces no instances of creak in the scene analyzed here. This scene involves the character's very explicit conversation about the qualities of the penis of a male stranger who she has just met in a bar. Notably, this exchange is accompanied by Bridget's manual inspection of the stranger's genitalia to assure herself of the veracity of his earlier statement.

There is, nevertheless, a rather crucial difference between our femme fatales and the female characters analyzed by Shaw and Crocker (2015). As Shaw and Crocker (2015: 26) suggest, creaky voice is associated with characters who show "deliberate provocativeness and flirtation as key components" of their persona. While femme fatales could be considered provocative by their very nature, they are not necessarily flirtatious and/or normatively playful. If we return to the example of Bridget in *The Last Seduction*, although she is undoubtedly in the process of seducing the male character, this is done in a manner that goes beyond the provocative. The male character, who embarks on making Bridget's acquaintance—no doubt to flirt and seduce her in a normative way—ironically ends up being on the seductee end of their constellation, where he is being almost clinically assessed by the extremely assertive Bridget. We would, therefore, like to propose that future research should distinguish between different types of seduction, as different types of seduction may utilize different phonatory profiles—including creaky voice, in different ways.

Do the fictional characters analyzed here reflect sociolinguistic changes in the real world?

The presence of creaky voice chronologically increases in the films analyzed, especially as we move to third-wave feminism, which corresponds to the non-academic and some academic (e.g., Dallaston and Docherty 2020 for an overview) reports of the rise in creak in the speech of female speakers of American English. There is, nevertheless, another interesting observation to be made about our femme fatales: They become more diverse in recent films, and it would be surprising if this diversity was not reflected also in the vocal variation of these characters. The more recent femme fatale characters are shown in a more dynamic, multifaceted manner, which includes various aspects of their persona, better understanding of their motivations, and a more varied range of situational contexts in which we find them. For instance, the post-war femme fatales analyzed here are less overt about their motivations and sexual desires, while the more recent ones present a spectrum. Similarly, the more recent femme fatales are also shown in a wider range of situations, including—in the scenes analyzed here—a bar, the house of their parents, an office, their own flat, their boyfriend's house, fighting practice ring and various combat environments, and more. In addition, the recent femme fatales also find themselves in economically more powerful positions, as is the case, for instance, with Meredith in *Disclosure*—a very successful businesswoman.

Finally, the discussion could not be complete here without a note on the quote that introduces this chapter. The voices of the femme fatales analyzed in this chapter are undoubtedly "deep," corresponding to f0 values that are below the average typically associated with female speakers of American English. One may, however, wonder what the "scratchiness" corresponds to exactly. In our examples, it is not creak. Instead, this label likely corresponds to non-pathological hoarseness (or huskiness), as observed in the femme fatales played by Johansson and Moore in this chapter. This phonatory profile is associated with, for example, higher values of jitter and shimmer, which do not, however, lead to the production of creak. Non-pathological hoarseness and creak should be distinguished in future research, as should different types of creak.

Carefully distinguishing different types of femme fatales in future research may shed further light on the social functions of creaky voice and other phonatory profiles. In addition, femme fatales should ideally be systematically contrasted with other female characters to further bolster the claim that low f0 and a high degree of breathiness are indeed the key components of the vocal aspects of the femme fatale trope, at least in American cinematography. Needless to say, the vocal means to construct the femme fatale likely vary across languages and cultures.

Conclusion

Our main findings indicate that creaky voice is not used to construct the femme fatale persona in our data. It increases in occurrence in the most recent films and most often signals a negative stance of the speaker and/or emotional turmoil. Second, pitch, pitch range, and breathiness are used to construct the trope fairly consistently across the years, with some character-specific variation.

Endnotes

1. Here, f0 stands for fundamental frequency, which is the rate at which the vocal folds vibrate per second. As such, f0 is different from pitch, which is the percept of f0. f0 is a physical phenomenon that can be measured through articulatory and acoustic analyses.
2. Fundamental frequency (f0) was extracted in VoiceSauce (Shue 2010; Shue et al. 2011) with MATLAB (2022) using the Straight algorithm. Cepstral Peak Prominence (CPP) was also extracted, bearing in mind that this measure is, first, extremely sensitive to noisiness, which is not necessarily vocal in nature (just like other measures quantifying friction in some way), and it does not distinguish between breathiness and creakiness (Fraile and Godino-Llorente 2014). Instead, it allows for assessment of differences between modality and other phonations (e.g., breathiness, creakiness, whisperiness). This being the case, we also quantified the amount of creak as percentage of the number of sentences produced by the character that contained creak. A simplistic aspect of this approach

is that it does not reflect whether the creak only affects one syllable or 10 syllables. Nevertheless, this is borne in mind in the qualitative analysis of the data. Segmentally caused creak (as in *Manhattan* [mæn'hæʔn]) and creak found at the beginning of vowel-initial words (as in *apple*) were not considered instances of what is labelled "creak" here. Both f0 and CPP were quantified through their mean value within each vowel token. f0 was normalized through semitone (ST) conversion as well. Since the visual results for Hz and ST are the same, only the former is provided. Graphs were produced in RStudio (RStudio 2022), using the basic packages as well as the lattice package (Sarkar et al. 2021).

3. Interestingly, the characters analyzed here are also more likely to smoke in the films from the second and third waves of feminism. Mrs. Robinson in *The Graduate* smokes, Matty in *Body Heat* smokes (although this seems to be primarily for visual and, therefore, more obviously symbolic purposes), Adrian in *The Crush* is a teenage femme fatale smoking a cigar in a very confident and skilled manner, Nola in *Matchpoint* smokes (and the camera captures this detail in a way that makes it difficult not to notice), and Bridget in *The Last Seduction* smokes. Smoking is also explicitly discussed in *Basic Instinct*. Any potential effects that smoking might have are, nevertheless, not reflected in the voices of these female characters.

Primary sources (Film clips)

ND. "I'm tired of what's right and wrong—Jessica Lange emerges in Postman (with Jack Nicholson)," YouTube, accessed June 19, 2022, www.youtube.com/watch?t=83&v=UXHjIfgZBD0&feature=youtu.be.

ND. "Top Gun (I think I'm falling for you)," YouTube, accessed June 19, 2022, www.youtube.com/watch?v=Ib8QahlcZNw.

2011. "Vertigo (4/11) Movie CLIP—What Happened? (1958)," YouTube, accessed June 19, 2022, www.youtube.com/watch?v=ZruKu2N6nQw.

2012. "Double Jeopardy (7/9) Movie CLIP—The Prosecution Rests (1999)," YouTube, accessed June 17, 2022, www.youtube.com/watch?v=m2REqMDEXNU.

2012. "Match point (2/8) Movie CLIP—An Aggressive Game (2005)," YouTube, accessed June 17, 2022, www.youtube.com/watch?v=Hj9WsioJbJw.

2014. "Eva Green Web—Eva Green Sin City Clip 1," YouTube, accessed June 17, 2022, www.youtube.com/watch?v=jGSf7vahC4I.

2014. "Linda Fiorentino The Last Seduction," YouTube, accessed June 18, 2022, www.youtube.com/watch?v=8mIl4pxMh4M.

2016. "Total Recall—What a Bitch," YouTube, accessed June 16, 2022, www.youtube.com/ watch?v=o9sA3JYmrvA.

2017. "La La Land—I'm Always Gonna Love You," YouTube, accessed June 16, 2022, www.youtube.com/watch?v=Gu08f15KnMI.

2017. "Splendor in the Grass—Natalie Wood," YouTube, accessed June 19, 2022, www.youtube.com/watch?v=-4KfXkEYbQQ&t=178s.

2017. "To Die For—Kidman Monologue," YouTube, accessed June 17, 2022, www.youtube.com/watch?v=DC88UajiTnM.

2018. "Marilyn Monroe in 'Niagara' – 'Kiss'," YouTube, accessed June 19, 2022, www.youtube.com/watch?v=aDIvP2nyb2E&t=232s.

2019. "THE ELEVATOR SCENE (Top Gun)," YouTube, accessed June 19, 2022, www.youtube.com/watch?v=bt8FqaMsYDY.

2020. "Marilyn Monroe in "Gentlemen Prefer Blondes"—"A Wonderful Moon Out Tonight," YouTube, accessed June 19, 2022, www. youtube.com/ watch?v=njn1QfASvCA&t=34s.

2020. "Sultry Kathleen Turner sizzles in 'Body Heat' (1981)," YouTube, accessed June 19, 2022, www.youtube.com/watch?v=ynzlZVlxLzk.

2020. "Disclosure—morning meeting," YouTube, accessed June 17, 2022, www.youtube.com/watch?t=30&v=O3wc_E9TBV8&feature=youtu. be.

2020. "The Graduate (1967) | Iconic Scene | Dir. by Mike Nichols & starring Dustin Hoffman," YouTube, accessed June 19, 2022, www. youtube.com/ watch?t=182&v=CjuPkYiDTzw&feature=youtu.be.

2021. "Alicia Silverstone—The Crush (1993)," YouTube, accessed June 18, 2022, www.youtube.com/watch?v=EkO0IjbZStU.

2021. "Basic Instinct (1992)—Trying to Resist," YouTube, accessed June 16, 2022, www.youtube.com/watch?t=131&v=CBkuTVArMWw&fe ature=youtu.be.

2021. "Best of Scarlett Johansson HOT COMPILATION !! Best fight scenes & kissing scenes," YouTube, accessed June 17, 2022, www.youtube. com/watch?v= UWHOjmIPlHo.

2021. "Lauren Bacall Voice Style," YouTube, accessed June 19, 2022, www.youtube.com/watch?v=elXijEbb_ao.

References

Anderson, R. C., Klofstad, C. A., Mayew, W. J., and Venkatachalam, M. (2014). "Vocal fry may undermine the success of young women in the labor market." *PLOS ONE*, **9**(5): 1–8.

Andreetti, S. (2019). "Xayide, enchantress or Femme Fatale? Magic and empathy at the service of manipulation in Michael Ende's *The Neverending Story*." In D. Farnell, R. G. da Noiva, and K. Smith (Eds.), *Perceiving Evil: Evil Women and the Feminine*, 69–76. Leiden: Brill.

Anolli, L. and Ciceri, R. (2002). "Analysis of the vocal profiles of male seduction: From exhibition to self-disclosure." *The Journal of General Psychology*, **129**(2): 149–169.

Babel, M., McGuire, G., and King, J. (2014). "Towards a more nuanced view of vocal attractiveness." *PLOS ONE*, **9**(2): 1–10.

Bell, A. (2011). "Falling in love again and again: Marlene Dietrich and the iconization of non-native English." *Journal of Sociolinguistics*, **15**(5): 627–656.

Blum, H. (2016). "Totally fried: What do you need to know about vocal fry?" *The ASHA Leader* (February 2022): 50–56.

Boersma, P. and Weenink, D. (2022). *Praat: Doing Phonetics by Computer*. Retrieved from https://www.fon.hum.uva.nl/praat/ (accessed March 1, 2022).

Boozer, J. (1999/2000). "The lethal femme fatale in the noir tradition." *Journal of Film and Video*, **51**(3/4): 20–35.

Bronfen, E. (2004). "Femme Fatale: Negotiations of tragic desire." *New Literary History*, **35**(1): 103–116.

Cameron, I. (Ed.). (1993). *The Book of Film Noir*. New York: Continuum.

Chiari, S. (2019). "Overflowing the measure: Cleopatra unbound." *Actes des congrès de la Société française Shakespeare*, **37**.

Cook, P. (1978). "Duplicity in Mildred Pierce." In E. A. Kaplan (Ed.), *Women in Film Noir*, 69–80. London: British Film Institute.

Copjec, J. (Ed.). (1993). *Shades of Noir: A Reader*. London: Verso.

Cowie, E. (1993). "Film noir and women." In J. Copjec (Ed.), *Shades of Noir: A Reader*, 121–166. London: Verso.

Dallaston, K. and Docherty, G. (2020). "The quantitative prevalence of creaky voice (vocal fry) in varieties of English." *PLOS ONE*, **11**.

Dean, J. (n.d.). "Vocal fry: Why women should avoid speaking with 'creaky' voice." *Psyblog*. Retrieved from https://www.spring.org.uk/2012/05/get-your-creak-on-is-vocal-fry-a-female-fad.php (accessed May 5, 2022).

Edwards, K. L. (2010). "The mother of all femme fatales: Eve as temptress in Genesis 3." In H. Hanson and C. O'Rawe (Eds.), *The Femme Fatale: Images, Histories, Contexts*, 35–45. Palgrave Macmillan.

ErkenBrack, E., Pardo, R., and Jackson, J. L. (2012). "Popular culture" (Oxford Bibliographies). *Oxford Bibliographies*. Retrieved from https://www.oxfordbibliographies.com/view/document/obo-9780199766567/obo-9780199766567-0010.xml (accessed June 3, 2022).

Esling, J. H., Moisik, S. R., Benner, A., and Crevier-Buchman, L. (2019). *Voice Quality: The Laryngeal Articulator Model*. Cambridge: Cambridge University Press.

Esposito, L. (2016). "'I am a perpetual underdog': Lady Gaga's use of creaky voice in the construction of a sincere pop star persona." BA thesis, Swarthmore College.

Fraile, R. and Godino-Llorente, J. I. (2014). "Cepstral peak prominence: A comprehensive analysis." *Biomedical Signal Processing and Control*, **14**: 42–54.

Gledhill, C. (1978). "Klute 2: Feminism and Kluke." In E. A. Kaplan (Ed.), *Women in Film Noir*, 99–114. London: British Film Institute.

"Gloaming" (n.d.). Definition in the *Cambridge Dictionary*. Retrieved from https://dictionary.cambridge.org/dictionary/english/gloaming.

Greer, S. D. F. and Winters, S. J. (2015). "The perception of coolness: Difference in evaluating voice quality in male and female speakers." *18th International Congress of Phonetic Sciences, Glasgow*.

Grivicic, T. and Nilep, C. (2004). "When phonation matters: The use and function of yeah and creaky voice." *Colorado Research in Linguistics*, **17**(1): 1–11.

Grossman, J. (2020). *The Femme Fatale*. Rutgers University Press.

Grossman, J. (2009). Rethinking the Femme Fatale in Film Noir: Ready for Her Close-Up. New York: Palgrave Macmillan.

Guins, R. and Cruz, O. Z. (2005). *Popular Culture: A Reader*. London: Sage.

Hejná, M. and Jespersen, A. (2022). "Ageing well: Social but also biological reasons for age-grading." *Language & Linguistics Compass*, **16**(5–6): e12450.

Henton, C. and Bladon, A. (1988). "Creak as a sociophonetic marker," in V. Fromkin, L. M. Hyman, and C. N. Li (eds.), *Language, Speech, and Mind: Studies in Honour of Victoria A. Fromkin*, London; New York: Routledge. Published in the USA in association with Routledge, Chapman and Hall: 3–29.

Hillenbrand, J., Getty, L. A., Clark, M. J., and Wheeler, K. (1995). "Acoustic characteristics of American English vowels." *The Journal of the Acoustical Society of America*, **97**: 3099–3111.

Hornibrook, J., Ormond, T., and Maclagen, M. (2018). "Creaky voice or extreme vocal fry in young women." *New Zealand Medical Journal*, **131**(1486): 36–40.

Hunt, S. (2009). "The femme fatale." *Calgary Herald*, February 28, 2022.

Jancovich, M. (2011). "'Vicious womanhood': Genre, the femme fatale and postwar America." *Canadian Journal of Film Studies*, **20**(1): 100–114.

Jilliand. (2020). "Marlene Dietrich and the Independent Femme Fatale" *Cine 345 Midterm Blog: Monroe, Dietrich and Davis Reviews*. Retrieved from https://blogs.uoregon.edu/jilliand/2020/04/30/marlene-dietrich-and-the-independent-femme-fatale (accessed April 30, 2020).

Kaplan, E. A. (Ed.) (1978). *Women in Film Noir*. London: British Film Institute.

Keating, P., Garellek, M., and Kreiman, J. (2015). "Acoustic properties of different kinds of creaky voice." *18th International Congress of Phonetic Sciences, Glasgow*.

Kiesling, S. F. (2022). "Stance and stancetaking." *Annual Review of Linguistics*, **8**: 409–426.

Krahé, B. and Papakonstantinou, L. (2020). "Speaking like a man: Women's pitch as a cue for gender stereotyping." *Sex Roles: A Journal of Research*, **82**(1–2): 94–101.

Lacasse, S. (2010). "Slave to the supradiegetic rhythm: A microrhythmic analysis of creaky voice in Sia's 'Breathe Me'." In A. Danielsen (Ed.), *Musical Rhythm in the Age of Digital Reproduction*, 141–158. London: Routledge.

Laver, J. (1980[2009]). *The Phonetic Description of Voice Quality*. Cambridge: Cambridge University Press.

Luke, C. (1994). "Women in the academy: The politics of speech and silence." *British Journal of Sociology and Education*, **15**(2): 211–230.

Mendoza-Denton, N. (2011). "The semiotic hitchhiker's guide to creaky voice: Circulation and gendered hardcore in a Chicana/o gang persona." *Journal of Linguistic Anthropology*, **21**(2): 261–280.

Nachbar, J. and Lause, K. (1992). *Popular Culture: An Introductory Text.* Chicago: Popular Press.

Pisanski, K. and Feinberg, D. R. (2018). "Vocal attractiveness." In S. Frühholz and P. Belin (Eds.), *The Oxford Handbook of Voice Perception*, 1–24. Oxford: Oxford University Press.

Prasad, M. M. (2018). "Popular culture and cultural studies." *Oxford Research Encyclopedias: Communication*. https://oxfordre.com/communication/view/10.1093/acrefore/ 9780190228613.001.0001/acrefore-9780190228613-e-561 (Accessed June 5, 2022).

Richards, J. M. (1997). "'To promote a woman to Beare rule': Talking of queens in Mid-Tudor England." *The Sixteenth Century Journal*, **28**(1): 101–121.

Rogue Amoeba. (2022). *Audio Hijack*, accessed from https://rogueamoeba.com/audiohijack/ (accessed June 17, 2022).

RStudio Team. (2022). *RStudio: Integrated development for R*. Accessed February 1, 2022, http://www.rstudio.com/.

Sarkar, D., Andrews, F., Wright, K., Klepeis, N., Larsson, J., and Murrell, P. (2021). "Package 'lattice.' Version 0.20-45." Retrieved from https://cran.r-project.org/web/packages/lattice/lattice.pdf.

"Second Wave Feminism" (n.d.). *Oxford Reference*. Retrieved from www.oxfordreference.com/view/10.1093/oi/authority.20110803100451509.

Shaw, F. and Crocker, V. (2015). "Creaky voice as a stylistic feature of young American female speech: An intraspeaker variation study of Scarlett Johansson." *Lifespans and Styles*, 1, 21–27.

Shue, Y.-L. (2010). *The Voice Source in Speech Production: Data, Analysis and Models*. PhD Thesis, UCLA.

Shue, Y.-L., Keating, P., Vicenik, C., and Yu, K. (2011). "VoiceSauce: A program for voice analysis." *17th International Congress of Phonetic Sciences*: 1846–1849.

Smith, K. (2019). "'Keep mum, she's definitely not dumb': The complex and cunning femme fatale in espionage fiction and history." In D. Farnell, R. Noiva, and K. Smith (Eds.), *Perceiving Evil: Evil Women and the Feminine*, 129–138. Leiden: Brill.

Stathopoulos, E. T., Huber, J. E., and Sussman, J. E. (2011). "Changes in acoustic characteristics of the voice across the life span: Measures from individuals 4–93 years of age." *Journal of Speech and Language Hearing Research*, **54**: 1011–1021.

Steinmetz, K. (2011). "Get your creak on: Is 'vocal fry' a female fad?" *Time*.

"Third Wave Feminism" (n.d.) *Oxford Reference*. Retrieved from www.oxfordreference.com/display/10.1093/acref/9780199532919.001.0001/acref-9780199532919-e-708.

Tyler May, E. (1988). *Homeward Bound: American Families in the Cold War Era*. New York: Basic Books.

"The Femme Fatale Trope, Explained" (n.d.). YouTube video. Retrieved from https://www.youtube.com/watch?v=XjAKd-Lfmt0.

van Borsel, J., Janssens, J., and de Bodt, M. (2009). "Breathiness as a feminine voice characteristic: A perceptual approach." *Journal of Voice*, **23**(3): 291–294.

Wolk, L., Abdelli-Beruh, N. B., and Slavin, D. (2012). "Habitual use of vocal fry in young adult female speakers." *Journal of Voice*, **26**(3): 111–116.

Wolf, N. (2015). "Young women, give up the vocal fry and reclaim your strong female voice." *The Guardian*.

Xu, A., Leung, S.-S., and Lee, A. (2016). "Universal vs. language-specific aspects in human vocal attractiveness: An investigation towards Japanese native listeners' perceptual pattern." *172nd Meeting of the Acoustical Society of America*, 29.

Xu, A. and Lee, A. (2018). "Perception of vocal attractiveness by Mandarin native listeners." *Proceedings of the 9th International Conference on Speech Prosody, Poznań*, 344–348.

Yuasa, I. P. (2010). "Creaky voice: A new feminine quality for young urban-oriented upwardly mobile American women?" *American Speech*, **85**(3): 315–337.

Zimman, L. (2017). "Gender as stylistic bricolage: Transmasculine voices and the relationship between fundamental frequency and /s/." *Language in Society*, **46**(3): 339–370.

Chapter Nine

The role of creaky voice in gender attribution in Scottish English

Joe Pearce

Introduction

How does creaky voice affect how listeners perceive a speaker's gender? As discussed throughout this volume, creaky voice has a variety of functions, and research on the perception of creaky voice has found that creaky voice can take on gendered sociocultural meanings, from "non-aggressive" and "hesitant" in the speech of young American women (Yuasa 2010), or as gay-sounding among American men (Zimman 2013). However, while research looking at production has found gender differences in creaky voice usage in many accents of English (Henton and Bladon 1988; Syrdal 1996; Stuart-Smith 1999; Wolk *et al.* 2012; Abdelli-Beruh *et al.* 2014; Szakay and Torgersen 2019), only limited preliminary research

Creak: Theories and practices of pulse phonation
Edited by Francesco Venturi

ISBN 978-981-5129-29-8 (Hardcover), 978-1-003-61606-1 (eBook)
www.jennystanford.com

has considered the idea that creaky voice may affect gender perception (Foulkes *et al.* 2011; Greer 2015; Lee 2016). Considering whether and how creaky voice may influence gender perception may allow us to understand the relationship between the varied indexical meanings of creaky voice, bridging research on gender differences in creaky voice with research on listener perceptions of creak. Understanding the role of creaky voice in gender perception may also have practical implications for gender-affirming voice therapy, where a change in voice quality often forms part of voice therapy protocols (Davies *et al.* 2015).

The present study considers how creaky voice affects how listeners perceive speaker gender. To investigate this question, I consider the role of creaky voice in comparison to two other voice qualities: modal voice, to serve as a reference point, and breathy voice, which previous research has found to be associated with femininity and female speakers (e.g., van Borsel *et al.* 2009; Porter 2012; Skuk and Schweinberger 2014). These questions were considered in two experiments where Scottish listeners (Experiment 1) and North American listeners (Experiment 2) heard Scottish English stimuli in different voice qualities (modal, breathy, creaky) and at different f0 levels (120 Hz, 165 Hz, 210 Hz), and made judgments about speaker gender, rating masculinity and femininity, and categorizing the gender of the speaker. As I explore below, listeners appeared to use creaky voice alongside other information to make judgments about speaker gender. Listeners rated creaky voices as less masculine than modal voices and were also less likely to categorize the speaker of creaky voice as a woman if other cues to speaker gender were ambiguous. Listeners also reported using their perceptions of other speaker characteristics to make judgments about the speaker's gender, suggesting that listeners bring their wider listening experience with them as they use creak to make judgments about gender in voices.

Background

Fundamental frequency (f0), the acoustic correlate of pitch, plays a major role in how listeners attribute gender to a voice. In spoken

English, female speakers tend to use higher-sounding voices with an f0 of 200–210 Hz, while male speakers tend to use lower-sounding voices with an f0 of around 100–120 Hz (Simpson 2009). This difference, in turn, affects how listeners attribute gender to a voice, with f0 being the main acoustic cue contributing to their perceptions of the speaker's gender (Leung *et al.* 2018). However, little research has examined the role of creaky voice on gender perception, despite much research suggesting potential differences in creaky voice usage in production. Both Stuart-Smith (1999) and Beck and Schaeffler (2015) have looked at gender differences in voice quality in Scotland and found male speakers to be creakier than female speakers. However, in research looking at American English, female speakers have been suggested to use more creaky voice (Syrdal 1996; Wolk *et al.* 2012; Podesva 2013; Abdelli-Beruh *et al.* 2016). The present study considers how the gender differences in creaky voice, which may exist in production, might affect the listener's perception of gender when hearing voices with different voice qualities. I also consider whether f0, an established cue to gender, affects the role of creaky voice in gender perception. Furthermore, I consider whether the listener's sociolinguistic background plays a role, asking whether creaky voice affects the listener's perception of gender differently for North American and Scottish listeners, who may have been exposed to different culturally specific meanings of creaky voice.

The role of creaky voice in gender perception is relevant for discussions about a potential iconic link between creaky voice and masculinity. Podesva and Kajino (2014) suggest that there may be an iconic link between creaky voice and masculinity, proposing that listeners hear a resemblance between the low f0 of creak and the low f0 that male speakers tend to use and thus interpret creak as masculine, an idea that reflects Ohala's (1994) discussion of the frequency code. Podesva and Kajino (2014) also stress that the link between creaky voice and masculinity is still mediated by ideology, allowing it to become associated with femininity through use in women's speech, which they suggest has occurred in the United States.

This idea of an iconic link between creaky voice and masculinity is, to an extent, validated by previous work on the perception of creak for American English listeners. Greer (2015) asked listeners to judge which voice was more masculine (for male speakers) or more feminine (for female speakers) when presented with two voice samples produced with different voice qualities. She found that for male speakers, creakier voices were perceived as more masculine than non-creaky or less creaky voices. Similarly, Lee (2016) investigated perceptual judgments of creaky voice in a study where listeners rated the masculinity and femininity of creaky and modal voices on Likert scales and found that listeners rated creaky voices as more masculine and less feminine than modal voices. Yuasa (2010: 318) suggests that the use of creak by female speakers of American English may also recruit some of these masculine associations, saying that female newscasters and politicians may make use of creaky voice and its low f0 to project a "male-like authoritative image." Ultimately, Yuasa's (2010) study, which looks at how American English listeners perceive women's use of creaky voice, finds that listeners perceive creaky voice in female speech as hesitant and non-aggressive, indicating that the use of creaky voice by female speakers of American English may have led to the sociocultural meanings of creaky voice shifting away from masculine associations.

In this study, I consider how creaky voice affects gender perception in two experiments where listeners heard creaky, modal, and breathy voices and made judgments about the gender of the speaker. This allows consideration of whether listeners associate creaky voice with masculinity. I expected that listeners might rate creaky voices as more masculine and less feminine, which would have supported the idea of an iconic link between creaky voice and masculinity based on creaky voice's low f0. However, in this study I found that listeners perceive creaky voice as less masculine from the modal voice of a Scottish English female speaker. In the Results section, I present these results and return to discuss their implications in the Discussion section.

As discussed by Leung *et al.* (2018), different studies measure "perceived gender" differently: Some studies ask listeners to categorize the gender of a speaker, while others ask listeners to rate traits such as masculinity and femininity. While these different dimensions of gender are often treated as equivalent in research, it is possible that listeners understand them differently, leading creaky voice to affect their perceptions of these different dimensions in different ways. To consider whether the types of responses captured affected whether creaky voice plays a role in gender perception, participants were asked to rate the gendered attributes of a voice (masculinity and femininity on scales of 1–100), or to categorize the speaker's gender (whether they thought the voice belonged to a "man," "woman," or "neither/can't tell"). I found different factors affected masculinity ratings, femininity ratings, and gender categorizations differently, suggesting that these terms are not necessarily equivalent for listeners. I present these results in the Results section, examine this issue in more detail by considering the qualitative feedback of the participants on the experiment, and then return to discuss this issue in the Discussion section.

I also ask whether the f0 of a voice, a well-established cue for gender perception, affects how creaky voice affects gender perception. I used stimuli that fell into f0 ranges that were expected to cue the perception of either a male voice (120 Hz) or a female voice (210 Hz), or which were designed to be ambiguous (165 Hz). I expected that while creaky voice might affect how listeners perceived gender attributes (masculinity and femininity) at all f0 levels, it might only affect how listeners categorized the speaker's gender (man, woman, or neither/can't tell) at the ambiguous f0 level. At this ambiguous f0 level, f0 would be less informative as a cue to gender, which could potentially lead listeners to rely more on other cues, such as creaky voice. However, I found that 120 Hz functioned as an ambiguous f0 level rather than 165 Hz, and that at this f0 level was less likely to categorize voices as belonging to men or as neither/can't tell, in comparison to how they categorized creaky voices in relation to modal voices at higher f0 levels.

I focus here on Scottish English and compare perceptions between Scottish listeners (Experiment 1) and North American listeners (Experiment 2). I expected that the culturally specific meanings of creaky voice in North America might lead to North American listeners reinterpreting the link between creaky voice and masculinity: rating creaky voices as more feminine and be more likely to categorize them as women, compared to Scottish listeners. However, I did not find a difference between Scottish and North American listeners in terms of how creaky voice affected their perception of gender. In the third section, I return to this finding alongside other ways that listeners' sociolinguistic background affected gender perception, and discuss this issue in more detail concluding the fourth section.

Methods

To investigate how creaky voice affects gender perception, two experiments were carried out. In each experiment, listeners heard stimuli produced by a Scottish English female speaker. Each stimulus was manipulated to have a median f0 of either 120 Hz, 165 Hz, or 210 Hz, and either produced with a modal, breathy, or creaky voice quality, or manipulated synthetically to have a modal, breathy, or creaky voice quality. Listeners heard these stimuli and were asked to either say whether they thought the voice sounded like that of a man or a woman, or rate how masculine and feminine the voices sounded on scales of 1–100.

In Experiment 1, participants consisted of listeners from around the Central Belt of Scotland who had lived in the Greater Glasgow area for at least a year. In Experiment 2, participants consisted of listeners who were born in and currently resided in North America and were recruited online using Prolific (www.prolific.co) and completed the experiment online. This allowed consideration into whether there was any difference in how listeners of two different sociolinguistic contexts—who may have different indexical associations for creaky voice—perceive creaky voice in terms of gender.

Stimuli

Stimuli consisted of 108 short phrases and sentences containing only sonorants, such as "We were running" or "A long year," which enabled differences in voice quality to be synthesized using KlattGrid (Klatt and Klatt 1990; Boersma and Weenink 2019) based on sentences produced by the speaker. The stimuli were produced by a cisgender female L1 English speaker from Glasgow, Scotland. She was selected to produce the stimuli because she was trained in theater and so was able to control the quality of her voice.

The stimuli were recorded in a sound-attenuated booth with a Sennheiser MKH 40 P48 cardioid condenser microphone in Audacity (Mazzoni 2018) in monoaural sound at 44,100 Hz. The microphone was positioned approximately 30 cm away from the lips. The speaker produced stimuli for the "natural" condition in modal voice, creaky voice, and breathy voice, modeling each quality from the examples that accompany Laver (1980). For the breathy voice condition, she heard examples of both breathy and whispery voices and produced a voice quality with audible glottal friction that lay on the auditory continuum of whispery–breathy voice. The overlap-add pitch synthesis in Praat was used to manipulate the median pitch of each stimulus to one of three levels: 120 Hz to cue the perception of a male voice, 165 Hz to create ambiguous stimuli where the effect of voice quality on gender perception could be maximized, or 210 Hz to cue the perception of a female voice.

The speaker produced stimuli for the "synthesized" condition in modal voice to allow voice quality differences to be synthesized using KlattGrid (Weenink 2009) in Praat (Boersma and Weenink 2019) to create highly controlled voice quality differences. The values for each parameter were chosen based on the values given in Klatt and Klatt (1990) and Gobl and Ní Chasaide (2003) for the synthesis of different voice quality and altered where this seemed to improve the resulting auditory quality. The settings for each parameter are detailed in Pearce (2020). Stimuli were processed after manipulation to remove artifacts of the copy synthesis process as far as possible.

Participants

Experiment 1: Scottish participants

Thirty-three Scottish participants from the wider Central Belt of Scotland took part in Experiment 1, in person, in Glasgow. The participants were aged between 19 and 45. All reported being native English speakers. Of the 33 participants, nine were male and 17 were female, and seven self-reported their gender identity. All participants reported having normal hearing.

Experiment 2: North American participants

Forty participants from the United States and Canada took part in Experiment 2. All reported being native speakers of English. The participants were aged between 18 and 62. Of the 40 participants, there were 17 female participants and 23 male participants. All the participants reported having normal hearing. They were recruited using Prolific (www.prolific.co) [2019], an online tool for recruiting participants to take part in online experiments and compensate them for their time. They all completed the experiment in a quiet space using headphones.

Procedure

The study was created using PsyToolkit, an online tool for creating and running experiments online (Stoet 2017). After answering demographic questions, the participants began the experiment.

For each stimulus, the participants rated how natural the voice sounded on a scale of 1–100, as a distractor question. Then, they were asked one of the following questions:

- Is the voice you hear that of a man or a woman? (Response: "Man," "Woman," or "Neither or can't tell")
- Please rate how masculine and feminine this voice sounds to you. (Response: Two sliding scales from 1–100)

The procedure was the same for both sets of participants, with slight differences in the demographic questions. An archived version of Experiment 2 can be accessed here: https://www.psytoolkit.org/cgi-bin/psy2.5.4/survey?s=zsxzS. At the end of the experiment, the participants were also asked the following questions:

- What did you pay attention to when deciding whether you thought the voices sounded like a man or a woman?
- What did you pay attention to when rating how masculine/feminine a voice sounded?
- Do you have any further comments about your experience completing this experiment?

Experimental design

The independent variables were the naturalness of the stimuli (either natural, where the speaker had produced them, or synthesized, where voice quality differences were introduced with KlattGrid), the median f0 of the stimuli (210 Hz, 165 Hz, or 120 Hz), and the voice quality of the stimuli (modal, breathy, or creaky). The study involved a 2×3×3 within-subjects design where naturalness was crossed with voice quality and f0 to create 18 conditions. The conditions are summarized in Table 9.1.

Table 9.1 Conditions of the experiment, where stimuli naturalness is crossed with f0 level and voice quality

	Natural (n = 54)			**Synthesized (n = 54)**		
	Breathy	**Modal**	**Creaky**	**Breathy**	**Modal**	**Creaky**
210 Hz	Breathy 210 Hz $n = 6$	Modal 210 Hz $n = 6$	Creaky 210 Hz $n = 6$	Breathy 210 Hz $n = 6$	Modal 210 Hz $n = 6$	Creaky 210 Hz $n = 6$
165 Hz	Breathy 165 Hz $n = 6$	Modal 165 Hz $n = 6$	Creaky 165 Hz $n = 6$	Breathy 165 Hz $n = 6$	Modal 165 Hz $n = 6$	Creaky 165 Hz $n = 6$
120 Hz	Breathy 120 Hz $n = 6$	Modal 120 Hz $n = 6$	Creaky 120 Hz $n = 6$	Breathy 120 Hz $n = 6$	Modal 120 Hz $n = 6$	Creaky 120 Hz $n = 6$
Total				$n = 108$		

Participants heard a total of 108 stimuli, with 54 in the natural condition and 54 in the synthesized condition. Within these two conditions, participants heard each sub-condition six times in total, responding to each question on three occasions for each condition. Stimuli order was randomized.

The dependent variables were gender categorization (man, woman, or neither/can't tell), perceived masculinity, and perceived femininity (each rated on scales of 1–100).

Analysis

All statistical analysis was conducted using R v.4.1.1 (R Core Team 2020). Listener ratings of masculinity and femininity were analyzed in mixed-effect linear regression models using the lmer() function of lme4 (Bates *et al.* 2015). Each model was stepped down from the full model to select an optimal model using the step() function of the stats package.[1]

The categorizations of the speaker's gender by the listeners were analyzed in a mixed binomial generalized linear mixed-effects regression model using the glmer() function of lme4 (Bates *et al.* 2015), where man and neither/can't tell responses were grouped together into a "not a woman" category. This model was stepped down manually by removing predictors and comparing the AIC of each model in a log likelihood ratio test using anova().[2] Plots showing the effects of these models were created using the effects package (Fox and Weisberg 2019) and ggplot2 (Wickham 2016).

I also made use of tidyverse (Wickham 2016) for data manipulation and exploration and was assisted by stargazer (Hlavac 2018) in formatting the output regression table.

Qualitative participant feedback

Participant responses to feedback questions were analyzed using thematic analysis. A set of initial codes were identified prior to analysis. These were then revised iteratively during coding, with additional codes added to account for unexpected points that emerged in participant responses. These codes were then collapsed into several themes to connect the ideas that arose across different participant responses to the three questions.

Results

Table 9.2 summarizes the results of three mixed-effects models predicting listener ratings of femininity, masculinity, and categorization of gender as a function of different independent

Table 9.2 Summary of the mixed-effects models predicting listener ratings of masculinity, femininity, and categorization of gender as a function of independent variables

	Dependent Variable		
	Femininity (1)	**Masculinity (2)**	**Gender Categorization (3)**
Constant	78.301*** (2.300)	28.099*** (2.102)	−2.672*** (0.370)
165 Hz *Ref = 210 Hz*	−9.623*** (1.322)	7.238*** (1.661)	1.634** (0.498)
120 Hz *Ref = Mean of 210 Hz & 165 Hz*	−28.719*** (1.248)	35.752*** (1.436)	5.835*** (0.410)
Breathy voice *Ref = Modal voice*	4.312*** (1.145)	−2.625** (0.832)	−0.916*** (0.260)
Creaky voice *Ref = Modal voice*	1.233 (1.257)	−2.363** (0.859)	−0.051 (0.227)
N. American *Ref = Scotland*	0.381 (2.771)	−8.837** (2.701)	0.531 (0.423)
Synthesized *Ref = Natural*	−20.405*** (0.827)	12.019*** (0.646)	2.657*** (0.184)
165 Hz: Breathy voice		−2.063 (1.917)	−1.731* (0.679)
120 Hz: Breathy voice		−4.688** (1.817)	−1.071* (0.439)
165 Hz: Creaky voice		1.957 (2.104)	1.112 (0.612)
120 Hz: Creaky voice		−2.130 (1.816)	−1.579** (0.507)
165 Hz: N. American	4.230** (1.335)	−4.535*** (1.349)	−1.621*** (0.295)
120 Hz: N. American	3.117** (1.149)	−11.583*** (1.166)	−1.187*** (0.234)
Log Likelihood	−16,887.490	−16,926.600	−1,287.036

Note: *p<0.05; **p<0.01; ***p<0.001

F0 level is coded using Helmert coding, so that 165 Hz is compared to 210 Hz and 120 Hz is compared to the mean of 165 Hz and 210 Hz. I used the codingMatrices package (Venables 2021) to facilitate Helmert coding.

variables. The first column of the table shows the different independent variables included in each model: voice quality, f0, listener origin/experiment, naturalness, and the interaction between listener origin/experiment and f0 for all three models, as well as the interaction between voice quality and f0 for the masculinity and gender categorization models. The other columns give information about what effect the model predicts a change in each independent variable will have on the outcome variable. The first number in each cell, called the estimate, shows the predicted size and direction of the effect of each level of the independent variable. For masculinity and femininity, the unit of the estimate is the 1–100 rating scale that listeners used to rate voices. However, for the model predicting gender categorizations, the unit estimate is in log(odds): This is the odds of listeners categorizing a voice as not a woman rather than as a woman, transformed to exist on a scale from $-\infty$ to $+\infty$. The second number in each cell, shown in parentheses, shows the standard error of the estimate. Asterisks indicate that the effect was significant. Each model showed variance according to random effects of participant and sentence.[3]

I had expected that creaky voice might increase listener ratings of masculinity and decrease listener ratings of femininity. However, the results of the two models predicting listener ratings of masculinity and femininity did not support these initial predictions. While creaky voice did not have a significant effect on listener perceptions of femininity, it did affect listener ratings of masculinity: Listeners rated creaky voice stimuli as significantly less masculine than modal voice stimuli ($B = -2.625$, $SE(B) = 0.859$, $p < 0.01$, $t = 2.750$). The predicted effect of voice quality on listener ratings of masculinity is shown in Fig. 9.1.

I had expected that creaky voice might affect listener perceptions of speaker gender differently for North American and Scottish listeners. The interaction between voice quality and listener origin did not improve the fit of any of the models, so was taken out when models were stepped down. Because of this, it is unclear whether the sociolinguistic backgrounds of the listeners affect the role of creaky voice in gender perception.

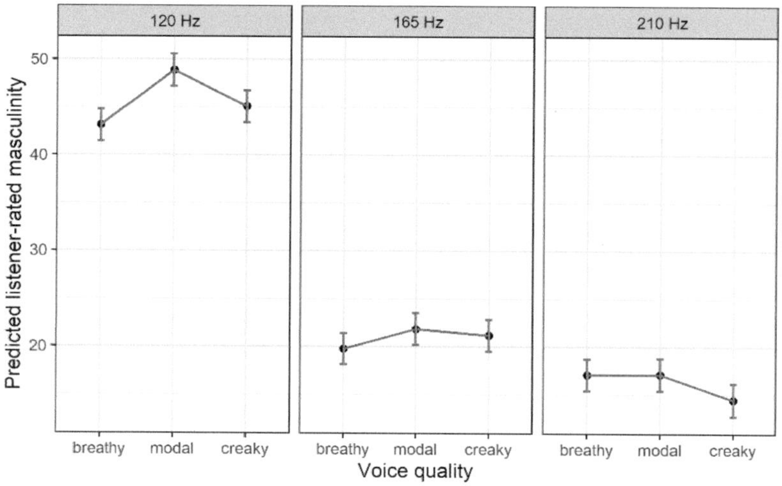

Figure 9.1 Effect plot showing predicted listener-rated masculinity as a function of voice quality of stimuli at different f0 levels.

However, the origin of listeners did affect their perceptions of gender in other respects: The log(odds) of a North American listener categorizing a voice as not belonging to a woman decreased by −1.621 at 165 Hz, compared to the log(odds) of Scottish listeners categorizing a voice as not belonging to a woman at 210 Hz (SE(B) = 0.295, $p < 0.001$, $t = 5.486$).

The log(odds) of a North American listener categorizing a voice as not belonging to a woman further decreased by −1.187 at 120 Hz, compared to how North American listeners categorized voices in relation to Scottish listeners categorized voices at higher f0 levels (SE(B) = 0.234, $p < 0.001$, $t = -5.081$). The proportions of man, neither/can't tell, and woman responses by f0 level and listener origin are shown in Fig. 9.2.

The proportions of listener categorizations of the speaker's gender as man, woman, or neither/can't tell at different f0 levels are shown in Fig. 9.3. There was no significant difference between how listeners categorized the speaker's gender in creaky voices compared to how they categorized gender in modal voices overall. There was also no difference in how they categorized creaky voices compared to modal voices at 165 Hz, in comparison to how the listeners categorized the speaker's gender in creaky voices

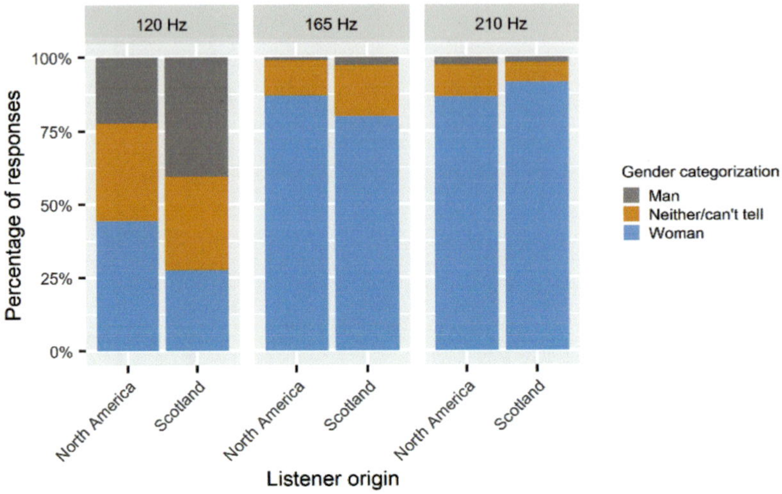

Figure 9.2 Stacked bar plot showing the percentage of Man, Neither/Can't tell, and Woman responses by f0 level and listener origin.

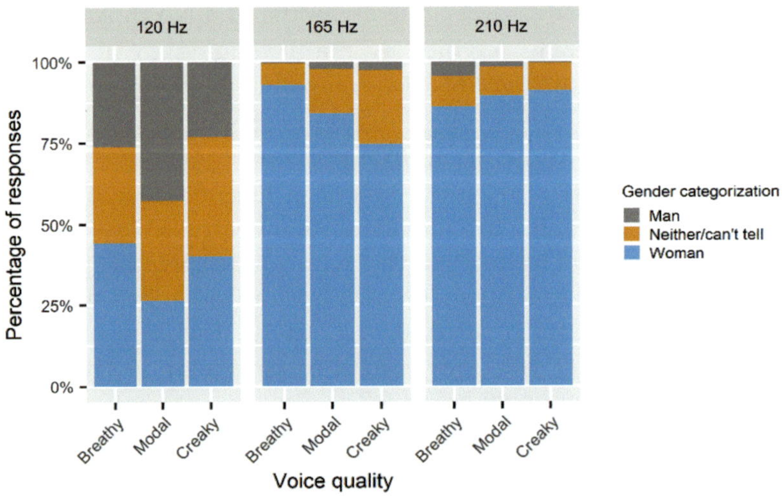

Figure 9.3 Stacked bar plot showing the percentage of Man, Neither/Can't tell, and Woman responses by f0 level and voice quality.

in relation to modal voices at 210 Hz. However, the listeners were significantly less likely to categorize a voice as man or neither/can't tell at 120 Hz, in comparison to how they categorized creaky voices in relation to modal voices at higher f0 levels ($B = -1.579$, $SE(B) = 0.507$, $p < 0.01$, $t = -3.114$).

Qualitative analysis of participant feedback

Pitch was the most mentioned cue that listeners mentioned paying attention to when categorizing gender and rating masculinity and femininity. Respondents reflected an awareness of how pitch affects gender perception, saying, for example, "If voice was deeper, it was more likely a man." The participants also reported paying attention to a wide range of other cues in the voice, among them "tone," "harshness," "breathiness," "vocal fry," "nasality," "intonation," "timbre," "raspiness," "warmth," "smoothness," "speech patterns," and "clarity."

Many participants also reported that rather than thinking about the speaker's gender in isolation, they imagined what the speaker might look like, compared them to real people they knew, and used this other information about the speaker to help them rate gender. One participant, for example, reported they "could often imagine a man or a woman [they] knew saying the words." Another participant reported they tried to "distinguish the age of the speaker," because "what might have been young female could well have been late adolescent male." Another listener reported paying attention to perceived sexuality, saying: "A lot of the ones I put as 'masculine' were because I could imagine gay men I know saying it in that tone/pitch/whatever it's called." Two participants noted specifically that some of the voices sounded like transgender people that they knew.

Some listeners commented specifically on the two different tasks given to them in the experiment, rating masculinity and femininity or categorizing the gender of the speaker, in their feedback. Some listeners noted that they did not see the different terms as equivalent, with one saying, "Sometimes a more masculine voice meant a less feminine one but not always so it was good to have the two sliders." Another participant said that "There should have been an option for 'women [*sic*] speaker but with a masculine voice'." However, for others these terms were more similar, with another participant saying, "I paid attention to the same things when deciding whether man or woman, masculine or feminine," a sentiment echoed by other participants.

Discussion

Creaky voice and perceptions of masculinity and femininity

I expected that Scottish listeners would perceive creaky voice as more masculine than modal voice. However, I found that listeners perceived creaky voice as less masculine than modal voice, and this was true for listeners overall, not just for North American listeners. However, listeners did not rate creaky voice any differently from modal voice in terms of femininity.

This suggests that for Scottish and North American listeners of Scottish English, the iconic link between creaky voice and masculinity has been reworked through indexical meanings of creaky voice. This is contrary to previous work where creaky voice was rated as more masculine than modal voice (Greer 2015; Lee 2016). In contrast to the previous studies of Greer (2015) and Lee (2016), the present study used only a female speaker to produce stimuli, and despite attempts to alter listener perception of her gender through manipulating the pitch of her voice, residual cues to gender in her voice meant that listeners overwhelmingly perceived the speaker as female. However, my findings support Yuasa's (2010) suggestion that the use of creaky voice by female speakers has led to a shift in the meanings of creaky voice, at least in the voices of speakers perceived as female, as I find that listeners rate creaky voices as less masculine than modal voices in a female speaker who was overwhelming perceived as female by listeners. It is possible that listeners would perceive creaky voice differently when used by male speakers. Although Lee (2016) did compare the perception of creaky voice in male and female speakers, her findings that creaky voice is perceived as masculine in female speakers are at odds with the findings of the present study, so future research should revisit this issue and directly compare how creaky voice affects perceptions of masculinity between male and female speakers.

Because Scottish listeners also perceived creaky voice as less masculine than modal voice, this could also suggest a change in progress in the gender differences in creaky voice in Scottish English. This prediction was in part based on production data from Scottish English suggesting that creaky voice is more common among male speakers. However, research conducted in London (Szakay and Torgersen 2019) suggests that the usage of creaky voice is increasing among young female speakers. If this trend has spread to Scotland, this may explain why creaky voice was perceived as less masculine than modal voice, at least by Scottish listeners. However, this is only tentative evidence and there are other potential explanations for this finding too. One possibility is that even if male speakers are still creakier in Scotland, Scottish listeners may not associate creaky voice in a female speaker with masculinity because of hearing creaky voice in female speakers in other varieties, or because of cultural conversations that have drawn attention to young women's use of creaky voice. As reviewed by Dallaston and Docherty (2020) and Davidson (2021), evidence that female speakers use more creaky voice than male speakers in American English is not consistent across different studies, so listener ratings of creaky voices may not necessarily relate to actual differences in production, but rather to ideologies about creaky voice circulated in the media that associate creaky voice with young, middle-class American white girls (e.g., Slobe 2018).

Listener ratings of creaky voice as less masculine than modal could also be connected to other indexical meanings of creak identified in previous studies, rather than directly to gender. Instead of investigating ratings of masculinity directly, Yuasa (2010) asked listeners to rate creaky voices on scales such as "confident" to "hesitant" and "aggressive" to "compliant" and found that listeners perceived creaky voice in women's speech as "non-aggressive" and "hesitant." This might help to explain why creaky voice only affected ratings of masculinity, but not femininity: Listeners may be associating creaky voice with other speaker traits, which they connect back to a lack of masculinity, rather than to femininity.

This finding that creaky voice leads listeners to rate voices as less masculine but not more feminine also suggests that these terms are not equivalent to all listeners. This was echoed in the participants' feedback on the experiment, where some participants appeared to see different dimensions of gender they were asked about as equivalent while others saw them as more separate. Future research should consider the ways in which the participants' understanding of gender may play a role in their responses in gender perception experiments in more detail.

Creaky voice as a cue to gender when other cues are ambiguous

As well as rating masculinity and femininity, listeners also categorized the gender of the voice they heard, as man, woman, or neither/can't tell. I expected that while creaky voice might affect how listeners perceived gender attributes (masculinity and femininity) at all f0 levels, it might only affect how listeners categorized the speaker's gender (man, woman, or neither/can't tell) at the "ambiguous" f0 level.

There was no overall effect of creaky voice on how listeners categorized the speaker's gender, nor did creaky voice affect how listeners categorized gender at the intended ambiguous f0 level, 165 Hz, in comparison to how they categorized gender at 210 Hz. However, listeners were significantly less likely to categorize a creaky voice as belonging to a man or as neither/can't tell than as belonging to a woman at the 120 Hz level, in comparison to how they categorized the speaker's gender in creaky voices at higher f0 levels. While this 120 Hz level was initially the intended cue for the perception of a male voice, it appears that because of residual cues to gender in the speaker's voice, 120 Hz actually functioned as the ambiguous level.

This finding has potential implications for voice therapy for transgender individuals, as it suggests that creaky voice can affect how listeners attribute gender to a speaker when other cues might be ambiguous. While these findings are only preliminary, voice therapists working with transgender clients, and transgender individuals working independently to alter the way their voices are perceived, might want to reflect on the use of creaky voice and consider trialing the increase or decrease in creak usage. The exact way in which, if at all, this affects how a speaker's voice is perceived may depend on how creak is used alongside other features of a speaker's voice, as well as the specific indexical meanings of creaky voice that exist in the speaker's linguistic and cultural context.

Sociolinguistic background and perception of other speaker attributes

One other factor that may affect the role of creak in gender perception is the sociolinguistic background of listeners. I had expected culturally specific gender differences in creak usage between North American and Scottish contexts might lead to creak playing a different role in gender perception. However, there was no evidence that this was the case. This could potentially be attributed to the fact that studies on creaky voice in American English and Scottish English are not directly comparable because they tend to use different methodologies for considering creak—studies looking at Scotland have used Vocal Profile Analysis, which is not typically used in studies on creaky voice in American English. Furthermore, as mentioned earlier, evidence of female speakers using more creaky voice than male speakers in North American varieties of English is not consistent; this may mean that listeners may be drawing on a wider range of indexical meanings of creaky voice than simply the amount it is used by male and female speakers in their own variety when answering questions about the gender of a speaker using creaky voice.

However, North American and Scottish listeners did perceive gender differently overall, with North American listeners being less likely to categorize voices as man or neither/can't tell than woman—compared to Scottish listeners, an effect that was stronger at lower f0 levels, and similar effects in terms of masculinity and femininity. This suggests that speakers of different sociolinguistic backgrounds can and do perceive gender differently in voices.

Other aspects of the sociolinguistic backgrounds of listeners may have also affected their responses. Participants reported a comparison of the voices to those of real people they knew, suggesting they brought their experiences as listeners with them. Some participants also commented specifically that the voices sounded like trans people that they knew; how much exposure a participant has had to trans voices may well influence the way that they think about gender in the voice, and thus how they categorize gender and rate gender attributes in voices. Recent works by Brown *et al.* (2021) and Hope and Lilley (2022) consider diverse gender samples and consider how transgender and non-binary participants rated gender in voices in comparison to cisgender listeners. While Brown *et al.* (2021) did not find any difference in how their cisgender as well as transgender and non-binary participants rated gender in voices, Hope and Lilley (2022) found that transgender and non-binary listeners were more likely to categorize voices as non-binary and used a wider expanse of rating scales. Hope and Lilley (2022) attribute this to participants having a more flexible conceptualization of gender, which in turn leads to increased perceptual flexibility. It is unclear, however, whether all participants in that category would have had a more flexible understanding of gender, and how this might have influenced their ratings independently of their familiarity with voices that did not conform to gender norms. A potential avenue for future work could be to compare the strategies of transgender and non-binary participants for rating voices with that of cisgender participants, comparing not only their ratings but also building on the qualitative aspect of this study to examine the reasons behind any differences. Asking the participants about, for example, their familiarity with voices belonging to transgender people, as well as asking about their understandings of gender, may help to disentangle how these different factors contribute to the identities of participants affecting or not affecting their ratings and categorizations of gender in an experiment.

The participants also commented that their responses relied on how they perceived characteristics such as the age and sexuality of the speaker. As creak can be associated with gay-sounding speech (Podesva 2007; Zimman 2013), as well as with young women specifically rather than older women, the way that listeners took other perceived characteristics into account when making judgments about gender may have affected how creak affected their judgments. It is clear from the comments of these participants that listeners do not perceive gender in a vacuum, but in tandem with other characteristics, bringing their experience as listeners with them. This should be taken into consideration in the design of future research on the role of creaky voice in gender perception.

Conclusion

The results of this study suggest that creaky voice can influence gender perception. Specifically, I found that creaky voice decreases the listener's perception of masculinity among Scottish and North American listeners and that listeners use creaky voice as a cue to the speaker's gender where other cues to gender in the voice are ambiguous. These findings suggest that the meaning of creaky voice for listeners is mediated by sociocultural ideologies about creaky voice, rather than being intrinsically linked to masculinity through its low f0.

Qualitative participant feedback suggests that listeners brought their understandings of gender and experiences of hearing real voices with them to the experiment. Taken together, these findings highlight the role of the participant's exposure to sociocultural ideologies and voices outside of the experimental setting when making judgments in gender perception experiments.

This was supported by findings that sociolinguistic background affected how listeners categorized gender across all voices and that creaky voice affected the listener ratings of masculinity, ratings of femininity, and categorization of gender in different ways. Together, these findings suggest that gender perception is a complex process and that the role of creaky voice within gender perception cannot be easily separated from the other speaker characteristics that creaky voice might index. Future research on the perception of creak might find value in taking a bottom-up approach, where listeners are able to respond freely to voices in the first instance, creating qualitative data that could be used to decide on rating scales and categorical response options for a second experiment. Taking this approach would forefront the role of listeners in the perception of creak, which may be key to understanding perceptions of creak if, as the results of this study suggest, the experiences of listeners are integral to the role creak plays in gender perception.

Endnotes

1. The full models for listener ratings of masculinity and femininity predicted masculinity and femininity as a function of the fixed effects of f0, voice quality, listener origin and naturalness of the interactions between f0, voice quality, and origin; and the random intercepts of sentence and participant.

   The full models were then stepped down automatically using the step() function of the stats package, which adds and removes predictors and compares each model using Akaike's information criterion (AIC) in a Stepwise Algorithm to choose an optimal model. The final models selected were as follows: 1. lmer(masculinity ~ voice quality + origin + f0 + naturalness + (1 | sentence) + (1 | participant) + voice quality:f0 + origin:f0); 2. lmer(femininity ~ voice quality + origin + f0 + naturalness + (1 | sentence) + (1 | participant) + origin:f0).
2. The full model predicted listener gender response as a function of the fixed effects of f0, voice quality, origin and naturalness; the interactions between f0, voice quality, and origin; and the random intercepts of sentence and participant. The full model was then stepped down manually by removing predictors and comparing the AIC of each model in a log likelihood ratio test using anova(). The final model selected was as follows: 1. glmer(Gender categorization ~ voice quality + origin + f0 + naturalness + (1 | sentence) + (1 | participant) + voice quality:f0 + origin:f0).
3. Listener ratings of femininity showed variance according to participant (Variance = 133.52, SD = 11.555) and sentence (Variance = 13.59, SD = 3.687). Listener rating of masculinity showed variance according to participant (Variance = 126.422 , SD = 11.244) and sentence (Variance = 2.792, SD = 1.671). Gender categorizations also showed variance according to participant (Variance = 2.9745, SD = 1.7247) and sentence (Variance = 0.4136, SD = 0.6431).

References

Abdelli-Beruh, N. B., Wolk, L., and Slavin, D. (2014). "Prevalence of vocal fry in young adult male American English speakers," *Journal of Voice*, **28**(2): 185–190.

Bates, D., Mächler, M., Bolker, B., and Walker, S. (2015). "Fitting linear mixed-effects models using lme4," *Journal of Statistical Software*, **67**(1): 1–48.

Beck, J. and Schaeffler, F. (2015). "Voice quality variation in Scottish adolescents: Gender versus geography," *Proceedings of the 18th International Congress of Phonetic Sciences (ICPhS 2015)*, Glasgow, Scotland.

Boersma, P. and Weenink, D. (2019). *Praat: Doing Phonetics by Computer*. [Computer program]. Version 6.0.49, Amsterdam.

Brown, K. M., Dahl, K. L., Cler, G. J., and Stepp, C. E. (2021). "Listener age and gender diversity: Effects on voice-based perception of gender," *Journal of Voice*, **35**(5): 739–745.

Dallaston, K. and Docherty, G. (2020). "The quantitative prevalence of creaky voice (vocal fry) in varieties of English: A systematic review of the literature," *PLOS ONE*, **15**(3): e0229960.

Davidson, L. (2021). "The versatility of creaky phonation: Segmental, prosodic, and sociolinguistic uses in the world's languages," *WIREs Cognitive Science*, **12**: e1547.

Davies, S., Papp, V., and Antoni, C. (2015). "Voice and communication change for gender nonconforming individuals: Giving voice to the person inside," *International Journal of Transgenderism*, **16**(3): 117–159.

Foulkes, P., Docherty, G., Khattab, G., and Yaeger-Dror, M. (2011). "Sound judgements: Perception of indexical features in children's speech," in D. R. Preston and N. A. Niedzielski (Eds.), *A Reader in Sociophonetics*, New York, N.Y.: De Gruyter Mouton.

Fox, J. and Weisberg, S. (2019). *An R Companion to Applied Regression*. Thousand Oaks, CA: Sage.

Gobl, C. and Ní Chasaide, A. (2003). "The role of voice quality in communicating emotion, mood and attitude," *Speech Communication*, **40**(1–2): 189–212.

Greer, S. (2015). "The perception of coolness: Voice quality and its social uses and interpretations," MA thesis, University of Calgary.

Henton, C. and Bladon, A. (1988). "Creak as a sociophonetic marker," in L. M. Hyman and C. N. Li (Eds.), *Language, Speech and Mind: Studies in Honour of Victoria A. Fromkin*, 3–29. Routledge.

Hlavac, M. (2018). "Stargazer: Well-formatted regression and summary statistics tables," R Package version 5.2.1. Rerteived from https://CRAN.R-project.org/package=stargazer.

Hope, M. and Lilley, J. (2022). "Gender expansive listeners utilize a non-binary, multidimensional conception of gender to inform voice gender perception," *Brain and Language,* **224**: 105049.

Klatt, D. and Klatt, L. (1990). "Analysis, synthesis, and perception of voice quality variations among female and male talkers," *Journal of the Acoustical Society of America,* **87**(2): 820–857.

Laver, J. (1980). *The Phonetic Description of Voice Quality.* Cambridge: Cambridge University Press.

Lee, K. (2016). "The perception of creaky voice: Does speaker gender affect our judgments?" MA thesis, University of Kentucky.

Leung, Y., Oates, J., and Chan, S. P. (2018). "Voice, articulation, and prosody contribute to listener perceptions of speaker gender: A systematic review and meta-analysis," *Journal of Speech, Language and Hearing Research,* **61**(2): 266–297.

Mazzoni, D. (2018). *Audacity(R): Free Audio Editor and Recorder* [Computer application]. Version 2.3.2. Audacity Team.

Ohala, J. J. (1994). "The frequency code underlies sound-symbolic use of voice pitch," in L. Hinton and J. Nichols (Eds.), *Sound Symbolism,* 325–347, Cambridge: Cambridge University Press.

Pearce, J. (2020). "Exploring the contribution of voice quality to the perception of gender in Scottish English," MPhil(R) thesis, University of Glasgow.

Podesva, R. (2007). "Phonation type as a stylistic variable: The use of falsetto in constructing a persona," *Journal of Sociolinguistics,* **11**(4): 478–501, doi: 10.1111/j.1467- 9841.2007. 00334.x.

Podesva, R. and Kajino, S. (2014). "Sociophonetics, gender, and sexuality," in *The Handbook of Language, Gender, and Sexuality,* 103–122, John Wiley Sons, Ltd.

Porter, C. (2012). "Voice quality and gender identification: Acoustic and perceptual analysis," PhD thesis, Dalhousie University.

Prolific. (2019). Prolific. Version July 2019. Oxford.

R Core Team. (2020). *R: A Language and Environment for Statistical Computing.* v.4.1.1. Vienna, Austria: R Foundation for Statistical Computing.

Skuk, V. G. and Schweinberger, S. R. (2014). "Influences of fundamental frequency, formant frequencies, aperiodicity, and spectrum level on the perception of voice gender," *Journal of Speech Language and Hearing Research,* **57**(1): 285.

Slobe, T. (2018). "Style, stance, and social meaning in mock white girl," *Language in Society*, **47**(4): 541–567.

Stoet, G. (2017). "PsyToolkit: A novel web-based method for running online questionnaires and reaction-time experiments," *Teaching of Psychology*, **44**(1): 24–31.

Stuart-Smith, J. (1999). "Voice quality in Glaswegian," *Proceedings of the 14th International Congress of Phonetic Sciences*, 2553–2556, San Francisco, United States.

Syrdal, A. K. (1996). "Acoustic variability in spontaneous conversational speech of American English talkers," *Proceeding of the Fourth International Conference on Spoken Language Processing, ICSLP '96*, 438–441, Philadelphia, United States.

Szakay, A. and Torgersen, E. (2019). "A re-analysis of f0 in ethnic varieties of London English using REAPER," *Proceedings of the 19th International Congress of Phonetic Sciences*, 1675–1678, Melbourne, Australia.

Van Borsel, J., Janssens, J., and De Bodt, M. (2009). "Breathiness as a feminine voice characteristic: A perceptual approach," *Journal of Voice*, **23**(3): 291–294.

Venables, B. (2021). codingMatrices: Alternative Factor Coding Matrices for Linear Model Formulae. R Package version 0.3.3. https://CRAN.R-project.org/package=codingMatrices.

Weenink, D. (2009). *The KlattGrid Speech Synthesizer*. Amsterdam.

Wickham, H. (2016). *ggplot2: Elegant Graphics for Data Analysis*. New York: Springer-Verlag.

Wolk, L., Abdelli-Beruh, N. B., and Slavin, D. (2012). "Habitual use of vocal fry in young adult female speakers." *Journal of Voice*, **26**(3): 111–116, doi: 10.1016/j.jvoice.2011.04.007.

Yuasa, I. P. (2010). "Creaky voice: A new feminine voice quality for young urban-oriented upwardly mobile American women?" *American Speech*, **85**(3): 315–337.

Zimman, L. (2013). "Hegemonic masculinity and the variability of gay-sounding speech: The perceived sexuality of transgender men," *Journal of Language and Sexuality*, **2**(1): 1–39.

Chapter Ten

Gilmore Girls, cowboys, *bon viveurs*, and Lacanian psychoanalysis

Vocal fry's significance as symptom of object *a* (capitalist semioblitz, masculine menace, and ineradicable excess)

Tristam Vivian Adams

Introduction

This chapter argues the unbearable silent object *a* voice is at the core of vocal fry; attendance to vocal fry is symptomatic of object *a* voice. To begin, vocal fry is examined in two registers. Initially, the significance creaky voice (vocal fry) plays in verbal velocity and rhythms in the context of highly technologized "semioblitz" of capitalism is explored (*Gilmore Girls* being a key example).

Creak: Theories and practices of pulse phonation
Edited by Francesco Venturi

ISBN 978-981-5129-29-8 (Hardcover), 978-1-003-61606-1 (eBook)
www.jennystanford.com

This initial step is a development stemming from an earlier paper "Accelerations and speed limits: An essay on the vocal limits of semiocapitalism" (Adams 2018), which explored vocal fry in the context of Franco Berardi's notion of "semiocapitalism" in *Precarious Rhapsody: Semiocapitalism and the Pathologies of the Post–Alpha Generation* (Berardi 2009). Second, the role of vocal fry in male characters is explored in films such as *No Country for Old Men* and *The Talented Mr. Ripley*. Third, the text explores, by way of Mladen Dolar's examination of spirit and voice in terms of the statue of Memnon, and Adriana Cavarero's survey of the status of voice in Plato and Aristotle, how and why we seek significances in the creaking edges of voice. This leads to the argument that the preceding two readings are symptomatic of the problem of thinking the Lacanian object *a* voice—a voice of ineradicable excess, a terrible remainder, of which one strategy of denial, one obfuscatory comfort blanket, involves clothing the creaking horizon of voice with layers of significance. Arguments from parts one and two are thus reframed as the want to wrap voice in meaning rather than confront its silent asignifying core: object *a* voice, the extimate of body and meaning.

Stethoscopes and canaries: I *know* that sound

Several theorists, thinkers, and writers in many disciplines have attended to voice in terms of significances and meanings of vocal characteristics and traits. One trend across many literatures is the presupposition that vocal characteristics divulge something hitherto unknown, hidden, or unspoken. Voice is frequently framed as a betraying "tell": those that make deductions and/or draw out inferences operate in degrees of an empirical "stethoscopic" vein. The ear is pressed up against a sound, and characteristics of the sound are suggested to tell something of the history and emotional or physical state of the voicer. The tether between sound and what it tells us, what one presupposes the voice to betray, runs the spectrum of being well founded at times (as many rigorous studies demonstrate) and fantastic—of fantasy—in other turns (as many presuppositions around voice are). What is at stake in the question of attending to the significances of voice is not so much *what* voice is divulging, but *what and how we ascribe causalities to the unseen cavern that produces voice*. This is, on many levels, the crux of voice explored in questions Chion frames on the relation of the *acousmatic* and the *acousmêtre* (1999: 9; 169). *"What's that sound?"* is a rich seam explored in film and literature countless times (e.g., the opening pages of Wyndham's *The Day of the Triffids* (2000), Calvino's "A King Listens" (2009), or Schwob's "The King in the Golden Mask" (2017). Beyond the veil of labyrinth partitions, beyond the Pythagorean curtain (Dolar 2006: 61), the source of voice is forever occluded. The "origin" of voice is *always acousmatic; it has no specific origin*—nonetheless, as one attends to voice, we seek to fathom the causes and effects behind its chimerical oscillating qualities. Listen to the spilt dominoes: discern consequences; assume causalities; speculate; mythopoiesthesize; fantasize; dream. Such is it to wonder about voice.

Anne Karpf, in *The Human Voice: The Story of a Remarkable Talent*, catalogues several stethoscopic presuppositions whereby voice is taken as a canary-in-the-cave forewarning of malady. Karpf writes: "voice is a stethoscope, and transmits information not only about anatomical abnormalities but even illnesses" (2007: 10). Much later in the book, Karpf lists a number of ways in which stress, trauma, or illness impacts vocal and speech production:

> When we're stressed or excited [...] our laryngeal muscles tense up, making the vocal folds tauter, so that the speaker has to produce more pressure to force the air through. The vocal musculature is a highly sensitive instrument. [...] Emotional states like deception, conflict and anxiety can change breathing, which in turn influences subglottal pressure and so impacts voice. [...] Simply remembering an emotion – a happy event, the shock of an accident – affects the movement of the diaphragm. [...] Changes in facial expression also affect the pharyngeal muscles. In a grimace, the corners of the mouth are turned down, the vocal tract shortened and its walls tensed. This helps makes the voice higher, more nasal, and narrower. When a person is apprehensive or fearful, their voice 'shrinks' as the mucous membranes become dry. (Karpf 2007: 134–135)

This is not without merit, for the production of voice is a precarious harmony: It is a fragile culmination (an orchestration easily disturbed or affected) of complex synchronicity. Robin Dunbar, the British anthropologist and evolutionary psychologist, in *Grooming, Gossip and the Evolution of Language*, describes the precarious delicacy of vocalization in terms of sequence, rhythm, and temporal coordination:

> Speaking requires very fine motor control of the lips, tongue, vocal chords and chest, all of which have to be *integrated in just the right sequence* to produce a particular sound. […] In addition to fine motor control, speech also requires precise control of breathing so that air is released from the lungs in just the right quantity and with just the right amount of force (1996: 133, emphasis added).

It is notable just how many component parts work together under the umbrella term "vocal tract" to produce voice. Salted speech marks are added to *vocal tract* here because voice does not have an organ of its own. French composer, film and sound theorist Michel Chion touches on this briefly and powerfully in *The Voice in Cinema* when he writes: "it paradoxically appears that the human body does not have a specific organ for phonation" (1999: 127). Adriana Cavarero, in *For More Than One Voice: Toward a Philosophy of Vocal Expression*, details with more sustain the volumes of body components co-opted for voice: "lips, mouth, palate, tongue, teeth, […] larynx, nasal cavities, lungs, diaphragm—come together for acoustic purposes" (2005: 65). Each of these parts has a more vital primary function. Lips, mouth, palate, tongue, and teeth specialize in consumption and nourishing activities. The larynx, commonly called the "voice box," guards the trachea from food aspiration, preventing foreign matter entering the windpipe and thus protecting us from choking. Nasal cavities, the lungs, and diaphragm take care of the ever-necessary requirement of respiration for the body to live. Voice co-opts this array, interrupting their primary functions, in service of its own sound. Many parts of the corpus are employed briefly in a momentary secondment (from functions crucial to maintain life) to crucial to contribute to the precarious harmony of producing voice.

Dunbar's line, concerning the timing and synchronicity of voice's contributing component parts, not only impresses a render of voice as a finely tuned motor-neurological exercise but also underscores the temporality of vocal production—if one player is off-rhythm, the whole delivery is ruined. Vocal production must occur in just the right sequence, with just the right timing, to produce intended sounds. Contingency to sequence, timing, and productive unison plays into a number of inferences whereby vocal qualities betray malady. Karpf's citation concerning the connection between heart disease and vocal speed ("risk of coronary heart disease can be predicted, it's been claimed, on the basis of voice characteristics like volume and speed alone") (2007: 10) is one of many causal tethers running along this vector. Beyond physical malady, the same connection, the same stethoscopic eavesdropping, is at work in studies of mental illnesses such as depression. Deceleration in speaking delivery is a general symptom of depression. Karpf writes of how "[d]octors have even maintained that picking up on changes to the sound of the voice can be life–saving in cases of throat cancer and people contemplating suicide" (2007: 10). Karpf also references a paper by France *et al.* "Acoustical properties of speech as indicators of depression and suicidal risk" (2000: 829–837) that examines "suicidal speech." Ellgring and Scherer write in the *Journal of Nonverbal Behavior*: "rate or tempo of speech goes down" (1996: 84) in depressed patents. The monitoring of speed, the speaker's ability to articulate words at a certain frequency, is also cited in terms of the maternal voice by Karpf, who writes: "The voices of depressed mothers are more likely to be flat, and they usually talk less than non–depressed mothers, using shorter utterances as well as longer pauses" (2007: 77). The effects of depression on speech and voice have established causal connections with long histories. Buyukdura *et al.* state that "[p]sychomotor retardation has been characterized as a major feature of depression since antiquity," before referring to how both Hippocrates and Aretaeus of Cappadocia "described psychomotor retardation as a characteristic of depression" and Darwin who "also discussed visible psychomotor symptoms and depressed patients" (Buyukdura *et al.* 2011: 2).

In much the same casual structuration, there are studies concerning "pressured speech" as a symptom of anxiety, bipolar disorder, or types of schizophrenia. Pressured speech is characterized generally by rapid speech that does not pause for others to speak; disorganized argumentative articulation; "above normal volume for the situation"; jokes; and urgency, among other characteristics (Fletcher 2017). Although there is some credence to the contingency of vocal expression to physical and/or psychological illnesses, the context of such inferences needs to be considered with caution. What is a "normal" volume for a situation; what is too slow or too fast. What are the ideal speaking cadence, pitch, volume, and tone for a situation? Such questions highlight the risk of reductive approaches to finding significances in vocal characteristics—they also illustrate how, for every well-meaning inference concerning voice, there is also a judgment of what standard is, what the ideal voice is, by a certain group of specialists working in disciplines that have their own existing challenges of homogeneity, access, Western-centric, and Anglophone inequalities. And this is before one touches on the additional shifting of speech and voice by conscious affect or unconscious mirroring. This is all to say that as much as voice and speech qualities may point to certain maladies, psychological and/or physical disturbances, the attention and attendance of said qualities are contingent, by varying degrees, to an ethics of what another's voice ought to be. Too fast, too slow: *for whom*?

This is not a criticism of the empirical sciences by a humanities theorist—the work to better understand maladies such as depression is necessary and urgent. And, perhaps in some corners of interdisciplinary fields, where the rhizomes grow, credit is seldom afforded to the caution and rigor science papers exercise in presenting data. The—perhaps dissatisfying for the lay reader—anticlimactic caveat of "more studies required" is testament to such caution. Furthermore, this care of the empirical sciences does not translate well to popular media representations of the studies: what *might* point to or *may* signal on one peer-reviewed page *proves* on a baiting platform's garish regurgitation.

There is an established causal connection between voice and the statuses of the body and mind. Many studies of voice pursue the correlations of vocal attributes to departures from standard norm, vocal "disruptions," and aberrations. The premise of such correlations is that vocal characteristics are a register of environmental impacts upon the corporeal and psychological health and stability of the human animal.

In this turn, rather than ripping the voice–mind–body causality up in the service of a papier-mâché humanities narrative, I propose the opposite: that an appraisal of voice's significance in regard to the vessel it sounds from always runs into the fundamental philosophical tension of lack and excess. Adriana Cavarero's analysis of Alcibades' likening of Socrates to a *Silenus figure* and Mladen Dolar's recourse to puppets, inanimate vessels, and a singing stone (e.g., Kempelen's *die Sprech–Maschine* (2006: 6–11) or the *Statue of Memnon* in Egypt (2008: 3–7)) both explore this tension: insofar as setting a thought of voice upon its germinal ambiguity, by turns wondrous and unnerving: a voice coming from only *somewhere*, a lacuna, a void, is never enough and one must strive to divulge, unveil, understand or fathom the mysteries of voice. It is through an unmasking of significance, the unveiling of epistemological precarity, that one can begin thinking the nature of object *a* voice. But before all of these, let us consider what voice requires of its vessel.

Asphyxiation, DOOM, and the torture house of language

Steven Pinker notes: "Syntax overrides carbon dioxide: we suppress the delicately tuned feedback loop that controls our breathing rate to regulate oxygen intake, and instead we time our exhalations to the length of the phrase or sentence we intend to utter" (1994: 164). Our respiratory rhythm, the automatic inhale and exhale that sound softly against milky flocculent pillows aglow in twilight as we dream, does not take precedence over the pneumatic requirements of the lilt of language. Syntactical asphyxiation.

Speaking voice (and singing voice indeed) is *always already* steeped in doom—the sounding of a body employed for semiotic and/or aesthetic sonic transmissions over vital biological necessities. The hyperbole of "All this talking is killing me!" is apposite on two levels. First, the aforementioned orchestra of seconded organs and biological processes neglect primary life-support mechanisms in the service of voice. This is the irony of the so-called "voice box"—ironic in the odd nomenclature in that it is not an organ with a primary function of producing voice; irony in a visceral register: metallic irony, the taste of blood wrought from hurt. Second, the rhythmic syntactical requirements of language and song impose an operation that is a degree of asphyxiation—interrupting breathing.

Let us move, now, from physics to metaphysics. In regard to the violence of vocalization, one can regard voice as a *corp–aural* register, a sounding out of both metaphysical and physical distress, of the human animal subjected to, ensnared in, what Jacques Lacan touches on as man subjected to the torture of language. In his famous commentary on Freud—"An Address: Freud in the Century" from May, 9, 1956; printed in the Alain-Miller-edited *The Psychoses 1955–1956: The Seminar of Jacques Lacan Book III*—Lacan—"writes": "Psychoanalysis should be the science of language inhabited by the subject. From the Freudian point of view man is the subject captured and tortured by language." (1997: 243). Žižek refers to this moment of Lacan's thought using the term "the torture house of language" (2008: 4). Žižek elaborates this torture of the subject imprisoned by language:

> the entire psychopathology deployed by Freud, from conversion–symptoms inscribed into the body up to total psychotic breakdowns, are scars of this permanent torture, so many signs of an original and irremediable gap between subject and language, so many signs that man cannot ever be at home in his own home. This is what Heidegger ignores: this dark torturing other side of our dwelling in language. (2008: 4)
>
> Throughout his work, Lacan varies Heidegger's motif of language as the house of being: language is not man's creation and instrument, it is man who 'dwells' in language: 'Psychoanalysis should be the science of language inhabited by the subject' Lacan's 'paranoid' twist, his additional Freudian turn of the screw, comes with his characterization of this house as a torture–house: 'In the light of the Freudian experience, man is a subject caught in and tortured by language.' (Žižek 2012: 870)

Žižek's focus on this line is erudite, for the notion of the human animal, the flabby scarred body from which desire-vectors lash, bound, restrain, and strangle—like the butcher-bound meat from the butchers (*"I've just been to the butchers"*)—is a recurring theme in Lacan's animations of Freud. (See Lacan 1997, 48–50.) The notion of language as a torture the subject is ensnared within,

Lacan's paranoid twist, is something the Italian philosopher Lorenzo Magnani emphasizes in particularly physical terms:

> [A]cknowledging Lacan's ideas, Žižek tells us that it is man that 'dwells' in language in a sort of constitutive passivity, and not language that dwells in humans: 'man is a subject caught and tortured in language', language is the 'big Other', and, again, human animal does not 'fit' language: [...] An eloquent explanation of this kind of violent effect of language, which constitutively affects the individual psyche, is illustrated by the case of mutilations, where parts and/or aspects of the real body suffer—so to say—from the 'signifier' effected by the language activity. The semiotic pressure established by language constrains the body [...]. (2011: 51)

In Magnani's implementation of Žižek's writing of Lacan's "torture house of language," the physical constraint, suffering—the brutalizing facet of language—is elaborated. The body suffers as semiotic forces constrain and bind the body: recall the imposition of speech on rhythms of autonomic respiration—the constraints of language imposed on breathing are a low-level asphyxiation: semiotic strangulation. Is the vocal fry, then, to be heard, attended to as a creak, as a sounding out of the pressures and forces the speaking animal is subject to?

Creaks of semiocapitalism

Although Lacan is not merely referring to vocalization or speech when he uses the term language (he is speaking of language in the wider sense of human's ensnared in a signifying matrix), speech and voice are, I propose, symptoms of living amid the pressures and forces of our semiotic maelstrom. That is to say voice and speech can betray the scars, ruptures, fissures, and maladies of the human animal trapped in a signifying matrix. More specifically, by way of an application of Berardi's articulation of the tensions and torsions between the human animal and communicative technologies in the context of semiocapitalism, voice—in this turn—is posed as symptomatic of exposure to intensifying signifying forces. This is an acute take, to consider how being subject to language in turn affects and registers as symptom in the very execution of anthropoid vocalization. Does overexposure to language affect the biological protocols of human vocalization? If humans are trapped in the torture house of language, what symptoms develop in our voices as language intensifies?

Franco Berardi, in *Precarious Rhapsody,* introduces the notion of semiocapitalism—that is, capitalism as defined by semiotic pressures upon the organism, a world or excessive significance, signs, and communicative pressures. He writes that "the kind of overproduction manifest in semiocapitalism is specifically semiotic; an infinite excess of signs circulates" (2009: 108). And that semiocapitalism is "characterized by an excess of speed of the signifiers" (2009: 111). Perhaps the most productive method of framing this notion is by drawing out the relationship of four of Berardi's key terms he employs to draw out two concepts: Infosphere/Cyberspace and Psychosphere/Cybertime. The Infosphere and Cyberspace are expanding, getting faster, denser, more complex, and detailed; it is the world of technologically afforded information transmission and dissemination. It could be radio, televisual, or Internet based—any communicative technology.

A purpose of Berardi presenting the reader with this conceptual juxtaposition is to highlight a key difference of capacity between Infosphere/Cyberspace and Psychosphere/Cybertime. The former never ceases expansion; it is a curious facet of online media, social media, torrents, streaming pornography, e-commerce, and Internet archives and encyclopedia (and much else). Saturation never arrives; these realms have certainly *felt* saturated ever since one can remember, but saturation never comes. There is *always* Moore—a never-ending exponential growth in data capacities. Limitless. The latter, however, is contingent to the human animals' collective psychological capacities and emotional and cognitive attention time. Exposure to the infinite scroll of online media is an experience of at once excess and lack—insofar as there is always already an excess of stimuli one frantically consuming while always already being behind, late, second, third, nth hand. The symptom of logistic commitment apathy in chronic social media users resonates with online life. One is always late. Cybertime is not only inadequate for the content of the Infosphere/Cyberspace, but engagement is necessarily one of lack and lag—which is to say the temporalities of Cyberspace and Cybertime are not only asymmetrical in capacity but not parallel in temporal terms. Anecdotally, when I first started using the Internet, it was I who waited for it. Waiting through the metallic Rickenbacker twang of dial up, waiting for images to load top-down segment, waiting for text to load or videos to buffer. All that is gone. Now I am the one that cannot keep up, not matter how much I scroll down. Although the mediascape of the Infosphere has always felt saturated, there was at least some speed limit to consumption. Today, gratis of 5G and superfast fiberoptic broadband with great speed, it is beyond our abilities to consume. One is no longer inconvenienced by lag or buffering on desktop devices, seldom on portable oblong portals. We are the lagging agent in this relationship. In this reversal of roles, our waiting, our hopeful anticipation for technology, has been inverted into a hopeless doom. FOMO Sisyphus.

What happens as we accelerate to keep up? Voice is one register of our exposure to the vertiginous velocities of the exploding Infosphere. Berardi references a study by Richard Robin:

> Evidence suggests that globalisation has produced faster speech emission rates in areas of the world where the Western mode of transmission of signs has come to replace traditional and authoritarian ones. For instance, in the ex-Soviet Union the speed of transmission measured in syllables per second has almost doubled since the fall of the communist regime: from three to almost six syllables per second; similar findings reached the same conclusions in the Middle East and China. (Robin in Berardi 2009: 112)

Robin's study suggests there is a relationship of syllabic speed to capitalism; that the pressures of competition and production are reflected in the speed of speech. The accelerating demands of semiocapitalism that exploit our communicative and cognitive capabilities are reflected in syllabic speed. We are in the business of talking faster. Increased syllabic intensity is a symptom of the pressures and demands of semiocapitalism.

Television, in particular, plays a double role acceleration; characters speak at increasing rates for a number of reasons. The diegetic glossocelerants act out symptoms of semiocapitalism and, in turn, play a role in the catalytic feedback loop of exposing viewers to increasing dialogue tempos. Accelerating dialogue and semiotic intensity have been employed to "make [...] characters appear more intelligent and appeal to younger viewers" (Plunkett 2002). Accelerated syllabic delivery is also transmitted for exactly the same purpose as Berardi describes as characterizing semiocapitalism: *increasing communicative transmissions. Gilmore Girls* (2000–07) is an example of this and has gained notoriety and, in turn, traded on its fast-paced dialogue. As Ken Honeywell comments, this is partly due to episode script length: "while scripts for most TV shows of that length are 40–50 pages long, Gilmore Girls scripts were known to hit 80 pages" (2011). *Gilmore Girls* fits almost twice the semiotic verbal content into its timeframe compared to its chronological equivalent. A greater semiotic blitz is compressed, intensified to fit a finite timeframe. The show's promotional tagline attests to this: *Life's short. Talk fast.* Lauren Graham, who plays Lorelai, the central character in *Gilmore Girls,* references an increase in dialogue per show in her autobiography. Emphasis is added to this line to resonate with Moore's Law:

> For you speed nerds, note how over the course of this year, the pace of the show *increased exponentially,* and everyone began talking a whole lot faster. This was around the time when we started to be known for that. Watch the first and last episodes back–to–back, and it's super–evident.[1] (Graham 2017: 75–76)

Humorous delivery harbors a proclivity for increased speech rates. *Gilmore Girls* contains highly accelerated syllabic frequencies and a frenetic ricocheting dialogue between characters; the show's necessity to compress and intensify dialogue into a standard-length show is at once a symptom of increasing rates of semiotic transmissions in televisual media and a diegetic character trait serving much of the comedy. Kreiman and Siditis (2013: 300) have noted:

> One exception to the dearth of scientific study of prosody in humour comes from a study analysing speech from a well–known television sitcom, *Friends* (Purandare and Litman, 2006). Speech samples were selected immediately preceding artificial laughter (designated as 'humourous') and compared with other ('nonhumourous') speech in various measures of tempo, pitch and energy. Even when gender and speaker differences were taken into account, humourous turns were faster. (2013: 300)

Frasier (1993–2004) also contains accelerated speech as a direct result of increasing communicative content, but the accelerating impetus is not by virtue of scriptwriters, directors, or in the service of characterization or comedy. It is symptomatic of greater semiotic content being crammed into a fixed timeslot, but not the semiotic content of the show. Anne Karpf references an article by Aaron Barnhart, "Speeded–up Frasier gives KSMO extra ad–time." The article describes how a Kansas TV station sped up Frasier episodes to fit more adverts into the slot:

> American television now routinely speeds up sitcoms and compresses speech in order to fit in more ads. If "Frasier" seems to talk faster on one affiliate station than on another that's probably because he is talking faster – the station has accelerated the episode. (Karpf 2007, 44)

Gilmore Girls, Frasier, Friends, and many other shows are manifestations of the Richard Robin study cited by Berardi in a dramatic microcosm. Syllabic speed and vocal tempos are intensified and accelerated as a result of pressures and requirements to fit ever more communicative and semiotic content within a fixed broadcast window (be they diegetic, as in the case of *Gilmore Girls,* or extra-diegetic and symptomatic of a wider media platform reliant on squeezing more advertising revenue from a given timeslot, i.e., *Frasier*). The same trend is evident in advertising. Kreiman and Siditis write that "[f]aster speaking rates generated more emotional responses from the listeners than did slower rates [...] Ads that combined a high syllable speed with low pitch were better liked, and enhanced attitudes toward the product" (2013: 380). Television shows are accelerated for effect, dialogue compression, and freeing up airtime for additional advertisement placements. The advertisements themselves quicken speech rates not just to communicate greater product information but also to elicit more emotional responses in the receiver.

Expanding on Burroughs' line, "The word is now a virus (...) It is now a parasitic organism that invades and damages the central nervous system" (2014: 56). Speech speeds accelerate in a feedback loop; in interaction, speech rhythms drift toward a synchronicity (Karpf 2007: 277–281). Such syllabic rhythmic convergences are mostly a subconscious (Warren et al 2006:

13074) symptom of speech's rooting in gesture: We unconsciously match our speech rates to the other as other animals such as primates match lip-smacking rhythms and tempos (Morill *et al.* 2021). This subconscious contagion of accelerating speech gains greater significance in the din and glare of high-velocity media; to return to Berardi, we can consider a Rose Golden line from 1975: "For the first time in human history, there is a generation that has learned more words and heard more stories from the televisual machine than from its mother" (Berardi 2009: 9). Jukes, McCain, and Crockett also comment on the difference in online exposure between generations:

> by the time they're 21, the digital generation will have played more than 10,000 hours of video games, sent and received 250,000 emails and text/instant messages, spent 10,000 hours talking on phones, and watched more than 20,000 hours of television and 500,000 commercials (and most assuredly these estimates are on the extreme low side). Almost none of these experiences our parents or we had while we were growing up. (2010: 29)

We live in a morass of influencing exposures—audition alone constitutes being part of a conversation, a conversation that affects our speech. As speech accelerates, the delicate cognitive and neuro-motor protocols of vocal production must be orchestrated with increasing precision (imagine an orchestra repeating a phrase of "The Birdie Song" with ever-increasing tempo). Humans have limits (we are not limitless like technologies we are subjected to). The vocal fry often serves as a hesitancy in North American and European English—playing much the same pausing role as "y'know," "um," or "like." Significantly, the vocal fry is used *between words,* as a phonic pause, often at "the end of sentences" (Wolk *et al.* 2012: 114). It "serves as a marker of syntactic boundaries" (Ibid: 112). The vocal fry interrupts as a temporal glitch, a "gravelly hesitation" in rapid speech delivery (Luckwaldt 2013).

In the context of the accelerating demands of semiocapitalism on the human, the vocal fry is proposed here as a creaking register of the human speed limit being reached. It is the sounding of the disjunct between the finite and human animal and the limitless and accelerating Infosphere of semiocapitalism: human buffering. Berardi writes of a "hyper–excitement [...] that leads to depression" (2009: 116). The creak sounds precisely at the limit of speech, psychomotor speed, and cognitive capacities. As syntax overrides respiration, as we struggle with a cognitive–motor–syllabic pile up, we croak. The creaking vocal fry: the human-phonic equivalent of a grinding Window hourglass or Apple rainbow spinning wheel.

Creaky menace

The vocal fry does not always occur as hesitancy between words but can constitute a large proportion of the tenor of speech. Particularly in the burgeoning Neo-Western style of films, the overriding characteristic of an actor's dialogue or voiceover's commentary is the presence of a gravelly and hoarse tone. An example of this is evident in *No Country for Old Men* (Coen Brothers 2007), based on Cormac McCarthy's 2005 novel of the same name. The lines of the film's opening monologue—"I was Sheriff of this county when I was 25 years old, hard to believe. My grandfather was a lawman, father too..."—the words of Ed Tom Bell (played by Tommy Lee Jones), are delivered as a voiceover. The screen is dark, and slowly images of the American West vista at dawn (or is it dusk?) impart some color and light to the screen. The load of such a scene needs little cultural capital to establish a preliminary level of import. Ed Tom Bell is an old and hardened lawman, weathered by a harsh world—there is a patina, a drawl: a stoic fatigue. We are on the cusp of invoking Barthes' *grain of the voice* here. We are also tracing traditional American masculinity tropes—but again, the discussion on the seam between Cowboys in Hollywood Western films vis-à-vis gender and masculinity requires no further commentary in this chapter, except to note that these tropes, especially in Neo-Western films like those by the Coen Brothers, who are among the most notable specialists, undergo a deliberate theatricization. The leathery creak of the world-weary Cowboy or Lawman is exaggerated just enough to be registered as theatrical fantasy rather than a pretention of realism. Later in the film, the affectless doe-eyed antagonist, Anton Chigurh, played by Javier Bardem, stops at a roadside

Texaco "gas station." The cashier enquires if he might be "getting any rain up your way?" Chigurh, bristling, replies "What way would that be?" The line, and many subsequent, is croaked out almost exclusively in the vocal fry register. Were one to watch the film with a sound system with a sub-woofer or have the opportunity to view it at a cinema, Anton's voice would pierce the sleepy silence of the gas station with a flutter of sub-bass—a raven, amplified to Chinook, effect. In contrast to the opening monologue of Ed Tom Bell, vocal fry is employed here to different effect: menace and the simmering threat of lethal violence. From Chigurh's initial upset at the cashier's innocuous small talk concerning the weather, the scene descends via a series of offenses Chigurh sustains from small talk inexactitudes, each one—"'now' is not a time"—smarting Chigurh till he asks "What's the most you ever lost in a coin toss?" The cost of the bet is plain; we have already witnessed Chigurh glibly and apathetically murdering all he meets with a captive bolt stunner.

The presence of vocal fry in scenes of menace and intimidation is not novel to the Western or Neo-Western film genres. Its prevalence as a deliberate performative affect to impart sober severity into a sentiment is partly to the need of lowering both tone and volume; a vocal fry is seldom heard at higher volumes. Rob Brydon and Steve Coogan's attempt to chart the deterioration of Michael Caine's voice in various roles as he ages is an effective performance of this. As they try to mimic his grave tenor in scenes across his career, both succumb to vocalizations that are almost exclusively in the vocal fry register (BBC 2010; Madman Films 2014). Coogan refers one mode of Caine impression as a "broken voice," and Brydon refers to the croakiness of late-Caine tenors as "coming down several octaves" due to all the "brandy and cigars." There is, in *The Talented Mr. Ripley* (Minghella 1999), an instance of creaky vocal patina manifesting at the thematic confluence of traditional male intimidation and class superiority. Vocal fry imparts at once low-tone menace and a patina of the privileges of a bon-viveur. Tom Ripley, played by Matt Damon, is aboard Dickie Greenleaf's boat with Marge Sherwood and Dickie's friend Freddie Miles (played by Philip Seymour Hoffman). When Dickie (Jude Law) ducks below deck for an amorous interlude with Marge (Gwyneth Paltrow), Freddie catches Tom glancing down into the cabin. Mingella provides the viewer with Tom's vantage; we see, reflected in a mirror, the bare entwined legs of Dickie and Marge, swim trunks toed off feet. Tom is *perving*. At this Freddie, who has long suspected Tom of being a clingy impostor (infatuated with Dickie, Dickie's life, and his privilege), takes his first opportunity in the film to call him out. "Tommy! How's the peeping?" Then again, croaked: "Tommy… How's the peeping? Tommy, Tommy, Tommy, Tommy…" The second line is entirely in the vocal fry register; it is at once menacing and taunting. The obnoxious and entitled Freddie is exposing Tom on a number of painful levels—his desire for *an idea* of Dickie, his curiosity, but also, on a more fundamental level, his desire to be a part of that world: "the repeated sidelong glances he casts at Dickie betray not a desire to have him, but to be like him" (Žižek 2003).

Later in the film, Freddie finds Tom staying in what Tom claims to be Dickie's apartment. Freddie throws his coat onto the edge of the open piano; the action cuts at the difference between Dickie and Tom—the former carefree; the latter prim, precious, uptight, and materialistic. He begins stabbing a key of the piano, riling, goading, Tom's preciousness, his uptight reverence for materials that signal his vision of a status beyond his own. It is a *Crime and Punishment/Colombo* reverse whodunit. The viewer knows what Tom did and also knows both parties know one another's game—the relish is in the cat "n" mouse; the squirm and strafe of their orbiting, their sparring. Freddie asks, in a much lower vocal fry tone, if the apartment came furnished: "it doesn't look like Dickies'… it's horrible isn't it?" He croaks, "it's so… *bourgeois…*"

Interpreting interpretation

Dolar, in "The Voice and The Stone: From Hegel to Beckett" (the title's first part would lend itself well to weird horror fiction), describes, by way of Hegel's *der Werkmeister* in *Phenomenology of Spirit*, the statue of Memnon. The statue, "as the story goes, produced a sound when the first rays of light touched it in the morning. The stone emitted a voice, it was a vociferous stone, a stone made vocal." (2008: 3). In the next paragraph, he goes on to frame this as a marvel, astounding "spectators"—notably, in this short passage, he uses the term uncanny twice, the second time referring to the sonic phenomenon as "an uncanny inhuman voice" (2008: 4). Dolar continues:

> But what does it say, this voice? What message does it convey? For Hegel, it says that even the inert stone has an interiority, that it is not simply equal to itself, it appears as the inert matter split in itself, it is the birth of the voice out of the spirit of the stone, or rather, more to the point, the opposite: *the birth of the spirit out of the voice of the stone*. The emission of the voice as if spiritualizes the stone. This voice has no content, what it says has no meaning—except for the meaning as such, the birth of meaning at its minimal, a minimal movement of interiority being exteriorized, a selfless thing producing the empty form of a self. There is nothing more selfless than a stone, yet it produces a selfless self, an unconscious self of the thing, as it were. Not out of its own accord, not from its inner nature, it needs the light to produce a voice, an external agency, so as to yield animation, an anima, from its entrails. The moment is iconic, it is the moment when the thing, so to speak, speaks, it gives birth to meaning. (2008: 4, original emphasis.)

The passage is sandwiched by subtle shifts across a fundamental metaphysical delineation: voice and meaning. Dolar begins the passage by asking what does the voice say, what message? And closes with the same equivalence presupposing that a sounding thing "speaks" and "gives birth to meaning." "Does voice *always* have meaning?" and "Do sounding things always have a meaning?" are two questions one should level at this juncture. Yet, nestled in the middle of this section is the proclamation that "what it says has no meaning" except for meaning at its most minimal. The opacity, the seeming reluctance to a categorical verdict, is what is most "telling" here. Continuing the word play, one might say it is in the spirit of what is at stake—that is to say, it is in the question of spirit vis-à-vis voice that we may better understand the significance of Dolar's presentation of meaning, voice, spirit, and stone.

Dolar, after some words around the history of the statue, summarizes its role in Hegel's text:

> it was for him emblematic for a certain theory of the sign, of signification, of art and the movement of spirit. The rays of sun elicit the voice out of the stone, the externality of light produces the interiority of the inanimate, it gives life to the dead, it endows thinghood with an incipient reflexivity. (Dolar 2008: 5)

This passage courts similar vectors as the introduction text in *A Voice and Nothing More* whereby Dolar explores chess "playing" automatons; his entrance is by way of Walter Benjamin's "Theses on the Philosophy of History." He is clear that this is to serve as "a prolegomenon to a theory of the voice" (2006: 5). He provides a brief survey of Johann Neopmuk Mälzel's chess automaton (that toured in the 1830s) and Wolfgang von Kempelen's previous construction from 1769 (2006: 6). Dolar details Kempelen's "other big obsession [...] to construct a speaking machine, a machine that would imitate human speech." (2006: 6) Dolar, again, touches on the nature of this voice as "uncanny" before elaborating: "It is as if the voice could subjectify the machine, as if there were an affect of exposure—something becomes exposed, an unfathomable interiority of the machine irreducible to its mechanical functioning." (2006: 8) The point of presenting these two marvels, writes Dolar, is that this speaking machine was used as an introduction to the chess-playing automaton. Dolar highlights that the speaking machine appeared mechanistic, whereas the subsequent presentation, the chess automaton, appeared "as human–like as possible—it made the pretense of being engrossed in thought, it rolled its eyes, and so on—while the speaking machine was as mechanical as possible." (2006: 9) Dolar's reading is to pursue "a teleology in the link between the two machines [...] that the speaking machine was presented as the introduction to the thinking machine, thus the former made the latter plausible [but] also teleology in a stronger sense: the second machine appeared as the fulfillment of the promise given by the first." (2006: 9) Dolar expands on this as a Hegelian transition from "in itself" to "for itself" and writes that the point of sequence "could be read in such as way that speech and voice present the hidden mechanism of thought." (2006: 10) In the penultimate paragraph, he invokes Agamben's line: "The search for the voice in language, this is what is called thought." (2006: 11)

Rather than afford Dolar's ultimate aim in the introduction of making a proposition concerning object *a* voice and historical materialism (the main point of his introduction, his riff on Benjamin's essay—which curiously does not attract a great deal of commentary) time here, let us instead follow the theme of interiority, voice, and thought (or *spirit*... Hegel with it!).

Let us go back to a foundation of the relationship of voice with spirit/thought/interiority, specifically Alcibiades' comments in Plato's *Symposium* (diegetically Alcibiades' lines come shortly after the "flute girl" is expelled from the Symposium—flutes have a history associated with Pan, nature, Dionysian Dithyramb, and the corrupting force of sensuous aesthetics). We shall draw heavily from Adriana Cavarero's chapter "Some Irresistible (and Somewhat Dangerous) Flute Playing" in *For More Than One Voice*. Alcibiades likens Socrates, his and Plato's *teacher*, to a Silenus figure:

> I think he's very like one of those Silenus–figures sculptors have on their shelves. They're made with flutes or pipes. You can open them up, and when you do you find little figures of the gods inside. I also think Socrates is like the satyr Marsyas. (Plato 2000: 73)

Alcibiades is likening his teacher to a figurine concealing a flute—an object where voice issues from within; in another sense, he is suggesting that not only Socrates' voice is intoxicating and spellbinding (like the flute) but also his external appearance is at odds with his inner worth. To put it bluntly, Socrates' "face and body are ugly," but his soul is beautiful—and it is through his voice that his soul is conveyed and transported to others. Socrates is an ugly flute with a beautiful voice. He continues his *vino veritas* monologue:

> You may not play the pipes, like Marsyas, but what you do is much more amazing. He had only to open his mouth to delight men, but he needed a musical instrument to do it. The same goes for anyone nowadays who plays music – I count what Olympus played as really Marsyas', since he learned it from him. His is the only music which carries people away (...) Such is its divine power, and it makes no difference whether it's played by an expert, or by a mere flute girl. (Plato 2000: 73)

There are several themes nestled within this passage. First, the shadow of pederast homosexuality "as a model of cultural and political formation in the Greek aristocracy" is alluded to (in part by double entendres) (Cavarero 2005: 68). The context is Alcibiades' unrequited love for Socrates. His woe stems from his military service in Potidaea where they were messmates. Alcibiades hoped that Socrates would "improve him" and wined him frequently with this hope. One evening, after plying Socrates with wine, Alcibiades suggested he stay, arguing it to be too late for him to leave. Socrates agreed. Alcibiades then details what happened after lights out:

> I decided it was time to abandon all subtlety, and say plainly what I was after. So I nudged him. "Socrates, are you asleep?" "No." "Do you know what I've decided?" "What?" "I think you're the ideal person to be my lover, but you seem to be a bit shy about suggesting it." (Plato, 2000: 77–78)

Alcibiades hoped to satisfy Socrates for his own improvement. Socrates, however, declined, and Alcibiades *sleeps and nothing more* next to him. Alcibiades admits that he "slept with Socrates all night, but absolutely nothing had happened. It was just like sleeping with one's father or elder brother." (2000: 79) Alcibiades, rejected but still full of admiration for Socrates, declares how he "went around infatuated with the man, No one's ever been so infatuated." (Ibid)

This monologue of Alcibiades is a duet. Alcibiades not only speaks for himself but also voices Plato's admiration for his teacher. Plato is ventriloquizing here. This is Cavarero's point on the opening page of the chapter—a sentiment she traces from Martha Nussbaum's more visual comment in *The Fragility of Goodness*: "The enamoured Alcibiades speaks for himself, but he also speaks for the author: "the Symposium's portrait of Alcibiades is in some sense Plato's own self–portrait" (2005: 68; see note 2). "Alcibiades'" riffs on the story of Marsyas and Olympus (or Zeus). Marsyas challenged Olympus to a musical duel. Challenging the Gods is doomed to failure. He lost. His punishment: flayed alive, leaving only his screaming mouth. "Alcibiades'" twist here is that Olympus was playing Marsyas' music, not his own. Alcibiades feels he is just as transported and beguiled by the voice of Socrates even when others are poorly parroting his arguments:

> You have the same effect on people. The only difference is that you do it with word alone, without the aid of any instrument. We can all listen to anyone else talking, and it has virtually no effect on us, no matter what he is talking about, or how good a speaker he is. But when we listen to you, or to someone else using your arguments, even if he's a hopeless speaker, we're overwhelmed and carried away. (2000: 73)

Alcibiades, in likening Socrates to Marsyas and in the echo of his confession of his unrequited love, is plagued by Socrates's voice—even when it comes from others' mouths recounting Socrates' arguments and wisdom. Socrates voice is a ubiquitous and inescapable torment for the spurned Alcibiades who is, nevertheless, fascinated and captivated:

> When I hear him, it's like the worst kind of religious hysteria. My heart pounds, and I find myself in floods of tears, such is the effect of his words. I used to listen to Pericles and other powerful speakers, and I thought they spoke well. But they never had the effect on me of turning all my beliefs upside down, with the disturbing realization that my whole life is that of a slave. Whereas this Marsyas here has often made me feel that, and decide that the kind of life I lead is just not worth living. (2000: 74)

Socrates voice, Alcibiades argues, torments, infects, plagues, and is disseminated by others, through the bodies of others. There is a theme of contagion, insidious voice, in this love story. In this textual duet to Socrates, Alcibiades-cum-Plato rehearse symptoms of lovesickness: "*When we listen to you, or to someone else using your arguments, even if he's a hopeless speaker, we're overwhelmed and carried away.*"

Alcibiades "feels his heart flutter in his breast, and he becomes a slave to what he hears." (Cavarero 2005: 70) He becomes tearful at the echo of Socrates' voice in others.[2] Alcibiades' solution is to resort to the same aural evasion as Odysseus does when his voyage takes him within earshot of the Sirens. He mentions the Homeric poem and references the irresistibility of voice:

> Even now I know in my heart of hearts that if I were to listen to him, I couldn't resist him. (...) So I tear myself away, as if stopping my ears against the Sirens; otherwise I would spend my whole life

> sitting there at his feet. (...) So I run for my life, and avoid him, and when I see him. (Plato 2000: 74)

Plato's response is very different to that of Alcibiades. Plato, unlike Alcibiades, stayed to listen to Socrates. Alcibiades, qua Plato's ventriloquized dummy in *Symposium*, can be interpreted as an alternative representation of how Plato feels (the love for his master's spellbinding voice) and what he does not (unlike Alcibiades) give in to (the *avoidance* of the beguiling voice). Capricious and handsome Alcibiades flees the spellbinding voice of Socrates, whereas Plato does not flee; he stays and takes notes. Alcibiades is the confident student who is first to speak to his teacher, glass in hand, at a *soirée*. Plato, by contrast, is the shy and studious pupil who takes copious notes that are revisited, poured over, and absorbed later on, in silence, in private.

Plato, ventriloquizing through Alcibiades, is describing by proxy what he cannot bear to confess under his own name: his love for his teacher's speech. He is also, as a veiled humble-brag, declaring his fealty, his suffering for love, that Alcibiades could not bear and fled from. Plato, were he to play *Second Life*, would be a self-congratulatory and satisfied character. Cavarero details:

> Alcibiades not only functions as the double face of himself and Plato, but also as the one from whom Plato wishes to distinguish himself. Alcibiades – who is beautiful, famous and charming – is an ambiguous figure on whom hangs, among other things, an accusation of sacrilege. (...) Alcibiades turned his back on Socrates; he stopped his ears in order to extract himself from the teaching of Socratic logoi. On the other hand Plato, the legitimate heir, stayed to listen – all the while worrying about to eliminate precisely the enchanting aspect of this listening. Plato, in other words, found his own way to stop his ears – namely, through the devocalizing of logos. (2005: 70)

"Socrates does not write; he speaks. *Logos* comes out of his mouth and effectively enters the ears of his interlocutors." (Cavarero 2005: 70) Plato, by scribing Socrates' verbalization, by setting voice to text, cleaves logos, language, soul, and meaning away from voice as all its messy indeterminate corporeality. Cavarero writes: "The devocalization of logos inaugurated by Plato, in addition to establishing the ontological primacy of speech over thought, above all tends to liberate speech from the corporeality of breath and voice." (2005: 62) Plato finds a way of having his teacher's words without having to bear his voice. The essence of Socratic *logoi* is recorded and silenced—devocalized—by Plato. *Logos* sans *Phone*.

But this is also Plato's bifurcation *of himself* and Socrates' speech, a pathology of love and denial. On one hand Plato must, albeit via the textual proxy of Alcibiades, admit Socrates' voice is wonderful, enchanting, and unbearably captivating (like love). Yet he does not wish to be so enchanted, because for him, or at least the political side, it is a dishonorable pleasure—a sentiment ventriloquized through Eryximachus: "I've another suggestion to make. I don't think we need this flute girl who's just started playing. She can play for herself, or to the women upstairs, if she feels like it, but for this evening I suggest we stick to conversation." (Plato 2000: 13) The Silenus figure, with their flute-like outside and god figures housed within, is Plato's analogy of Socrates' speech. Plato's operation of devocalizing *logos* allows him to access the god figures within the Silenus statue without bearing the enchanting and corrupting flute, the voice. An insight Cavarero draws out:

> [the] rhetorical surface of the Socratic *logoi* is ugly, but the content is beautiful. Alcibiades stops here. His drunken discourse has reached its end. The whole platonic philosophy, however, without a drop of wine in its veins, goes well beyond this. It devocalizes logos. It cuts off its sonorous skin and founds in its place an order rooted in the videocentric and noetic sphere of thought. It makes speech an acoustic shell of an idea. (2005: 70)

The likening of Socrates to Marsyas (flayed as punishment for losing the *Gods & Satyrs–Got Talent* knockout round) is violent symbolism for Plato's metaphysical operation of separating thought, *logos,* from voice *phone*. Plato, by virtue of this devocalizing procedure of flaying of his master's voice, sought to isolate knowledge, *logos,* and meaning from its libidinal corporeality and capture it in the silent textual realm of language.

Later, Aristotle would continue the devocalization project in terms of the time's materiality of meaning. In Hugh Lawson-Tancred's translation of *De Anima,* Aristotle writes:

> Now voice is a kind of sound of an ensouled thing. For none of the things without soul gives voice, though some are said by analogy to give voice, such as the flute and the lyre and whatever other of the things without soul have the production of a sustained, varied and articulate sound. (1986: 178)

One of the distinctions in this passage is that voice is contingent to soul—only soulful things give voice. Aristotle goes on:

> But many of the animals do not have voice, such as the bloodless ones and, of the ones with blood, the fish. [...] everything gives sound by something striking and against something and in something, and as this last is air it is reasonable that only those things that admit air should give voice. (1986: 178–179)

This is a *pneumacentric* take on voice; one draws out a definition of voice as the sonicization of air from within a body (which, in the case of the Statue of Memnon, accords with Dolar's proposition that it is a vociferous stone). But Aristotle goes further; there is one more clause he adds to his conception of voice:

> It is then the striking of this inhaled air by the soul in these parts of the body against the so-called windpipe that is voice. For it is not every sound of an animal that is voice, as we have said (for it is possible even with the tongue just to make a some sound and to make a sound like coughing), rather it is necessary that that which strikes be ensouled and have a kind of imagination, as voice is a kind of sound with meaning, and not, like a cough, just of the in breathed air, though it is this that it strikes the air in the windpipe against the windpipe. (1986: 179)

Aristotle moves from empiricism and observation to a synthetic, or instrumentalist, conception of voice; he goes from utilizing the biological pneumatics of vocalization as definitions to adding a supposed interaction this air has with soul within the bowels of the vessel, from that unknown voice-space within. In *Politics*, this distinction arrives in an ethical guise:

> Now, that man is more of a political animal than bees or any other gregarious animals is evident. Nature, as we often say, makes nothing in vain, and man is the only animal whom she has endowed with the gift of speech. And whereas mere voice (phone) is but an

> indication of pleasure of pain, and is therefore found in other animals (for their nature attains to the perception of pleasure and pain and the intimation of them to one another, and no further), the power of speech is intended to set forth the expedient and inexpedient, and therefore likewise the just and unjust. (2001: 1129)

The continuum of partitioning voice from meaning continues. The definitions of voice ascribe certain pneumatic requirements with a presupposition of their interactions "against" soul concealed within—speech is all this with the addition of a meaningfulness that can discern and articulate between right and wrong.

In Cavarero's brief chapter, "When Thinking Was Done with the Lungs…" she tracks the movement of "thought" through the corpus into its metaphysical status (contemporarily at stake under *eliminativist materialism)*. She writes that "prior to the triumph of metaphysics, the Greeks were thus convinced that thinking was done with the lungs, not the brain." (2005: 63) Tracking along with R. B. Onians' (1988: 44), she touches on several examples of how the "conscious 'I'" had "an aeriform character and is rooted in the lungs. This is true not only for *thumos*, which Plato himself links to the churning spirits of passions that are situated around the heartstrings." (2005: 64) And how beating the chest is a "direct gesture of the conscious 'I.'" (Ibid) After exploring the etymology that incorporates the sense of smell into the ancient concept of thought, and as mentioned at the beginning of this essay, highlighting that the speaking animal (man) "must rely on systems made for other vital necessities in order to reach speech. He must, as it were, distort their natural functions," Cavarero concedes that Plato's perspective—that thought originates in the head—aligns more closely with the scientific framework we recognize today (2005: 65, emphasis added). Nonetheless, this "scientifically correct" relocation of thought is decisive because it "moves the measure of the human being from the physicality of the body to the impalpability of the mind. (…) Thought, or the activity of the noetic soul (as it moves from the lungs to the head) gains autonomy from every corporeal cause and thus earns its metaphysical status." (2005: 65) This is the liquidation of *phone*, its ostracism from thought, soul, and spirit—the latter purified of sounding bodies; rendered mute. Curiously, in this Platonic turn, the soft matter of the brain, being regarded as the primary seat of the *psyche*, or soul, struggles with a neat divorce of the body. Etymology—*ah, what's in a name?*—betrays that 'psyche', "from the verb 'psycho', meaning 'to breathe', would undoubtedly sound to a Greek ear as a phenomenon linked to the emission of air. Similarly, the Latin term 'anima', deriving from the Greek 'anemos', meaning 'wind' or 'breath', conveys a comparable notion" (2005: 66). This passage then moves to a confusion of "air" with a pneumatic expulsion of sperm/cerebral material from the penis—which is not wholly thought of as "the soul" that such material houses in the cranium. "The soul is eternal, bodiless, and does not breathe (…) closer to a "mind" than a spermatic marrow, it is this soul that is identified with the work of thought." (2005: 67) Cavarero explains that while Plato's revision moves toward a modern scientific conception of thought, it also preserves some archaic implications vis-à-vis the "breath" as it relates to the mechanism of "procreation" and recuperating a spermatic functionality of *psyche* "which he has come to identify with the mind." The kin relations of *logos* as it passes through souls and Socrates as midwife—delivering the "true logoi with which their souls are pregnant"—exemplify this (2005: 67). Although on the surface Plato's devocalization of soul/thought/spirit is claimed clearly, the cut has not been clean. Extracting spirit and soul from the voice and body appears, on closer scrutiny, a botch.

Over 2000 years later, symptoms of the bungled devocalization of logos persist. Stones with sonic emissions are utilized as illustrations of soul, and for human voices, slight disturbances and "tells" of voice are framed as betraying both body and mind. Notably, Dolar's most direct address of this torsion between body and soul occurs as the culmination of an exploration of ventriloquism—because voice is always ventriloquized, and its "source" always acousmatic, disunderstood, veiled. He writes:

> [I]n many languages there is an etymological link between spirit and breath (breath being the "voiceless voice," the zero point of vocal emission); the voice carried by breath points to the soul irreducible to the body. One could use a French pun, and say that the voice

> is *plus–de–corps*: both the surplus of the body, a bodily excess, and the no–more–body, the end of the corporeal, the spirituality of the corporeal, so that it embodies the very coincidence of the quintessential corporeality and the soul. The voice is the flesh of the soul, its ineradicable materiality, by which the soul can never be rid of the body. (2006: 71)

In the following section, notably the section (perhaps along with "The Click") that deals most directly with desire and drives, Dolar presents this torsion as body–voice–language. He writes that voice "stands at a paradoxical and ambiguous topological spot, at the intersection of language and the body, but this intersection belongs to neither. *What language and the body have in common is the voice, but the voice is part neither of language nor of the body.*" He then states, "this is the topology of *object petit a*." (2006: 73) and provide Lacan's pet scheme:

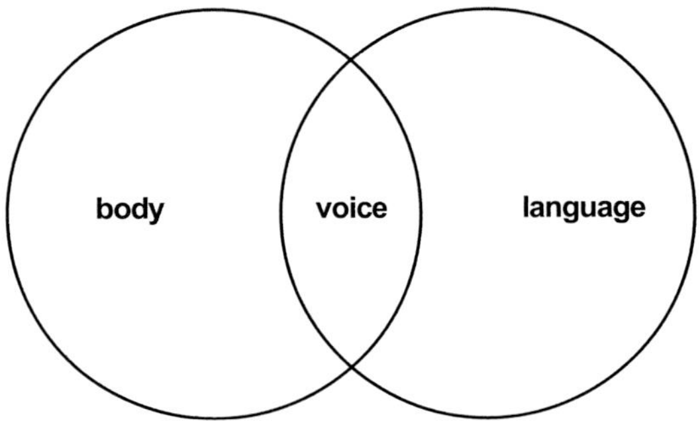

Figure 10.1 Dolar's Lacan scheme (2006: 73).

Dolar writes of the intersection as "being extimate to both" body and language (Ibid). This is an ineradicable ambiguity, but it is also ineradicable *excess*—excess of corporeality in something sonic and without flesh (something soulful, meaningful, thoughtful) and yet also something that "speaks" something that gives an excess of meaning to its meanings—thought, spirit—by virtue of its inextricability from metaphysically opposed materiality. The vocal fry, this creak from within the human vessel, cannot be attended to without running into the philosophical dilemma voice presents. The creak seemingly says *something* even as it says something. The creak betrays the body as it delivers thought—or *vice versa*. This is the wonderment of attending to voice, its chimerical nature of being impossibly at once metaphysical and corporeal—being at once spirit and flesh. To attend to voice purely in aesthetic terms, focusing only on the beauty or presence of the sonic body, is to miss some further meaning, yet to merely attend to its meanings—what the creak signifies—is to miss the voice's finer nature in the same way a music score misses the experience of live music—the swell, the weight, the flutter. Our problem with each of these is that they are symptoms of the unbearable ineradicable excess of voice.

Michel Poizat, a clear influence in Dolar's *A Voice and Nothing More* and also for Žižek's refrain of object *a* voice being "stuck in the throat" writes, in *The Angel's Cry: Beyond the Pleasure Principle in Opera*, of "rushing to the frenzy of applause while the last notes still hang in the air." And how "no one ever coughs during the orchestra's fortissimo (…) [f]or many spectators, coughing is not due to some chance conjuncture between a moment of silence and a tickling in the throat; more often than not, the larynx begins to tickle *because* there is silence." (1992: 89) This is the unbearability of the silent object *a* voice—the deafening silence must be punctured with noise, sound, or music. What Dolar describes as within voice the "inaudible and unbearable object voice." (1996: 16) This thought perhaps stemmed—to be floral—from "Jacques Lacan et la Voix" by Jacques-Alain Miller in *La Voix* (1989, colloquium proceedings from January 1988). An English translation is not at present available; however, Miller's sentiment—"If we make music and listen to it,… it is in order to silence what deserves to be called the voice as the object *a*."—is used in a number of Dolar's and Žižek's texts.[3] Žižek, in "I Hear You With My Eyes," writes, "Why do we listen to music?": *in order to avoid the horror of the encounter of the voice qua object.* […] it is a lure, a screen, the last curtain, which protects us from directly confronting the horror of the (vocal) object." (1996: 93–94). Yet, I ask, does one not *make* similar recourse with significance and meaning? When one is confronted with that genre-horror trope (the creaking doorway to the unknown), one staves off dread with meanings, sense, and reason. In terms of the interiority of the voice-giving vessel, where vocal fry creaks from, is there not also a defense of projecting, ascribing, smothering the unbearable inaudible core within with meaning? Is the ascription of meaning—to say "ah yes, this means such and such"—not the pareidolic, apophenic symptom of the unbearable object *a* voice nestled mutely within that creaky croak?

Endnotes

1. Graham gained a period of notoriety for her ability to deliver lines with *breathtaking* rapidity. This was showcased on *The Late Late Show with Craig Kilborn* (1999–2004) (DeniseMaus 2010) when she recited Lewis Carrol's 1871 nonsense poem, "Jabberwocky." Such a performance could be read as an absurdist caricature of a baroque—managerial jargon glossolalia. The viral proliferation and dominance of phrases such as "irregardless" and "daily–basis" over the more succinct "regardless" and "daily" to convey the same sentiment are perhaps registers of a tendency for conference call syllabic excess. Commonplace smatterings of filler terms used incorrectly but frequently (e.g., "like," "literally," and "physically": "like, literally died of embarrassment"; "physically log in," etc.) are also symptomatic of this trend toward syllabic excess.
2. Alcibiades' torment is echoed in contemporary popular music songs such as Britney Spears' *(You Drive Me) Crazy* (1999), *Toxic* (2004), *I'm a Slave 4 U* (2001) and Kylie Minogue's 2001 hit *Can't Get You Out of My Head* whereby voice and love are framed as pernicious agents of mania. Alcibiades is plagued by a love that strips him of agency and bodily control and equally plagued by a voice resulting in similar symptoms. The infectious and intoxicating spell of love is analogous to the compelling musical voice, the earworm that drives one "crazy."
3. See also "The Object Voice," in *Gaze and Voice as Love Objects* (Salecl and Žižek 1996: 10), and "The Linguistics of the Voice," in *The Sound Studies Reader* (Sterne 2012: 551).

References

Adams, T. V. (2018). "Accelerations and speed limits: An essay on the vocal limits of semiocapitalism," *Journal of Interdisciplinary Voice Studies*, 3(1): 21–43.

Aristotle (1986). *De Anima (On the Soul)*. Trans. H. Lawson–Tancred, London: Penguin.

Aristotle (2001). *The Basic Works of Aristotle (Modern Library)*. Ed. R. McKeon. New York: The Modern Library.

BBC (2010). "This Is How Michael Caine Speaks – The Trip – BBC Two," YouTube, https://www.youtube.com/watch?v=HFIQIpC5_wY (accessed January 10, 2022).

Buyukdura, J. S., McClintock, S. M., and Croarkin, P. E. (2011). "Psychomotor retardation in depression: biological underpinnings, measurement, and treatment." *Prog Neuropsychopharmacol Biol Psychiatry*, **35**: 395–409.

Berardi, F. (2009). *Precarious Rhapsody: Semiocapitalism and the Pathologies of the Post–Alpha Generation*. London: Minor Compositions.

Calvino, I. (2009). "A king listens," in: *Under the Jaguar Sun*. (Penguin Modern Classics). London: Penguin Books.

Cavarero, A. (2005). *For More Than One Voice: Toward a Philosophy of Vocal Expression*. Trans. P. A. Kottman, Stanford, CA: Stanford University Press.

Chion, M. (1999). *The Voice in Cinema*. Trans. C. Gorbman, New York: Columbia University Press.

Coen, J. and Coen, E. (Directors). (2007). No Country for Old Men [Film]. Paramount Vantage, Miramax Films.

DeniseMaus (2010). "Lauren Graham on Craig Kilborn, 31th January 2003 Part ½," YouTube, https://www.youtube.com/watch?v=77BgTTSAn14 (accessed January 10, 2018).

Dolar, M. (1996). "The object voice," in: Salecl R. and Žižek, S. (Eds.), *Gaze and Voice as Love Objects*, 90–126. Durham and London: Duke University Press.

Dolar, M. (2006). *A Voice and Nothing More*. Cambridge, MA: MIT Press.

Dolar, M. (2008). "The voice and the stone: From Hegel to Beckett," Eds. Gorlick, N. and Kerr, L. R. *Theory@Buffalo*, 12.

Dunbar, R. (1996). *Grooming, Gossip and the Evolution of Language*. London: Faber & Faber.

Ellgring, H. and Scherer, K. R. (1996). "Vocal indicators of mood change in depression," *Journal of Nonverbal Behavior*, **20**(2): 83–110.

Fletcher, L. (2017). "The everyday experience of engagement: What matters most?" *Human Resource Development Quarterly*, **28**(4): 451–479.

France, D. J., Shiavi, R. G., Silverman, S., Silverman, M., and Wilkes, D. M. (2000). "Acoustical properties of speech as indicators of depression and suicidal risk," *Transactions on Biomedical Engineering*, **47**(7): 829–837.

Graham, L. (2017), *Talking as Fast as I Can: From Gilmore Girls to Gilmore Girls, and Everything in Between*, London: Virago.

Jukes, I., McCain, T., and Crockett, L. (2010), *Understanding the Digital Generation: Teaching and Learning in the New Digital Landscape* (The 21st Century Fluency Series), Vancouver: Corwin.

Karpf, A. (2007). *The Human Voice: A Story of Remarkable Talent*. London: Bloomsbury.

Kreiman, J. and Sidtis, D. (2013). *Foundations of Voice Studies: An Interdisciplinary Approach to Voice Production and Perception*. Wiley-Blackwell.

Lacan, J. (1997). *The Psychoses 1955–1956: The Seminar of Jacques Lacan Book III*. Ed. Jacques-Alain Miller. New York: W. W. Norton & Company.

Luckwaldt, J. H. (2013). "Is vocal fry killing your career — or saving it?," Payscale, https://www.payscale.com/career-news/2013/08/is-vocal-fry-killing-your-career-or-saving-it (accessed 10 January 2018).

Madman Films (2014). "The Trip to Italy (Clip) Michael Caine," YouTube, Retrieved from https://www.youtube.com/watch?v=Vukpjl44Yo0 (accessed March 10, 2022).

Magnani, L. (2011). *Understanding Violence, the Intertwining of Morality, Religion and Violence: A Philosophical Stance*. Berlin and Heidelberg: Springer-Verlag.

Minghella, A. (Director). (1999). The Talented Mr. Ripley [Film]. Paramount Pictures; Miramax Films.

Morill, R. J., Paukner, A., Ferrari, P. F., and Ghazanfar, A. A. (2012). "Monkey lipsmacking develops like the human speech rhythm," *Developmental Science*, **15**(4): 557–568.

Onians, R. B. (1988). *The Origins of European Thought About the Body, the Mind, the Soul, the World, Time, and Fate: New Interpretations of Greek, Roman and Kindred Evidence Also of Some Basic Jewish and Christian Beliefs*. Cambridge: Cambridge University Press.

Pinker, S. (1994). *The Language Instinct*. London: Penguin.

Plato (2000). *Symposium and Phaedras*. Trans. Tom Griffith, New York: Alfred A Knopf/Random House.

Plunkett, J. (2002). "Speed thrills, says American TV producers," *The Guardian*, November 15, 2002. Retrieved from www.theguardian.com/media/2002/nov/15/usnews.television (accessed August 11, 2017).

Salecl, R. and Žižek, S. (Eds.). (1996). *Gaze and Voice as Love Objects* (Vol. 1). Duke University Press.

Schwob, M. (2017). "The King in the Golden Mask," *The King in the Golden Mask*. Trans. K. Schluter, Cambridge: Wakefield Press.

Sterne, J. (2012). *The Sound Studies Reader*. Oxford: Routledge.

Warren, J. E., Sauter, D. A., Eisner, F., Wiland, J., Dresner, M. A., Wise, R. J., Rosen, S., and Scott, S. K. (2006). "Positive emotions preferentially engage an auditory-motor 'mirror' system," *The Journal of Neuroscience*, **26**(50): 13067–13075.

Wolk, L., Abdelli-Beruh, N. B., and Slavin, D. (2012), "Habitual use of vocal fry in young adult female speakers," *Journal of Voice*, **26**(3): 111–116.

Wyndham, J. (2000). *The Day of the Triffids* (Penguin Modern Classics). London: Penguin.

Žižek, S. (1996). "I hear you with my eyes," in: Salecl R. and Žižek, S. (Eds.), *Gaze and Voice as Love Objects*. Durham and London: Duke University Press.

Žižek, S. (2003) "Not a desire to have him, but to be like him," *London Review of Books*, **25**(16). Retrieved from https://www.lrb.co.uk/the-paper/v25/n16/slavoj–zizek/not–a–desire–to–have–him–but–to–be–like-him.

Žižek, S. (2008). "Language, violence and non-violence," *International Journal of Žižek Studies*, **2**(3): 1–12.

Žižek, S. (2012). *Less Than Nothing: Hegel and the Shadow of Dialectical Materialism*, London: Verso.

PART III

Experiences

Chapter Eleven

Doing things with creak

Francesco Venturi

Introduction

What is "creak"? This is a question I have been asked on many occasions, and an answer without demonstration can be much longer. However, most people can recognize pulse phonation when they hear it, often adding that they use it in one way or another—in the morning, they say, or as interjection. The opinions vary, also depending on the geographic origin of the interlocutor, but they often include a question: "Do you mean the vocal fry?" It behooves me to clarify the difference between the physiological phenomenon and the popular media trope,[1] one reason why I prefer alternative terms, such as the other classic onomatopoeia "creak," or the biologically inclusive "pulse register/phonation," the more descriptive "pulsed voice," or even the more technical "M0" (mechanism zero), to remind us of this difference.[2]

Apparently, a chasm has opened up between this quiet vocal sound, ill-suited to a turbulent world, and the "Great vocal fry dustup of 2015" (Wolf 2015)—as loud as a montage of random

Creak: Theories and practices of pulse phonation
Edited by Francesco Venturi

ISBN 978-981-5129-29-8 (Hardcover), 978-1-003-61606-1 (eBook)
www.jennystanford.com

words spoken in a famously creaky voice, surrounded by a fanfare of Google-friendly articles, podcasts, YouTube videos, and social media coverage.[3] The rich phenomena involving pulse phonation seem overshadowed by a paradigm that is reminiscent of the "phenomenon of extreme vocal fry" (Hornibrook, Ormond, and Maclagan 2018). In short, the pulsed voice is figured in a rhetoric of the artifice (e.g., Habasque 2021); as contrived as prosthetics, but anthropic enough not to fall into Mori's uncanny valley—a mimicry, fad, or technique, filtered through references to popular media icons who extremely use one of its variants.[4]

My concern is with this paradigm as a background for considering four images that put forward a different way of listening to the creak. All images provide insight into a neglected aspect, and each propagates a way of thinking about pulse phonation that surpasses conventional interpretations. They are intermingled, as we shall see, but most importantly they all show something that is endogenous. In each image, a natural potential is explored.

Through writing my way into these images, through a variety of examples of practice, I will try to show not only some of the things that pulse phonation does, but I will "do things" with it, combining analysis and play—looking at places of experience in which we are all potentially involved. What I hope to make tangible, audible, and usable here is the power of pulse phonation to inspire qualitative changes in the way voice is produced, experienced, thought, and policed. All four images are inspired by such potential. They bring together the "voices" of early cavemen, modern infants, and children; contemporary musicians and artists; other mammals, birds, amphibians, and plants; fantasy beings that materialize through voice; objects such as wooden doors or chairs. These sonic materialities will help me illuminate what can be called the transformative power of creak, a recurring theme of this book and one of its initial inspirations (Venturi 2021).

What I hope to show is that creak is not something in itself. Species, people, and cultures may *use* pulse phonation, within certain practices and ecological niches, as these forms enter situated circles of meaning production. I will try to argue that pulse phonation can lead to qualitative changes in how voices are perceived and produced; that the very doing of pulse phonation for a human is driven by transformation and change, triggered by the emergence of a different "kind" of vocalization, and accompanied by an encounter with something other—an encounter which, however, does not generate a self/other dichotomy. In the following "images of creak," I will take up and elaborate on these elements of otherness and transformativity in different contexts: creaking in the first weeks of life and later in voice games; uses of the pulse register by Diana monkeys, toothed whales, goitered gazelles, and other animals; voice imitation, onomatopoeia, and creative invention of sounds; experimentation in music, art, and the moving image.

These images show some of the things that species do with the pulsed voice. The privilege of these images is that they are *partial*: not because they are inherently incomplete, but rather for the sake of the unforeseen connections and unexpected openings they make possible (Haraway 1988). What follows is in pursuit of such openings.

Endogenous sound

The first image is that of creak as a protophone. Protophones are the precursors of language: functionally flexible vocalizations, distinct from cries, that have occurred since the earliest days of human life. These sounds are not vegetative (e.g., coughing); they are not associated only with the expression of emotion, nor do they occur only in interactive situations. They arise from an "endogenous propensity for exploratory vocalization, perhaps the most fundamental ability underlying vocal language" (Oller *et al.* 2019: 1). The study of these protophones is crucial to the study of the origin of language.

In 2019, a group of researchers published a study on this "tendency" of children to produce sounds that do not express obvious immediate needs (Oller *et al.* 2019: 5). Their first goal was to challenge the notion that the human voice begins with the cry and that "the cry is the primary means of communication" (Zeifman 2001: 265), by examining how in early human life protophones occur. Another goal was to compare protophone and cry rates in preterm and full-term infants. Behind this was an attempt to test a promising hypothesis: The launch of protophone production might be associated with natural selection pressures. The study showed not only that protophones occur much more frequently than cries, but also that they occur as soon as infants vocalize.

These protophones include periods of creak. The research team analyzed full-day recordings of 41 infants from 8 and 4 weeks before birth (−2 months) to 1, 3, 6, 9, and 12 months of age. They grouped protophones into three categories:[5]

> (1) Vocants (or vowel-like sounds): By far the most frequent protophone category (>70% of all protophones at all ages) … Vocants are usually produced in comfort. (2) Squeals: Produced with very high pitch, indicated by periods of widely-spaced harmonics; squeals represented ~7% of protophones, occurring in all but 3 infants. (3) Growls: these exemplars include considerable interharmonic noise and pitch lower than squeals. Some growls also include periods of very low pitch in a phonatory pattern called vocal fry (or pulse). Growls represented ~15% of protophones, and occurred in all the infants. (Oller et al. 2019: 3)

This study tells us at least two remarkable things. First, that humans creak as early as they can breathe on their own. Pulse phonation patterns emerge naturally alongside cries and other protophones, starting in the first days of life. Second, that humans creak normally at this little age. Important evidence also comes from other sources, such as Buder's, Warlaumont's, and Oller's catalog of prelinguistic vocalizations, which reports that "growls that are not harsh or noisy are often associated with the pulse register" (2013: 108), or a study from the 1970s analyzing sounds produced by two infants between one and eight weeks of age, which shows that creak occurred at least as commonly as other protophones (Stark, Rose, and McLagen 1975).

Yet, once protophones give way to initial articulations (e.g., "baba") and these to speech, pulse phonation falls behind in most human languages. This raises a question: If creaking is common in our early days, why does it fail to survive the next stage of language development? One explanation is that children who explore and imitate the sounds around them focus on those that help them thrive. For a beginning speaker of an Indo-European language, for example, the pulse register has little application. All consonants and vowels are picked up by *phoné*, while creak remains in the prototype stage. As it does not participate in the development of protophones into phonemes, it remains in the realm of vocal exploration, kin to nonverbal expression and voice games. There it reappears later, in fact, in a silly voice, an imitation of a running engine, a frog croak, or a door opening slightly.[6]

We are thus presented with an image of creak as a protophone that never really developed into a phoneme. An endogenous, non-functional, and playful voicing that did not make it into the race for language. With relatively few exceptions (see Davidson 2020), most of the world's languages did not elevate creak to the status of *phoné*. At best, they left it with the status of *psophos*, a sound with non-anthropocentric traits. This image shows that creak, unlike other human activities such as speech, does not have to be taught. Nevertheless, it shows that something innate to us was cast away from *phoné*, and from the discourse that makes it possible. If the phoneme is a unity, the "atom" of language, creak resists becoming that measure. In this form of resistance lies a great opportunity for play, encounter, and change. This is of great interest to a human child who, increasingly framed within the rules of language, seeks alternatives. For an adult, the pulse register opens up the experience of being outside or beside language, from a ludic point of view: to return to voice games is to return to a world outside *phoné*. A world that we are nevertheless part of.

Voicing in communion

With the emergence of the first traces of a prelinguistic otherness, we reach the second image: that of pulse phonation as a more-than-human sound—where more-than-human refers to "a world that includes the emergent, permeable human and all of its animal and earth intra-active others" (Davies 2021: 1). It may be surprising to think of "vocal fry" in these terms, but there is no other vocalization that brings humans so close to their animal other. Many living things, each in their way, use a similar mode of sound production (Riede and Zuberbühler 2003). In other words, the pulse register is far from unique to humans.[7]

Creak-like dynamics are "a common production mechanism for low-frequency sounds in songbirds," meaning that no larynx is required to vocalize in pulses (see Chapter 5). The syrinx of crows "shows a vibration pattern of the labia similar to that of the human vocal fry register" (Jensen *et al.* 2007: 2703), and pulse tones "similar to vocal fry in human speech" have been observed in the calls of Australian magpies (Suthers, Wild, and Kaplan 2011: 45). Even crocodiles, the sister taxon of birds, use the pulse register. Vocalization is rare in non-avian reptiles, but crocodiles "have a complex vocal repertoire" that is used both above and below water, and they produce a "very consistent pulse tone pattern" at the end of contact calls (Riede, Tokuda, and Farmer 2011: 1). Pulse phonation is also known in amphibians: The voice of frogs and toads is well known to consist of pulses (Martin and Gans 1972). It is perhaps less well known that many nonhuman mammals phonate in pulses, including black lemurs (Gamba and Giacoma 2009), Squirrel monkeys (Winter, Ploog, and Latta 1966), Red howling monkeys (Schön Ybarra 1986), Chacma baboons (Fischer *et al.* 2002), Gelada baboons (Richman 1976), Pigtailed macaques (Grimm 1967), as well as koalas (Smith 1980), felidae (Peters 1981), and toothed whales such as dolphins and orcas (Au 1993; Møhl *et al.* 2003). Animals also do specific things with the pulse register.

Diana monkeys use pulse phonation to protect themselves from predation in their West African rainforest habitat. For a Diana monkey, creaking has a very specific meaning: "predator alarm!" (Riede and Zuberbühler 2003). In the BBC documentary *Life of Mammals*, David Attenborough and his team showed a pack of Diana monkeys a dummy leopard in the middle of the forest and filmed their reaction—the clip is easily found on the Internet. As soon as the monkeys see the predator, they cry out and begin to "creak" loudly. One by one, all members of the group join in the call to spread the alarm, producing a droning sound. The similarity between the leopard alarm call of Diana monkeys and the human creak is striking. Indeed, these "alarm calls are the product of the same source production mechanism that is responsible for the human pulse register" (ibid.: 2922). The fundamental unit is a pulse, lasting between 8 and 30 milliseconds. Each call comprises up to 30 pulses, and multiple calls form a bout. They use these alarm calls to signal to both leopards and eagles, and there are minute and fascinating ways in which the two are distinguished. These sounds can travel long distances, reaching up to a kilometer in dense forest habitats.

The pulse register is used by dolphins, orcas, sperm whales, and other toothed whales (odontocetes) to produce echolocation clicks that are essential for haunting in obscure marine environments. Although they hunt at great depths, toothed whales have an air-driven sound system that enables both complex social communication and echolocation (Madsen, Siebert, and Elemans 2023). Researchers have spent two decades solving the mystery of how these whales can produce loud sounds without air, since the volume of air at extreme depths is compressed (less than 1% of what is available at the surface). They recently demonstrated that whales divert air from their lungs into a pouch in their nose to bypass the lungs and vocal cords and then push it through membranes in the pouch called "phonic lips," producing a powerful, air-efficient pulsed voice. The study suggested that the phonic lips may also explain the typical "Mona Lisa smile" of orcas and dolphins, a trait that may have evolved in exchange for precise control over pulse register phonation. "While vocal fry may be controversial in humans," said Elemans on the *Science Magazine* (ibid.), "it doubtlessly made toothed whales an evolutionary success story."

We are thus presented with a fascinating image of pulse phonation—one of wild animal sounds; one of humans and nonhumans vocalizing in consonance (in the sense of *sounding together*). Not to get too carried away, an image that brings us into the realm of cross-species sound. This research field is rarely explored outside of bioacoustics. An interesting attempt is *Apotomē,* a short film by Allora and Calzadilla (2013), revolving around a singer and the bone remains of two elephants. The artists collaborated with Tim Storms, the vocalist who holds the world record for the lowest note produced by a human, an outstanding G-7 (0.189 Hz). This subsonic range is so low that only animals the size of elephants can hear it. While caressing the bones, Storms vocalizes in such extraordinary tessitura, and of course above that (one cannot sing melodies in the pulse register). What is to note is the film's attempt at a cross-species utopic vocalization. According to the accompanying notes, the short explores music "as a possible inter-species meta-language, a proto-linguistic, non-symbolic, and affective transhuman mode of communication" (Lisson Gallery 2014). What makes this possible, beyond utopias, is the pulsed voice: the sonic materiality where different life forms find a common vocal denominator.

Reminiscences of "the relationships existing in the pulsed phonation in humans and nonhumans" (Frey *et al.* 2011: 582) also mark the way goitered gazelles produce their rutting call. The analogy goes beyond the proportion between subglottal pressure and fundamental frequency, typical of the pulse register, and reaches a symbolical level related to body representation. During the rut, or the mating season, male goitered gazelles use vocalization to threaten rivals and attract potential mates. Body size plays a pivotal role in this endeavor (Fitch and Hauser 2002), given the tendency of females to favor larger males as mating partners (Birgersson and Ekvall 1997). The study of how vocalizations reflect body size is referred to as acoustic allometry. Calls provide information about the body size of the caller, and this information can be honest or dishonest (Fitch 2002; Pisanski, Anikin, and Reby 2022), since lower-pitched sounds can create the illusion that the source is larger than it actually is (Pisanski *et al.* 2017). The pulse register, therefore, with its ability to produce low frequencies, is an ally in these "allometry-cheating" scenarios. Male goitered gazelles leverage on the pulse register to bypass the boundaries set by evolution on their actual body size.

As with gazelles, the "low is large" principle (Pisanski *et al.* 2017) plays a role in human communication as well. Take the association between creak and dominance—according to which people can index authority by using pulse phonation (e.g., Hildebrand-Edgar 2016; Ward 2006; Lefkowitz and Sicoli 2007)—and consider the much-discussed negative responses to creak in female speech (e.g., Anderson *et al.* 2014; Ligon *et al.* 2019). In the pulse register, all human voices produce a similar fundamental frequency. At the acoustic level, then, gender differences vanish (Blomgren *et al.* 1998; Hollien and Wendahl 1968). What might be really at work here is the low-large association. Let us hypothesize on this basis. As long as "low is large" and "large is masculine," then creak will be foregrounded in relation to gender representation. In a chauvinistic society where masculinity is the reference for dominance, a longer larynx and deeper voice are associated with supposed greater stature or strength. In a communication where masculinity is the model, small bodies paired with deep voices may sound like "allometry-cheating."

This hypothesis is not so far removed from the recent reading of negative reactions to female creak put forward by Chao and Bursten (2021), who argued that these responses may be negative "due to female vocal fry's *violation* of the frequency code" (2021: 45, my emphasis). The frequency code is a concept introduced by John Ohala (1994) to describe how people associate higher pitched vocalities with femininity, while attributing lower-pitched vocalization to masculinity. The locus of this *violation* is the sole listener's mind. Therefore, as long as "creak is low," "low is large," and "large is masculine," then pulse phonation will continue to create the conditions for an act of violent listening, the violence being in the projection of such syllogism onto a sounding voice. It is not my intention to further elaborate on this hypothesis. It should only be noted that such nonverbal communication, hidden in a syllogism according to which "creak is low/large," occurs only as long as pulse phonation is listened to as a *deep* voice—and not, for example, as a *whole other* voice; one that sounds, in fact, gender-neutral (Venturi 2021).

In addition to gendered voices, echolocation clicks, and predator alarms, it is worth concluding this brief exploration of what species do with the pulse register by mentioning another specific use that concerns felids and crocodilians (humans do this too). Although these animal groups are far apart on the spectrum of evolution, both use pulse phonation for affectionate social communication. Lionesses and other feline mothers, for instance, use pulsed vocalization when interacting with their cubs.[8] On the other hand, juvenile crocodiles use pulse tone patterns in their "contact calls" to maintain social cohesion (Riede, Tokuda, and Farmer 2011). Indeed, the pulse register is used not only for feeding, self-defense, and role-playing, but also to express care and comfort. Such affectional use of pulse phonation reminds of the sound of primal longing that Katherine Meizel evokes when she listens to creak "in a bedroom moan, reaching for another's body, another's love" (this collection: 465).

The second image of the present tetraptych shows some of the things species do with the(ir) pulse register, among other things (see Chapter 5). It stages pulse phonation as a site where different vocalizing creatures find a common denominator for their vocalizations; one that is not merely the air, but a unique way of shaping it to "do things" (Austin 1962). The world is full of pulsed voices that are startlingly similar on a tonal and mechanical level, in species that are far apart on the evolutionary spectrum. On the other hand, it reveals that such a vibrant soundscape of pulsed voices has little to do with nature and much to do with nurture. Be it as a sign of authority in speech, as a predator alarm, or a song in communion with elephants; be it as a means of hunting, showing affection, or overcoming the limits set on body size.

Imitation and invention

In a recent article on the essence of writing about nature, Charles Foster reflected on the desire to express communion with the more-than-human world through words, writing, "Swifts should *screeeeeeeam*. Rain should *hissssssssssss*. Boughs should *creeeeeeeeak*" (2021, original emphasis). This leads us to the third image of the present tetraptych: the onomatopoeic power of creak. Indeed, creaking sounds are not exclusive to animals and humans; they are also commonly found in plants and their byproducts. The trees creak in the wind, the door creaks when it opens, the floorboards creak when you walk, and the chair creaks when you sit. The word "creak" reproduces these sounds in onomatopoeia, and we can *creak* (use the pulse register) to imitate these sounds, just as children do. In vocal imitation, the pulsed voice can evoke inanimate objects such as engines, wooden doors or metal gates, living organisms such as toads or tree branches, and fantasy creatures and monsters. In music ranging from extreme metal to radical improvisation; in contemporary art; through to video games, science fiction and horror films where pulse phonation evokes the other, the monstrous and the horrific. The "death rattle" from the *Ju-On/The Grudge* franchise, for example, is nothing more than director Takashi Shimizu's voice in the pulse register. Strictly speaking, pulse phonation is implicit in the title of the film series (13 films in all, encompassing both Japanese and American franchises).

Before we delve into the creative power of pulse phonation, it should first be noted that the expressions "vocal fry" and "creak" are in themselves onomatopoetic. Traces of verbal sound description are found throughout the literature, long before scientists agreed on the physiology of pulse phonation, let alone the terminology. Initially, speech pathologists had difficulty distinguishing hoarseness from creaking. Only with the help of onomatopoesis did they start to distinguish one from the other. Here are two of the early descriptions:

> The exact physiological mechanism of the creak is unknown… The auditory effect is that of a rapid series of tapping sounds, like a stick being passed over a railing. (Catford 1964: 32)
>
> You may recognize it as the sound many children make when imitating a motorboat, but to me it more resembles the sound of corn popping vigorously. (Moser 1942: 174)

The fact that the scientists who tried to identify creak relied on the sounds it could imitate is telling. Pulse phonation used to be known for what it does, not so much for what it is. This penchant for the imitative and performative is still an essential part of how we perceive pulse phonation. It is no accident that Catford's and Moser's descriptions are so frequently still cited in recent sources. When speech-language pathologists recognized to the pulse register the status of "natural," from the 1960s onward (see Chapter 1), a terminological and epistemic stamp of *mimesis* was put on it. Indeed, it is no accident that I refer to the onomatopoeic rather than, say, the evocative or imitative. Provided that "creak" and "fry" are onomatopoeias *sensu stricto*, although not universal, the pulsed voice has its own onomatopoeic force, and such creative force is universal to all people.

Like protophones, onomatopoeias reveal language universals.[9] The basic concept of onomatopoesis, or "the making of a name," is the production of a sound that resembles another sound (Bredin 1996; Assaneo, Nichols, and Trevisan 2011). These imitative sounds are found in all languages, and linguists believe they were the first words we spoke (e.g., Brown *et al.* 1955; Imai and Kita

2014; Donald 2016). There is consensus that imitative sounds have existed since the beginning of humankind and that they may have played a fundamental role in community formation and survival. Imitating the roar of a predatory cat or the sound of rushing water may have been how early humans initially referred to these phenomena (Edmiston, Perlman, and Lupyan 2018: 331). Here, onomatopoesis and speech conflate.

If protophones include the endogenous production of creak, we can assume that early humans too were given the pulse register—at least since the advent of the anatomically modern *Homo sapiens*, who walked upright and had a lowered larynx, resulting in an elongated and mobile vocal tract (Negus 1949). Our ancestors may have used imitative sounds to describe freshwaters, predators, and edible fauna. Frogs, for example, have occupied most types of waters on the earth for 7 million years (Fujita *et al.* 2016), and this abundance meant that "the predator Homo sapiens quickly noticed them" (Neveu 2009: 307). The earliest evidence of fishhooks comes from Sakitari in the Okinawa Islands and is dated to 20,000–23,000 years ago, and these faunal remains include frogs that "show signs of charring on a fire" (Gaffney 2021: 292). This leads to an intriguing hypothesis: Creaky sounds may have been part of our palette of imitative sounds since the dawn of humanity, and our ancestors may have communicated the presence of frogs by imitating their croak.

Imagine a *Homo sapiens* aged three in Sakitari near the end of the Paleolithic era. Her community has just moved near the water to gather resources. While the adults and grown-ups fish, she watches intently. They are on the lookout for those creatures that produce a croaking call—the very sound she uses to explain the others that she has just spotted a fat one. As she happily repeats this "proto-name," she sustains the noise, longer than any frog would. Here she is playing with this novel sound, which has evolved from a hitherto functional "name" to a non-functional "thing." What does it stand for? It is reminiscent of a bow being strung before the arrow is shot; of the sounds made by other cave occupants in the dark of night; of the rustling of branches in the wind; perhaps a combination of these with the frog croak. Imitation by creation has given way to creation by imitation.

This fantasy anecdote serves only to playfully introduce that vocalization fluctuates between *mimesis* and *poiesis*. The realm created through this fluctuation is the human imagination itself, which perfectly embodies this unique interplay between imitation and creation, between replicating and originating, or rather in between the innovative reuse of established elements (Moeller and Whitehead 2018). Imitation and production are integral parts of invention. In this sense, onomatopoesis is integral to creative vocalization, whether it is referential, as in the case of the frog croak, or non-referential, as in the case of the thing that our hominid child breathed life into with her voice.

To further refine this stretched notion of the onomatopoeic, let us dwell on the interplay between *mimesis* and *poiesis*.[10] Given that both are productive, and considering the creative function of the latter and the reproductive function of the former, should we regard them as antinomies? According to classical thought, yes: imitation and invention were "the traditionally opposed roles of the image" (Bachelard 2014: xx). It was postmodernism, with its emphasis on the "in-betweens" that helped overcome this Western antinomian thinking. The mirror of postmodernism, whose penchant for montage and reworking is well known (e.g., Bourriaud 2006), showed *mimesis* "at the heart of poiesis," marking the beginning of a new way of "thinking within the middle realm" (Levine 2009: 61, 63). Such shift in perspective requires a shift in logic: from the logic of identity to that of *resemblance*. According to the logic of identity, "imitation is reduced to identification," leading to a return to classical antinomies, while according to the logic of resemblance, we develop "a way of thinking within the middle realm 'between' the oppositions of traditional logic." Inhabiting such a space between invention and reproduction is the most important lesson the West has learned from postmodernism. In contemporary times, it increasingly engages with indigenous perspectives that eschew theoretical polarization (Magnat 2020: 163) and encounters philosophies rooted in community, emphasizing the concept of "participatory mimesis" (Nancy 1982: 64) through affective participation (*methexis*).[11] The onomatopoeic is rooted in *poetic mimesis*.

It is in the logic of resemblance that we find the real measure of the creative force of pulse phonation (as well as voice in general). In *poetic mimesis*, the voice has a phenomenology of poetic acts. Rully Shabara, the singer of the experimental duo Senyawa, for example, is a virtuoso of these poetic acts of voicing. Gigi Priadji's *Experience Series* (2016) shows Shabara producing a plethora of imagetic vocalizations, sung with either inhaled or exhaled creak, among other sounds. These intense voicings have the sound appearance of a life blossoming somewhere in between the animal, the vegetable, and the otherworldly. Indeed, the encounter between the human and the inhuman is a theme in Shabara's poetics, and also the name of the band (which means "compound") contains this reference. Another classic example is Joan La Barbara's *Q-/-uatre Petites Bêtes*[12]—written in 1978–79 and published in 1980. In this whimsical composition, four creatures are evoked by experimental vocalization: They "suddenly show up in this open space at the same time, and converse" (La Barbara, this collection). One of these creatures is made of inhaled creak, or "inspiratory vocal fry" (Paolillo *et al.* 2019). Note that she prefers the term "glottal click" because her use of this technique takes its cue from percussion, emphasizing the single pulse rather than the stream of pulses that produce a tone. When I interviewed La Barbara for this book (Chapter 18), I asked her to tell me more about this creaky creature. Who is it? How did she feel about this creation at the time it was recorded? Her answer is closely related to the notion of the onomatopoeic in pulse phonation and the nature of such a creative force. She explained that she was looking for "something different" and "less identifiable," so that people "could allow their minds to imagine different kinds of creatures" (this collection: 500).

In both Shabara's and La Barbara's creations, onomatopoesis defies any separation between referential and non-referential realms, as in the imitation of a sound that does not yet exist, or the sonic appearance of something that is yet to be imagined. Both artists make tangible the universal power of the pulsed voice to materialize entities, creatures, and things that are not anthropocentric. Not by chance, La Barbara has been hired by the film industries to give voice to monstrous characters, such as the Newborn from *Alien Resurrection* (1997). Yet what pulse phonation can bring to life in a poetic act of voicing is by no means universal, nor should it be, for we are alongside or outside the boundaries of *logos*. There is no identification involved because, in Bachelard's appropriate words, "the poetic act has no past, at least no recent past, in which its preparation and appearance could be followed" (1964: xi). As La Barbara quipped, "if I asked 10 people to draw the creatures, they would come up with 10 very different pictures" (this collection: 500).

A trait of what Jelena Novak calls "inhuman impurity" is typical of pulse phonation (2012: 63). The performativity of M0, either exhaled or inhaled, allows a voice to transform in a way that is not so easily achieved with other registers such as M2 (falsetto/head/etc.), or even M3 (whistle register). Those always preserve, to twist Novak's words, a certain human *purity*, or, in other words, a remnant of M1 (modal/usual). The term "modal" derives from a statistical "mode" or norm (Hollien 1974; Titze 2014). In contrast, nonmodal phonation (M0) shows remarkable *impurity*, because it literally erases the "modal" associated with the human voice. The stable vibrations of M1, M2, and M3, which all produce pitch, are lost in the irregular vibration of M0, which produces pulses.[13] These excitations can partially or completely disrupt the stability of the "mode" and involve different parts of the vocal folds. Pulse phonation patterns change the quality and affectivity of the vocal sound by mixing with other registrations and distorting them; see the "dist" tones of rock (Borch 2004), or the so-called "fry singing." The pulse register can integrate with breathy, whispered, and modal phonation to create sepulchral utterances in countless applications; listen, for example, to *FED UP* by rapper Ghostemane (2020), or the debut album of TikTok artist Kim Dracula (2023). Indeed, pulse phonation has affordances, but also effects. It can completely overturn the sound appearance of a voice in the many possible realizations of inhaled and exhaled creak. These transformations are processes that the voicer undergoes, but they are also in the ear of the perceiver, even more so in a recording than live on stage. The "inhuman impurity" is made evident in the absence of a singer's body to be looked at.

We are presented thus with an image of pulse phonation as a creative space, with an inclination for onomatopoesis, for voicing in communion with the world we are of, and a talent for poetry, or the poetic act of creating new worlds. Giving the pulsed voice the status of poetic space means, above all, tracing it back to presymbolic, sensual, and free vocalization. One can understand it as an inventive form or possibility of form that goes beyond the role of voice quality, interjection, or social marker, one that enables a meaningful reformation of the vocal matter in its human permeability.

In this image, pulse phonation has within itself what Bronwyn Davies calls "the power to exceed" (2021: 106); or that "ineradicable excess" already pointed out by Tristam Vivian Adams (this collection: 310). The idea of creak as primarily a linguistic tool or an affectation would be disastrous, for it would rip it from that excess. On the contrary, it is precisely here that creak becomes instigative of change.

Transformative power

The transformativity that belongs to the pulse register is encapsulated in Antony Aumann's reading of the Kierkegaardian project, which is to inspire transformations. These may be as dramatic as turning water into wine (Aumann 2020: 429). Aumann's reading draws on the influential account of Laurie Paul, a metaphysician who has paid particular attention to the topic of transformative experiences (e.g., Paul 2014). According to this view, transformations are twofold:

> [f]irst, they involve a shift in our desires, preferences, and values. In a transformation, we come to care about things we did not care about before. Second, transformations involve a shift in our epistemic standpoint. In a transformation, we enter a new way of seeing the world, one we did not and could not have understood beforehand. (Aumann 2021: 430)

This conceptualization captures two inherent aspects of creak: On the one hand, its ability to stimulate transformations that involve qualitative changes in how we express ourselves vocally, so how we negotiate our identities (see the ongoing discussions on gender performativity addressed in Chapters 6–10); on the other hand, the fact that creakiness itself is inspired by change. Beginning with a person's childhood, pulse phonation offers a glimpse of what it feels like to vocalize in a different way than we normally do; one we cannot comprehend in advance (see the testimonies from children in Chapter 17: 473). The creak offers a voicer the opportunity to "change voice" without the need for a physical technique (e.g., manipulating the throat) or a technical device (e.g., pitch-shifting). This has the potential to attract us to other ways of seeing the world, ones that we would normally dismiss. If the pulse register is a step into a prelinguistic otherness, then transformation is the step forward or the step that it makes possible.

Starting from ludic uses, passing by mimetic and poietic uses, through to gendered practices (that use voice quality to do things), pulse phonation resists assimilation by *phoné*. This opens up different scenarios of otherness, where "otherness" refers to deviation from the identical, and "identical" to what one gets used to. Indeed, if pulse phonation was bound to enter *phoné*, then it would have done it, perhaps at the time of our hominid child, and perhaps the world's onomatopoeias for "frog" would have been more similar to each other.[14] By resisting the program of *phoné*, the pulsed voice clarifies the otherness of language. If it is true that both infants and primates do a certain number of things with creak, then this suggest that we engage with a prelinguistic otherness. We are brought outside the tradition of *logos*, where we can meet with our presymbolic other. This encounter shows that we can do *something*, not *anything*, with the pulse register. It cannot be considered unconscious. On the other hand, its existence at the borders of phonation, at the boundaries of words and sentences (Redi and Shattuck-Hufnagel 2001), its occurrence in the voices of those who suffer from cognitive/physical effort in speech production (Cresti *et al.* 2015; Silvestro *et al.* 2021), all suggest that the pulse register does not require sophisticated (neural) control.[15] This encounter, then, shows that pulse phonation escapes control, first of all mechanically (for it is based on an irregular action). It is never fully deliberate and emerges spontaneously—while we look for a certain word and we sustain the voice (so not to be interrupted), or in a moan. Only in part can it be controlled. The pulse register puts the human voice on a threshold between the voluntary and the involuntary. This threshold is the very setting for an encounter with *another voice*, an opportunity to *become other* without generating a self-other dichotomy.

To delve into this connection between alterity, transformations, and control, let us recapitulate the sonic materialities evoked so far. The traces of otherness tangible in the phenomenology of creak, to use Radovanović's words, range "from extreme to everyday" (see Chapter 13). As we have seen, such phenomenology is threefold: first, the nonhuman, natural, as well as uncanny; second, the pathological, nonconformist, and uncontrollable; and third, the extraordinary, (cannily) artificial, and virtuosistic.

In the case of trained vocalists, the "otherly" quality of the resonant difference tones of skilled *Strohbass* lies not in production, but in perception.[16] Artistic applications of M0 in subharmonic singing arise from sophisticated vocal techniques, which are anything but "alien." These techniques are potentially accessible to anyone serious about learning the art, yet they sound "magical" to the listener, and those who can produce them marvel for their vocal abilities—it suffices to read the comments section of any YouTube video dedicated to these techniques. From the "fry scream" to the "creaking" technique of metal music to pseudo-Oktavism[17] and throat singing, pulse phonation is bound to produce special effects that seemingly go beyond the perceived limits of the human voice. In *Apotomē* by Allora and Calzadilla, this is taken to the extreme, when human perception is virtually zeroed. Aside from the already radical alterity implied by a man (Storms) vocalizing sounds he cannot hear himself, what should be noted here is the attempt at a transhuman sound production based on pulse phonation. Creaking has the potential to bring humans closer to their more-than-human other, whether in resemblance to a leopard alarm, in communion with a tree creaking in the wind, or in consonance with orcas, monkeys, and lions.

The ideal, the material, the poetic, and the mimetic come together when we move pulse phonation on a literal level. In the *Ju-on/The Grudge* film series by Takashi Shimizu, creak materializes the otherly itself. In these films, the creak is not only the voice of the ghosts, but also the sound of their emotions: the feeling of bitterness that keeps them in this world in the first place, materializing into a sounding voice. This encounter between creak and the feeling of "resentment" is crucial to the horror film series. As with many other monsters from literature and art, the inhuman is found within the human itself (Fisher 2016), and this narrative and emotional use of pulse phonation makes it palpable.

The potential to evoke the otherly is also central to onomatopoesis, in the stretched sense we have explored. Creaking derives creative force from its transformativity and transformative power from its otherness. In children's voice games and voice imitations at all ages, pulse phonation enables the speaker to transform their voice into another "thing," be it a piece of wood, a toad, a running engine, or a barking fantasy creature. This is also the case with vocal experimenters of the likes of Joan La Barbara, who gives substance-without-reference to a character because she knows that listeners will "imagine different kinds of creatures."

Although this recapitulation is far from exhaustive, it aims to make more tangible the many transformative experiences that creaking opens up. Rather than attempting to deconstruct the nature of such tangibility, if not the very idea of a "phenomenology of creak," I have attempted the opposite approach to the problem, asking whether particular places of experience can be made tangible with pulse phonation. I hope the reader has tried some of these sounds with their voice while reading.

If any transformation can be experienced at all, then it must involve an element of otherness, and otherness always implies the possibility of a transformation. By returning the voice to its presymbolic otherness, the creak triggers experiences that are distinct from language, from symbolic ascriptions of identity, and ultimately from control. I can deal with the pulse register but never fully control it, for the irregularity is integral. The force of M0 here lies halfway between *phoné* and a vegetative sound, where the unconscious borders on the conscious and both influence each other, allowing a voicer to *become other* without troubling classical antinomies. This transformative function is accessible and brings the other of the same to the surface. The otherness trait, then, gives the pulse register the potential to become political.

In light of the fourth image, then, the pulsed voice functions as the other of normal phonation, where "normal" means modal (M1) on the one hand and normative on the other, while "pulse" means nonmodal (M0) on the one hand and nonconformal on the other.[18] Only perhaps to make the "normal" voice sound as the actual, troubling "other."

Openings

The first three images—protophones in human infants, the pulse register between nature and culture, and onomatopoiesis in early hominids, modern children, and contemporary artists—meet in the fourth image, the transformative power that creak derives from its prelinguistic otherness. Retracing all the examples from the perspective of transformativity and otherness, we are presented with an image of pulse phonation as a force that bends the supposedly intrinsic properties of the voice. By making extrinsic the normative logic of any "normality," it implies other properties and enables new openings, making them immediately clear (and usable, vocally). On the other hand, it shows a vocal mechanism that is inherent. Creak, in this sense, can be inspired by and inspire a qualitative change. In the obvious and immediate change in one's voice agency, emergent from something that does not have to be learned and has not the characteristics of what it emerges from—a change that is accessible and, therefore, inclusive, for it empowers the voicer as the subject of their own change. The doing of pulse phonation becomes, thus, the search for sounds that lay silently apart from our habitual way of vocalization and derive from our very first voicings; sounds that are, to use an expression by Pikes and Campbell (2021), "our own," in a larger soundscape of more-than-human pulsed voices.

In some cases, nevertheless, people find it difficult to consciously use creak, at least at first. They may have been using the creak all their lives (at phrase boundaries, for instance) but find it difficult to do so intentionally. Indeed, the only physiological parameter that has a strong direct influence on the pulse register is subglottal pressure (Blomgren *et al.* 1998), and when pressure is higher, e.g., due to the stress of a task, someone might have difficulty "finding" the pulse register. Another person, in order to "find" it, might open the glottis and produce a breathy voice rather than close it and get a pulsed voice. However, the same two people might describe their difficulties a minute later with a strong creak pattern at the end of their sentences. The juxtaposition of very different modes of engagement with creak—from completely unconscious to increasingly conscious—is a window into the educational power of pulse phonation (the power to inspire transformations). For we do not learn creak, but we can learn about it. Yes, we can do things with creak, but what happens when creak does things back?

Having the pulse register "do things" opens a laboratory for vocal explorations, and this laboratory is greater than the exploration of the voice for exploration's sake. It is a laboratory for learning and unlearning structures that go beyond the vocal studio. First, because the pulse register is a natural ability, yet some find it difficult to use intentionally, so it creates an opportunity to learn our way through our bodies. Second, because collective uses of the creak in contemporary speech are controversial and associated with identities that have been historically othered. In other words, pulse phonation is neither innocuous nor innocent. Exploring the ways voices are supposed to sound and how they are able to sound holds great potential for meaningful change, and this change is not limited to voice art but concerns the lived experience.

The pragmatics of pulse phonation becomes thus a laboratory for learning scenarios. It is a matter of placing a vocal sound on the rails of a different epistemology. Conceptualized objects (such as "the creak") are the result of a cultural laboratory "where facts can only be made to run on the tracks laid down from the laboratory out" (Haraway 1988: 596; see also Latour 1984), and only diverging the rails do we change the object (which is why this book is titled "creak" and not "fry" for that matter). As the full choir of voice studies stands to indicate, exploring new ways in which the concept of "voice" is mediated may change the way voices are heard. These, too, are transformative experiences. All of these transformations have to do with different aspects of collective production, or how the pulse register is produced: physically (as a matter), societally (as a practice), and epistemically (as a resource). From this perspective, a critical approach to creak aims only partly at deciphering its sonic materiality; the primary aim is to understand the interplay between those three dimensions.[19] Their integration in lived experience is what makes invention (or *poetic mimesis*) possible. In so doing, the perceptual "tracks" on which pulse phonation "runs" may be changed and new possibilities opened. These are the "openings" that Donna Haraway advocated and from which this chapter began.

In place of a conclusion, I would like to draw attention one last time to the very first image mentioned, the "phenomenon of vocal fry" and the paradigm it portrays. We see that it now looks different. By examining different species, we have embedded pulse phonation in a living soundscape of more-than-human vocalizations, one that have multiple affordances and effects depending on the niche where they become performative. Furthermore, we have located the pulse register outside or alongside *phoné*, highlighting the transformative and creative experiences this enables through the creation of poetic, ludic, and less-than-human sounds, forming a whole area of our voices and those of our animal other. The pulse register eludes selection but remains at our disposal. In this sense, it can function as an escape and reconciliation. The world is full of creaky sounds. Things creak, creaking brings things about, and we can do things with creak. There are also things we may learn from it.

Endnotes

1. See Introduction and Chapters 1 and 6 in this volume. For further reading on the "phenomenon of vocal fry" in the United States, New Zealand, and England, see, e.g., Yuasa 2010; Wolk, Abdelli-Beruh, and Slain 2012; Anderson et al. 2014; Oliveira *et al.* 2016; Borrie and Delfino 2017; Cantor-Cutiva, Bottalico, and Hunter 2018; and Davidson 2020.
2. Each of the terms mentioned has its own *raison d'être* and should not be considered identical to "vocal fry." Although this term is often used interchangeably with "creak," there is a significant difference between them (Keating *et al.* 2015; see Chapter 4). Here, the focus is not on the minute acoustic and physiological differences between creak patterns across species, but rather on their familiarity at the perceptual level and their differences at the level of cultural practices and resources. "Creak," "pulse phonation," and "pulsed voice" encompass all variants. "Pulse register" and "M0" are interchangeable and point specifically to the mechanism by which all pulsed voices in the world are made possible, from bird calls and echolocation clicks to speech patterns and moans. "Pulse register" emphasizes the range (register) of the resulting sound (the term has been used to describe primates, humans, cats, and birds), while "M0" emphasizes the physiological reality involved (the term has been used for both land and sea mammals). The pulse register (M0) should be distinguished from the vocal-ventricular phonation mode (Fuks *et al.* 1998). Although less common, the term "pulsed

voice" has been used to describe certain subtypes of creak. Here, it is a descriptive synonym for "creak." The term "creaky voice" is used by some phoneticians as an umbrella term, while others use it to refer to a specific type (see Davidson 2019). People often do not distinguish between creak and creaky voice (Nix and Chappell, this collection), although the two have proved different (Meireles, this collection). The different specific circumstances of each of these sounds cannot be discussed here (see Chapters 1–5). This chapter acknowledges the cultural and contextual differences while recognizing the common qualities and emergent characteristics of the pulse register.

3. Kim Kardashian's "vocal fry" has been associated with her public persona, and her influence in popular culture helped raise awareness about pulse phonation. Several YouTube videos compile instances where her voice exhibits a distinct creaky quality. The editing process of these compilations results in a fragmented stream of words that lacks coherent meaning.
4. While in speech pulse phonation has been pejoratively associated with "younger (often white American) women" (Davidson 2020: 13), in singing it has been approvingly (yet in part erroneously) associated with the Tuvan *Kargyraa*, one of the two main variants of Zöömej, the traditional overtone singing of Tuva (Russia), which is based on the "vocal-ventricular phonation mode" instead. This mode is similar to the pulse register in terms of frequency range (under 100 Hz) but differs from the pulse register by involving the ventricular folds, also known as "false folds," in the process (Fuks *et al.* 1998; Lindestad *et al.* 2001). In both cases, the pulse register is made to coincide with the vocal techniques and the techniques of the self that it allows. The distinctions between creaking in singing (e.g., to emulate Oktavists), sustained creak in speech (e.g., whole sentences uttered in creak), occasional and nonverbal creaking (e.g., interjections such as "umm," end-of-phrase utterances, etc.), and creak with a linguistic function (e.g., in Japanese, see Chapter 4), all collapse into the speech pattern that has become famous, or infamous, under the name "vocal fry."
5. Elsewhere, protophones are grouped differently.
6. Other children's games include sustaining the pulsed voice and modulating it with vowels or trying to separate the pulses as much as possible.
7. In their groundbreaking research on animal vocalization, Tobias Riede and Klaus Zuberbühler were the first to adopt the term

"pulse register" for humans, cats, and primates alike, emphasizing the similarity of vocalizations across these species: "We used the term 'pulse register' to describe the Diana monkey calls, because these vocalizations strongly resemble the pulse register of humans" (Riede and Zuberbühler 2003: 2920).

8. For further reading, see Peters and Tonkin-Leyhausen (1999).
9. For a comparative analysis of onomatopoeias in languages around the world, see Tsoi and Chung (2004).
10. The ancient Greek word mimesis (μῖμησις) literally means "imitation," but scholars warn that understanding it this way may not be helpful (e.g., Woodruff 1992). The term gained importance with Plato and Aristotle. Depending on the context, there are English translations such as reproduction, representation, "functional deception" (Woodruff 1992: 91), and "iconicity" (Kosman 1992). In 20th century's literary criticism, the term referred to the representation of a sociocultural reality rendered as impersonally as possible (Auerbach 1946/2003). Contemporary views on mimesis move beyond its traditional association with representation, focusing instead on the act of imitating others as an anthropological phenomenon. In short, mimesis has to do with what exists. Poiesis (ποίησις), on the other hand, is more about the non-existent: bringing into existence something that did not exist before. The word literally means "to make" and is the root of "poetry," which in ancient Greek thought was conceptually close to the doing. In modern understanding, poiesis refers to the unfolding of a thing out of itself. Martin Heidegger theorizes the term in the sense of "bringing forth" (1977: 317), which he illustrates with the breaking open of a flower to blossom. For an account of more recent understandings of the theme of mimesis, see Lawtoo (2022).
11. On the connection between *mimesis* and *methexis*, see Nancy (2016).
12. The spelling of the word "Q-/-uatre" is an homage to Marcel Duchamp's four postcards (see Chapter 18: 445).
13. Sustained sound is the "other" of percussive sound, as much as, for example, a line would be the "other" of a dot.
14. If we compare the modern onomatopoeia for the frog croak in different world's languages, we will find many differences. The German "gribbit" and the English "ribbit" are similar, whereas the Swedish "ko ack ack ack kväk" is remarkably distant, as much as the Italian "cra cra" (or Giovanni Pascoli's famous "gre gre"). In Telugu, frogs go "beka beka," which is close to the Sinhalese's "baka

baka." Differently, in Marathi it is "darav-darav," while in Tagalog it is "kokak kokak," and in Vietnamese "ồm ộp, ộp ộp."

15. The fact that the pulse register does not need sophisticated neural control has also been discussed in birds (Jensen *et al.* 2007).
16. The term "*Strohbass*" derives from the German "stroh" ("straw"), and it often refers to difference tones in which a stable frequency is produced, but in certain languages (e.g., Russian), it becomes the equivalent of the English "vocal fry." While controversial, in music settings *Strohbass* names the loud and resonant subharmonic singing that surpasses basic creak in loudness (Wade 2019). Among the different creak patterns, the period-doubling pattern (Bailly, Bernardoni, and Pelorson 2010) is responsible for the loudest "subharmonics," an acoustic phenomenon where two frequencies are heard as one resultant sound, or difference tone. Similar to the harmonic series, but based on division and, therefore, reflected downward, difference tones follow the subharmonic series. When the frequency of two sources is in a relationship of fifth, the resultant tone sounds an octave lower (the first common subharmonic, or "octave subharmonic"). This technique is made possible either by the pulse register (thanks to "double pulse" pattern; see Chapter 2: 59) or the vocal-ventricular phonation mode (see note two) and can be handled perfectly with multiple tones, each produced with a combination of two sources (trains of pulses) to form a difference tone. While it is not possible to sing melodies, singers can learn to hit these tones with great precision.
17. The Oktavist is a deep bass choir singer, kin to the Italian *Basso profondo*, but none of the two would use M0 (only M1). The pulse register can be used to produce similarly low tones (below 100 Hz). See also note fifteen.
18. On the queer aesthetics of pulse phonation, see Venturi (2021).
19. On the relationships between lived matter, cultural practices, and cultural resources, see the "Introduction" to this book.

References

Alien Resurrection. (1997). [Film]. Directed by Jean-Pierre Jeunet. 20th Century Fox.

Allora, J. and Calzadilla, G. (2013). *Apotomē* [Film]. Super 16 mm film transferred to HD, Los Angeles: REDCAT.

Anderson, R. C., Klofstad, C. A., Mayew, W. J., and Venkatachalam, M. (2014). "Vocal fry may undermine the success of young women in the labor market." *PloS One*, **9**(5), e97506.

Assaneo, M. F., Nichols, J. I., and Trevisan, M. A. (2011). "The anatomy of onomatopoeia." *PloS One*, **6**(12), e28317.

Au, W. W. L. (1993). *The Sonar of the Dolphin*. Springer Verlag.

Auerbach, E. (2003). *Mimesis: The Representation of Reality in Western Literature*. (W. Trask, Trans.). Princeton University Press.

Aumann, A. (2021). "Kierkegaard on the transformative power of art." *History of European Ideas*, **47**(3), 429–442.

Austin, J. L. (1962). *How to Do Things with Words*. Oxford: Clarendon Press.

Bachelard, G. (1946/2014). *The Poetics of Space*. London: Penguin Books.

Bailly, L., Henrich Bernardoni, N., and Pelorson, X. (2010). "Vocal fold and ventricular fold vibration in period-doubling phonation: Physiological description and aerodynamic modeling." *Journal of the Acoustical Society of America*, **127**(5), 3212–3222.

BBC One. (2022). "Life of Mammals." Produced by David Attenborough. BBC One.

Birgersson, B. and Ekvall, K. (1997). "Early growth in male and female fallow deer fawns." *Behavioral Ecology*, **8**, 493–499.

Blomgren, M., Chen, Y., Ng, M., and Gilbert, H. (1998). "Acoustic, aerodynamic, physiologic, and perceptual properties of modal and vocal fry registers." *The Journal of the Acoustical Society of America*, **103**(5 Pt 1), 2649–2658.

Borch, D., Sundberg, J., Lindestad, P. A., and Thalén, M. (2004). "Vocal fold vibration and voice source aperiodicity in 'dist' tones: A study of a timbral ornament in rock singing." *Logopedics, Phoniatrics, Vocology*, **29**(4), 147–153.

Borrie, S. A. and Delfino, C. R. (2017). "Conversational entrainment of vocal fry in young adult female American English speakers." *Journal of Voice*, **31**(4), 513.e25–513.e32.

Bourriaud, N. (2006). *Postproduction*. (J. Herman, Trans.). Sternberg Press.

Bredin, H.T. (1996). "Onomatopoeia as a figure and a linguistic principle." *New Literary History*, **27**, 555–569.

Brown, R. W., Black, A. H., and Horowitz, A. E. (1955). "Phonetic symbolism in natural languages." *Journal of Abnormal and Social Psychology*, **50**, 388–393.

Buder, E. H., Warlaumont, A. S., and Oller, D. K. (2013). "An acoustic phonetic catalog of prespeech vocalizations from a developmental perspective." In B. Peter and A. N. MacLeod (Eds.), *Comprehensive Perspectives on Child Speech Development and Disorders: Pathways from Linguistic Theory to Clinical Practice* (pp. 103–134). Nova Science Publishers.

Cantor-Cutiva, L. C., Bottalico, P., and Hunter, E. (2018). "Factors associated with vocal fry among college students." *Logopedics, Phoniatrics, Vocology*, **43**(2), 73–79.

Catford, J. C. (1964). "Phonation types: The classification of some laryngeal components of speech production." In D. Abercrombie (Ed.), *In honor of Daniel Jones*. Longmans.

Chao, M. and Bursten, J. (2021). "Girl talk: Understanding negative reactions to female vocal fry." *Hypatia*, **36**(1), 42–59.

Cresti, C., Dovetto, F. M., and Rocha, B. (2015). "Schizophrenia and prosody. First investigations." In C. Manfredi (Ed.), *IX International Workshop Model and Analysis of Vocal Emissions for Biomedical Applications (MAVEBA)*, September 2–4, 2015. Proceedings (pp. 139–142). Firenze University Press.

Davidson, L. (2019). "Perceptual coherence of creaky voice qualities." *Paper presented at the International Congress of Phonetic Sciences* (ICPhS) 2019.

Davidson, L. (2020). "The versatility of creaky phonation: Segmental, prosodic, and sociolinguistic uses in the world's languages." *Wiley Interdisciplinary Reviews: Cognitive Science*, e1547.

Davies, B. (2021). *Entanglement in the World's Becoming and the Doing of New Materialist Inquiry*. Routledge.

Donald, M. (2016). "Key cognitive preconditions for the evolution of language." *Psychonomic Bulletin and Review*, 1–5.

Edmiston, P., Perlman, M., and Lupyan, G. (2017). "Creating words from iterated vocal imitation." *Proceedings of the 39th Annual Conference of the Cognitive Science Society* (CogSci 2017).

Fischer, J., Hammerschmidt, K., Cheney, D. L., and Seyfarth, R. M. (2002). "Acoustic features of male baboon loud calls: Influences of context, age, and individuality." *Journal of the Acoustical Society of America*, **111**, 1465–1474.

Fisher, M. (2016). *The Weird and the Eerie*. Repeater.

Fitch, W. T. (2002). "Comparative vocal production and the evolution of speech: Reinterpreting the descent of the larynx." In A. Wray (Ed.), *The Transition to Language* (pp. 21–45). Oxford University Press.

Fitch, W. T. and Hauser, M. D. (2002). "Unpacking 'honesty': Vertebrate vocal production and the evolution of acoustic signals." In A. Simmons, R. R. Fay, and A. N. Popper (Eds.), *Acoustic Communication, Springer Handbook of Auditory Research* (pp. 65–137). Springer.

Foster, C. (2021). "Against nature writing." *Emergence Magazine*. Retrieved from https://emergencemagazine.org/essay/against-nature-writing.

Frey, R., Volodin, I., Volodina, E., Soldatova, N. V., and Juldaschev, E. T. (2011). "Descended and mobile larynx, vocal tract elongation and rutting roars in male goitred gazelles (*Gazella subgutturosa* Güldenstaedt, 1780)." *Journal of Anatomy*, **218**(5), 566–585.

Fujita, M., Yamasaki, S., Katagiri, C., Oshiro, I., Sano, K., Kurozumi, T., Sugawara, H., Kunikita, D., Matsuzaki, H., Kano, A., Okumura, T., Sone, T., Fujita, H., Kobayashi, S., Naruse, T., Kondo, M., Matsu'ura, S., Suwa, G., and Kaifu, Y. (2016). "Advanced maritime adaptation in the western Pacific coastal region extends back to 35,000–30,000 years before present." *Proceedings of the National Academy of Sciences*, **113**, 11184–11189.

Fuks, L., Hammarberg, B., and Sundberg, J. (1998). "A self-sustained vocal-ventricular phonation mode: Acoustical, aerodynamic and glottographic evidence." *KTH Speech, Music and Hearing: Quarterly Progress and Status Report*, **3**, 49–59.

Gaffney, D. (2021). "Pleistocene water crossings and adaptive flexibility within the Homo genus." *Journal of Archaeological Research*, **29**, 255–326.

Gamba, M. and Giacoma, C. (2008). "Subspecific divergence in the black lemur's low-pitched vocalizations." *The Open Acoustics Journal*, **1**, 49–53.

Ghostemane. (2020). Anti-Icon [Album]. Blackmage Records.

Grimm, R. J. (1967). "Catalogue of sounds of the pigtailed macaque (Macaca nemestrina)." *Journal of Zoology*, **152**, 361–373.

Habasque, P. (2021). "Is creaky voice a valley girl feature? Stancetaking & evolution of a linguistic stereotype," *Anglophonia*, **32**.

Haraway, D. (1988). "Situated knowledges: The science question in feminism and the privilege of partial perspective." *Feminist Studies*, **14**(3), 575–599.

Heidegger, M. (1977). *The Question Concerning Technology and Other Essays* (W. Lovitt, Trans.). Garland Publishing, Inc.

Hildebrand-Edgar, N. (2016). *Creaky Voice: An Interactional Resource for Indexing Authority* (Master's thesis). University of Victoria.

Hollien, H. (1974). "On vocal registers." *J. Phon.*, **2**(2): 125–143.

Hollien, H. and Wendahl, R. W. (1968). "Perceptual study of vocal fry." *Journal of the Acoustical Society of America*, **43**(3), 506–509.

Hornibrook, J., Ormond, T., and Maclagan, M. (2018). "Creaky voice or extreme vocal fry in young women." *New Zealand Medical Journal*, **131**(1486), 36–40.

Imai, M. and Kita, S. (2014). "The sound symbolism bootstrapping hypothesis for language acquisition and language evolution." *Philosophical Transactions of the Royal Society of London. Series B, Biological Sciences*, **369**(1651), 20130298.

Jensen, K. K., Cooper, B. G., Larsen, O. N., and Goller, F. (2007). "Songbirds use pulse tone register in two voices to generate low-frequency sound." *Proceedings of the Royal Society B*, **274**, 2703–2710.

Keating, P., Garellek, M., and Kreiman, J. (2015). "Acoustic properties of different kinds of creaky voice." *Proceedings of the 18th International Congress of Phonetic Sciences*.

Kim Dracula. (2023). *A Gradual Decline in Morale* [Album]. Columbia Records and Sony Music Entertainment Australia.

Kosman, A. (1992). "Acting: Drama as the mimesis of praxis." In A. Rorty (Ed.), *Essays on Aristotle's Rhetoric* (pp. 51–72). Princeton University Press.

La Barbara, J. (1980). *Reluctant Gypsy* [Album]. Wizard Records.

Latour, B. (1984). *Les Microbes: Guerre et Paix, Suivi de Irréductions*. Paris: Métailié.

Lawtoo, N. (2022). *Homo Mimeticus: A New Theory of Imitation*. Leuven: Leuven University Press.

Lefkowitz, D. and Sicoli, M. (2007). "Creaky voice: Constructions of gender and authority in American English conversation." *Paper presented at the 106th annual meeting of the American Anthropological Association*, Washington, DC.

Levine, S. K. (2009). *Trauma, Tragedy, Therapy: The Arts and Human Suffering*. Jessica Kingsley Publishers.

Ligon, C., Rountrey, C., Rank, N. V., Hull, M., and Khidr, A. (2019). "Perceived desirability of vocal fry among female speech communication disorders graduate students." *Journal of Voice*, **33**(5), 805-e21.

Lindestad, P., Soedersten, M., Merker, B., and Granqvist, S. (2001). "Voice source characteristics in Mongolian 'throat singing' studied with high- speed imaging technique, acoustic spectra, and inverse filtering." *Journal of Voice*, **15**, 78–85.

Lisson Gallery. (2014). "Allora and Calzadilla: Apotomē." Retrieved from https://www.lissongallery.com/news/xxx.

Madsen, P. T., Siebert, U., and Elemans, C. P. H. (2023). "Toothed whales use distinct vocal registers for echolocation and communication." *Science*, **379**(6635), 928–933.

Magnat, V. (2020). *The Performative Power of Vocality*. Routledge.

Martin, W. F. and Gans, C. (1972). "Muscular control of the vocal tract during release signalling in the toad Bufo valliceps." *Journal of Morphological Sciences*, **137**, 1–27.

Moeller, H. G. and Whitehead, A. (Eds.). (2018). *Imagination: Cross-Cultural Philosophical Analyses*. Bloomsbury Academic.

Møhl, B., Wahlberg, M., Madsen, P. T., Heerfordt, A., and Lund, A. (2003). "The monopulsed nature of sperm whale clicks." *Journal of the Acoustical Society of America*, **114**, 1143–1154.

Moser, H. M. (1942). "Symposium on unique cases of speech disorders; Presentation of a case." *The Journal of Speech Disorders*, **7**(2), 173–174.

Nancy, J.-L. (1982). *Le Partage des Voix*. Paris: Galilée.

Nancy, J.-L. (2016). "The image: Mimesis and methexis." In C. Giunta (Ed.), *Nancy and Visual Culture*. Edinburgh: Edinburgh University Press.

Negus, V. E. (1949). *The Comparative Anatomy and Physiology of the Larynx*. Hafner.

Neveu, A. (2009). "Edible frogs." In P. Safran (Ed.), *The Encyclopedia of Life Support Systems, Fisheries and Aquaculture*, Vol. III. UNESCO Publishing.

Novak, J. (2012). *Singing Corporeality: Reinventing the Vocalic Body in Postopera* (Phd dissertation). Universiteit van Amsterdam, Amsterdam, Netherlands. Retrieved from https://hdl.handle.net/11245/1.378399.

Ohala, J. (1994). "The frequency code underlies the sound-symbolic use of voice pitch." In L. Hinton, J. Nichols, and J. J. Ohala (Eds.), *Sound Symbolism*. Cambridge University Press.

Oliveira, G., Davidson, A., Holczer, R., Kaplan, S., and Paretzky, A. (2016). "A comparison of the use of glottal fry in the spontaneous speech of young and middle-aged American Women." *Journal of Voice*, **30**(6), 684–687.

Oller, D. K., Caskey, M., Yoo, H., Bene, E. R., Jhang, Y., Lee, C.-C., Bowman, D. D., Long, H. L., Buder, E. H., and Vohr, B. (2019). "Preterm and full-term infant vocalization and the origin of language." *Scientific Reports*, **9**, 14734.

Paolillo, N. P., Carrozza, L., Oslo, M., Rosa, E., and Scalabrin, M. (2019), "Inspiratory vocal fry: Anatomical and physiological aspects, application in speech therapy, vocal pedagogy and singing. A pilot study." *Journal of Voice*, **35**(3): 394–399.

Paul, L. A. (2014). *Transformative Experience*. Oxford University Press.

Peters, G. (1981). "Das Schnurren der Katzen." *Säugetierkundliche Mitteilungen*, **29**, 30–37.

Peters, G. and Tonkin-Leyhausen, B. A. (1999). "Evolution of acoustic communication signals of mammals: Friendly close-range vocalizations in felidae (carnivora)." *Journal of Mammalian Evolution*, **6**, 129–159.

Pikes, M. and Campbell, P. (2021). *Owning Our Voices: Vocal Discovery in the Wolfsohn-Hart Tradition*. London: Routledge.

Pisanski, K. K., Anikin, A., and Reby, D. (2022). "Vocal size exaggeration may have contributed to the origins of vocalic complexity." *Philosophical Transactions of the Royal Society B: Biological Sciences*, **377**(1841).

Pisanski, K., Isenstein, S. G. E., Montano, K. J., and others (2017). "Low is large: Spatial location and pitch interact in voice-based body size estimation." *Attention, Perception, & Psychophysics*, **79**, 1239–1251.

Priadji, G. (2016). *The Experience Series: Rully Shabara* [Video]. YouTube. Retrieved from https://www.youtube.com/watch?v=AE6cYBvyFUI.

Redi, L. and Shattuck-Hufnagel, S. (2001). "Variation in the realization of glottalization in normal speakers." *Journal of Phonetics*, **29**(4), 407–429.

Richman, B. (1976). "Some vocal distinctive features used by gelada monkeys." *Journal of the Acoustical Society of America*, **60**, 718–724.

Riede, T. and Zuberbühler, K. (2003). "Pulse register phonation in diana monkey alarm calls." *The Journal of the Acoustical Society of America*, **113**(5), 2919–2926.

Riede, T., Tokuda, I. T., and Farmer, C. G. (2011). "Subglottal pressure and fundamental frequency control in contact calls of juvenile *Alligator mississippiensis*." *Journal of Experimental Biology*, **214**(18), 3082–3095.

Schön Ybarra, M. A. (1986). "Loud calls of adult male red howling monkeys (*Alouatta seniculus*)." *Folia Primatologica*, **47**(4), 204–216.

Silvestro, M., Dovetto, F. M., Corvino, V., Apisa, P., Malesci, R., Tessitore, A., Milizia, P., Tedeschi, G., Marciano, E., and Russo, A. (2021). "Enlarging the spectrum of cluster headache: Extracranial autonomic involvement revealed by voice analysis." *Headache. The Journal of Head and Face Pain*, **61**(9), 1452–1459.

Smith, M. (1980). "Behavior of the koala, *Phascolarctos cinereus* (Goldfuss), in captivity, III. Vocalizations." *Australian Wildlife Research*, **7**, 13–34.

Stark, R., Rose, S., and McLagen, M. (1975). "Features of infant sounds: The first eight weeks of life." *Journal of Child Language*, **2**(2), 205–221.

Suthers, R. A., Wild, J. M., and Kaplan, G. (2011). "Mechanisms of song production in the Australian magpie." *Journal of Comparative Physiology. A, Neuroethology, Sensory, Neural, and Behavioral Physiology*, **197**(1), 45–59.

Titze I. R. (2014). "Bi-stable vocal fold adduction: A mechanism of modal-falsetto register shifts and mixed registration." *The Journal of the Acoustical Society of America*, **135**: 2091–2101.

Tsoi, W. C. T. and Chung, H. W. C. (2004). *Characteristics of Onomatopoeia*. Unpublished study, Chinese University of Hong Kong.

Venturi, F. (2021). "Creaky voice gender." *Journal of Interdisciplinary Voice Studies*, **6**(2), 201–218.

Wade, E. (2019). "On the voice: Developing bass sections through extended techniques and auxiliary registers," *The Choral Journal*, **60**(2): 55–66.

Ward, N. (2006). "Non-lexical conversational sounds in American English." *Pragmatics & Cognition*, **14**(1), 129–182.

Winter, P., Ploog, D., and Latta, J. (1966). "Vocal repertoire of the squirrel monkey (*Saimiri sciureus*), its analysis and significance." *Experimental Brain Research*, **1**, 359–384.

Wolf, N. (2015). Young women, give up the vocal fry and reclaim your strong female voice. *The Guardian*. Retrieved from https://www.theguardian.com/commentisfree/2015/jul/24/vocal-fry-strong-femalevoice.

Wolk, L., Abdelli-Beruh, N. B., and Slavin, D. (2012). "Habitual use of vocal fry in young adult female speakers." *Journal of Voice*, **26**(3): e111–e116.

Woodruff, P. (1992). "Aristotle on mimesis." In A. Oksenberg Rorty (Ed.), *Essays on Aristotle's Poetics* (pp. 73–95). Princeton University Press.

Yuasa, I. P. (2010). "Creaky voice: A new feminine voice quality for young urban-oriented upwardly mobile American Women?" *American Speech*, **85**(3): 315–37.

Zeifman, D. M. (2001). "An ethological analysis of human infant crying: Answering Tinbergen's four questions." *Developmental Psychobiology*, **39**, 265–285.

Chapter Twelve

From extreme to everyday
Vocal fry in contemporary theory and practice of music

Bojana Radovanović

Introduction

Up until recently—and even today by numerous traditionalists—vocal fry, pulse phonation, or what otherwise goes by creak was considered an incorrect vocal gesture in music.[1] Notably, this was/is the stance of classical vocal pedagogy in Western music postulates. Regardless, more and more composers, performers, and pedagogues seem to be including this technique in their work.

Moreover, when looking at the broader picture, recent research demonstrates the prevalence of pulse phonation in contemporary speech patterns. Namely, in the last decade or so, researchers from different fields (but mainly linguistics) strived to examine the use, and listeners' recognition, of vocal fry in everyday speech. Their

Creak: Theories and practices of pulse phonation
Edited by Francesco Venturi

ISBN 978-981-5129-29-8 (Hardcover), 978-1-003-61606-1 (eBook)
www.jennystanford.com

research has indicated that monolingual and bilingual women and men speaking English (see Wolk *et al.* 2012; Abdelli-Beruh *et al.* 2014; Davidson 2019; Gibson *et al.* 2021) are particularly inclined to this, utilizing creak in place of punctuation marks or as a prosodic device. For instance, instead of lowering the intonation within a "normal speaking register"—which here signifies a modal voice register—people have been using vocal fry, the low and creaky phonation, to indicate the end of a thought or a sentence. These vocal gestures can be interpreted in different ways by the listeners, depending on the context and the speaker in question. Monica Chao and Julia R. S. Bursten, to take an example, look at this issue from the perspective of philosophy of language, feminist epistemology, and linguistics; they argue that the critics of vocal fry render this technique/register as an exclusively female trait in speech. As such, it serves as an inlet to criticizing and limiting female communicative autonomy: "[...] when fry is heard as annoying and distracting, it is because the hearer interprets the speaker as echoing an utterance from a position of authority to which she is not entitled. We show that this reaction encodes conscious or unconscious sexist attitudes toward women's voices" (Chao and Bursten 2021: 42).

The policing and criticism of vocal fry are not an exclusive and/or isolated occurrence. As Nassima B. Abdelli-Beruh shows in her review on literature on this phenomenon within the field of speech-language pathology (SLP), vocal fry "went from being viewed mainly as deviant (i.e., a sign of vocal pathology and/or affectation) to being viewed as 'normal', acquiring along the way the status of register in the late 1960s and therapeutic value in the 1990s" (Abdelli-Beruh, this collection: 25). This means that, before the second half of the 20th century in the United States, this vocal behavior was considered a bad habit that disturbed the purity of the "good American speech" (Abdelli-Beruh, this collection: 26).

In fact, throughout the history of the Western society, many vocal emissions are and have been considered undesirable, unfamiliar, unregulated, and non-standardized, in both everyday vocal behavior and in the performing arts such as music, theater, and spoken poetry. In this chapter, I will look into the phenomenon of vocal fry from the perspective of the "monstrous voice," a voice that simultaneously resists standardization, assumes the role of the Other, and teaches people in one society about themselves and their individual subjectivities mirrored in their voices. In that regard, I will address the gradual normalization of these extended vocal techniques with a particular attention given to creak and will do so looking into contemporary art, music, and science practices (composition, performance, theory/musicology, and vocal pedagogy). Furthermore, by focusing on the Yugoslav and Serbian context in this regard—a cultural space that is historically and artistically shaped primarily in line with European, i.e., Western cultural values—I will give voice to a peripheral scene and in turn contribute to a wider international discussion on the issue of vocal fry.

Vocal fry: From harmful to extended vocal technique

In order to define extreme vocalization, I refer to and broaden Jelena Novak's definition of the monstrous voice. Namely, in her book *Postopera: Reinventing the Voice-Body*, Novak describes the monstrous voice as a vocal expression/voice, "that voice that deviates from 'the natural,' and that has the sound appearance of a monster" (2015: 58). In the context of her research—the context of contemporary, post-opera production—this voice is aided by technological means and, as such, the "singing appears monstrous as a result of existing beyond the body that produces it" (Novak 2015: 15). Monstrous voice is built as a theoretical reflection on Bojana Kunst's theorization of the monstrous body—a body that is "other than human, but also a symptom of rethinking the human" (Novak 2015: 66; also see Kunst 2008).

This appearance of monstrosity in voice likewise intrigued my own considerations of voice in music. Namely, I argue that this notion of monstrous voice is applicable to the vocalizations that utilize extended vocal techniques, i.e., all those vocal techniques that are not considered a part of the musical vocal mainstream (see Radovanović 2021a; 2021b; 2023a; 2024 forthcoming). In addition to techniques such as growl, scream, shriek, gutturals, and others[2] that are commonly used in extreme metal music, creak can also be considered a monstrous vocalization, given its history and bad reputation during the history of Western vocal pedagogy.

So, what stands opposed to the monstrous voice? Opposed is the regulated, normalized, adapted, and cultivated voice of Western music, achieved through many years of learning and practice. Speaking in more general terms, the regulated voice is one with the traits of a civilized society—be it a singing or speaking voice. Extreme vocalizations are considered either a rare exception—in emotionally saturated situations—or unacceptable.

As for singing, looking back all the way to the Middle Ages—through sources that primarily documented vocal music in churches—there are several points that seem to have endured the test of time as the invariant norms of a good vocal performance. Those encompass precise intonation, clear pronunciation of the text, expressiveness, and durability (Conrad von Zabern, quoted in McGee 2012). In a shift from vocal poliphony to monody, and further on, in the establishment of what is now most commonly known as bel canto singing (see Stark 1999), those qualities—alongside the everlasting debate on vocal registers—remained the cornerstones of the singing practice of the Western cultural sphere.

With the development of contemporary vocal pedagogy, which is based on voice science ever since Manuel Patricio Rodríguez García (1805–1906) discovered the laryngoscope, vocal artists and pedagogues found efficient ways of regulating and training voice in classical music and, during the last several decades, music theater, jazz, contemporary, and popular music. The development of vocal science and phoniatric tools made it possible for many pedagogues and vocal performers to be better informed about their vocal health. In that respect, during the 20th century, vocal scientists claimed that creak was an abnormal type of vocal emission, which is dangerous and harmful for the one performing or speaking in that register (Moore and von Leden 1958; Wolfe *et al.* 1991). As previously mentioned, this attitude started changing in the SLP literature in the 1960s, which is the field that granted vocal fry the status of register.

More recent pedagogical contributions are more approving of the creak in music, as it is increasingly more prevalent in the domain of contemporary vocal art, especially when the branches of vocal pedagogy and phoniatry are successfully and expertly intertwined. This phenomenon, therefore, still causes different opinions from various sides of music pedagogy: in professional literature, some writers ignore it completely, some acknowledge it without further discussion on its nature or practicality, while there are also the ones who have strong positive or negative opinion on its use in singing training (Nix and Chappell , this collection: 68). An analogy to the therapeutic value of pulse phonation in SLP, as mentioned before with regards to Abdelli-Beruh's paper, can be seen in the works of music pedagogues such as William Vennard (1967), Frederick Swanson (1984), Kari Ragan (2020), and others, whereas they advocate for this technique as a valuable corrective or developmental tool (Nix and Chappell, this collection: 70). One of the examples of the second group coming from the Yugoslav/ Serbian context can be seen in a book *The Art of Singing* (Cvejić 1994; 2008), written by Serbian operatic soprano and vocal pedagogue Biserka Cvejić (1923–2021) and her husband, famous phoniatrist Dušan Cvejić (1923–1998).[3] In their study, Biserka and Dušan Cvejić classified vocal fry as a contraoctave register in male singers, without much detailing or acknowledging of the possibilities of female voices also performing creak.

Classifying vocal fry as an "extra-normal" vocal technique in his book *The 21st Century Voice: Contemporary and Traditional Extra-Normal Voice*, composer Michael Edward Edgerton further perpetuates the otherness/monstrosity discourse surrounding this vocal gesture (2015). However, together with his physiological analysis of this performing technique, Edgerton underlines that vocal fry is also an overused and stereotypical sound in contemporary music, and that it is usually a "vehicle or doorway toward finding more interesting asymmetries" (2015: 22).

In line with Edgerton's understanding of vocal fry, famous experimental vocalist and artist Joan La Barbara (b. 1947) views vocal fry as a type of extreme vocalization:

> Well, there's the vocal fry: it sounds like an egg frying. It's actually an exhaled subtone, a growly sound that be trained so that it gives a more musical effect. The inhaled glottal click is the exhaled version of the fry, which tends to be more percussive and one can control the rate and speed of the clicking sound somewhat. You make a low tone, so your vocal cords are in a relaxed configuration and then take in a little air and a clicking sound emerges (2017).[4]

One of the most prominent characteristics of La Barbara's poetic is the drive for experimentation and exploration of the capabilities and limits of her body and voice. The quote above tells us that she not only knew and wanted to produce a certain vocal gesture—in this case, the creak—but also was aware of the processes in the body that lead to the production of the said gesture. A deep awareness of one's own mechanisms is, for La Barbara, the source of inspiration and stimulus to study new techniques in the manner of an etude. This was sometimes done quite literally, as in the case of *Circular Song* (1975), where she works with the notion of circular breathing while vocalizing. In other cases, La Barbara uses many different techniques to conjure up a specific imagery, or *sound picture*. Inhaled creak and exhaled creak thus appear among other techniques in La Barbara's pieces for solo voice or voice with electronics such as *Voice Piece: One-Note Internal Resonance Investigation*. And in particular, *Twelvesong* (in the lowest layer of the vocal polyphony created for different tracks in studio) from her album *Voice is the Original Instrument* (1976).

In the context of expanding the expectation and possibilities of voice in music and theater, Roy Hart (1926–75) as well as his teacher Alfred Wolfsohn (1898–1962) should also be mentioned. Namely, as somewhat of a pioneer in a field of experimental voice in music, Wolfsohn was a German vocal pedagogue who, after World War I, suffered from PTSD and vivid auditory hallucination of screaming soldiers. Following the unsuccessful therapy prescribed to him as a cure, Wolfsohn left the bel canto approach behind and started exploring the range and possibilities of voice beyond gender boundaries and expectations (Newham 1997; Braggins 2011). As one of Wolfsohn's most well-known pupils, Roy Hart was a South African vocal artist and actor who introduced the cry, "a nonverbal, nonmusical form (or forms) of expression deriving from and conveying strong emotional states," and did not consider those "extended techniques" to actually be extended, but a neglected, natural part of human vocal repertoire (Salzman and Desi 2008: 275). Diversifying his vocal production to include, among others, belches, farts, growls, vocal fry, squeals, squeaks, and screams of all sorts (2008: 275) was a precondition for him to perform many contemporary vocal pieces that count on such an ability (i.e., Maxwell Davies' *Eight Songs for a Mad King*, 1969).

During the same time and in the similarly minded (i.e., experimentally oriented) scenes of other performing arts, vocal fry was also gaining more popularity as an expression tool. This is especially evident in the domain of sound poetry during the second half of the 20th century and in the first decades of the 21st—with Jaap Blonk (b. 1953) and The Four Horsemen featuring BpNichol (1944–88), Steve McCaffery (b. 1947), Paul Dutton (b. 1943), and Rafael Barreto-Rivera (b. 1944)—as a few examples of the artists who used vocal fry as one of many variants to perform vocally.

These experimental performing arts scenes served as a litmus test and a gateway to everyday and mainstream artistic usage of vocal fry. Vocal fry is on the spectrum of techniques in popular music (pop, r'n'b, blues, jazz), which is also recognized and noted in numerous news outlets and Internet portals,[5] and in its more extreme tendencies in metal music.[6] With this gradual adoption of vocal fry in various areas and genres of music and other performing arts, it is quite expected for this gesture to be completely included in vocal techniques repertoire in contemporary music. In the next part of the chapter, I will give a brief overview of Serbian history of vocal music, with special emphasis on the inclusion of new techniques, experimentation periods, and, finally, normalizations of vocal fry in music composition and performance.

Vocal fry in Yugoslav and contemporary Serbian music scene

Although there are many takes on using vocal fry in contemporary music, here I will take into account the composers and performers from Serbia. In doing so I intend to diversify knowledge in the field by giving space to less-known authors from peripheral music scenes. The choice of the examples I will present here is also a result of the specificity of the local context in which creak is—only recently and fairly carefully—used in the young authors' compositional process, as well as performers' repertoire of techniques.

The history of vocal music in the church and arts in Serbia is long and abundant, dating back to the Middle Ages—especially in the domain of choir and group singing in general. This line of development was empowered by the late awakening of Serbian art music in the 19th century, after centuries of slumber, when independence from the Ottoman Empire was regained. Some of the most important founders in that sense were Kornelije Stanković (1831–65), Stevan Stojanović Mokranjac (1856–1914), and Josif Marinković (1851–1931), who wrote extensively for church choirs and specific singing societies.

Perhaps this tardiness in the development of music institutions—i.e., places that stimulated composing, performing, and studying music—combined with the strong impetus of tradition and group musicking (partly influenced by the folk music) throughout the country and the region, influenced the general state of the field, impeding desire for the exploration and experimentation with voice. After World War II, in the context of socialist realism in the newly established Federal People's Republic of Yugoslavia (from 1963 to 1991, Socialist Federal Republic of Yugoslavia) and with the young or just established institutions, composers worked hard to make up for the times lost, and continued building a repertoire that could suit both contemporary styles and public requirements (Medić 2007, 2021; Mikić 2007, 2009). In that context, there could hardly be any place for vocal technique research. Even the most avant-garde figures of Serbian and Yugoslav music such as Vladan Radovanović (1932–2023) did not experiment much with the voice.

Musicological contributions to the study of art music with voice likewise reflect this condition, sometimes even working toward the goal of filling in the gaps and reading some compositional solutions as local reflections of the European avant-garde; such as detecting *Sprechgesang* or, more rarely, *Sprechstimme* in the compositions of expressionist provenance, which do not have that specific vocal requirement indicated (Masnikosa 2007). Namely, the culture of experimentation with voice was never completely embraced in 20th-century music in Serbia or Yugoslavia. Neither composers nor musical vocal performers strived to explore the possibilities of the vocal apparatus in a more significant way.

However, one notable exception should be mentioned here. Although her background is not in music but in acting and sound poetry, Katalin Ladik (b. 1942), a vocal artist, performer, and poet of Hungarian descent, born in Novi Sad, Yugoslavia, is a particularly captivating figure in that regard. As a vocal performer and a co-creator, she collaborated with several prominent Serbian, Croatian, and Hungarian composers during the second half of the 20th century. This list includes: Dušan Radić (1929–2010), Ernő Király (1919–2007),[7] Dubravko Detoni (b. 1937), Milko Keleman (1924–2018), Branimir Sakač (1919–79), Boris Kovač (b. 1955), Zsolt Kovács (b. 1969), and others. Ladik's musical endeavors encompassed electronic pieces, and experimental and improvisational music. As musicologist Milan Milojković pointed out, Ladik's career in music began in the late 1960s with her early interest in music with audiovisual work. Culminating during the 1970s, she intensely collaborated with Király, Detoni, Keleman, and Radić and was an integral part of the ACEZANTEZ ensemble (based in Zagreb) where she was responsible for vocal and dance elements (Milojković 2019: 123–124).

Regarding vocal experimentation where vocal fry can be incorporated (among other vocal gestures), during the 1970s, Katalin Ladik was active in the domain of film and phonopoetics.[8] The pieces created in that period, such as the sound layer of the film *O-pus* (1972) and her first solo record *Phonopoetica* (1976), in a way served as a polygon that demonstrated the various possibilities of Ladik's voice.

As noted before, artists in the domain of sound poetry were open to vocal experimentation, as a means of denouncing the domination of text and narrative in favor of non-semantic expression and composition/piece building. In line with that fact, theoretician Miško Šuvaković described Ladik's sound poetry/vocal art through the following words: "Katalin Ladik's phonic or oral—spoken or sonic—poetry rests on the asemantic use of human voice, speech acts and intersubjective situations capturing her body and extending it by her penetrative voice" (Šuvaković 2010: 109).

The way Katalin Ladik used her voice and body to serve her artistic persona was in complete synchronicity with foreign impulses in contemporary sound poetry, experimental music, and performance art. In such a context, *Phonopoetica* acted as an exhibition and an etude of her vocal possibilities. Embracing "inappropriate" and "unhealthy" vocal behaviors such as vocal fry, Katalin Ladik was one of the few artists who were not afraid to use and explore all the possibilities of her body. In *Phonopoetica*, Ladik also established electronic techniques as a prominent element in her work. During this decade, as previously noted, Ladik intensively collaborated with Király and the ensemble ACEZANTEZ. As a result of that collective endeavor—and in the light of vocal technique expansion—one should mention pieces such as *Apsurdna Priča / Abszurd Mese / Absurd Story* (name translated in Serbian, Hungarian, and English, respectively).[9] This piece was composed in 1971 and performed and recorded by Katalin Ladik as a lyricist and a soloist, with ACEZANTEZ and Dubravko Detoni as conductors.[10] Regarding vocal technique, this short work encompasses, on a small scale, everything Ladik was working with in this period: trills, ululations, vocal/glottal stops, creak, fast paced and intense pitch changes, screams and growls, non-semantic text, and other vocal sounds.

Interestingly enough, these vocal behavior breakthroughs did not happen in the domain of music—with the exception of Ladik's appearance in one of the most prominent roles in music events that were considered the avant-garde of their time, such as the performance of Dušan Radić's *Oratorio Profano* composed in 1972–74 and performed in 1979.[11] As the vocal soloist in front of three narrators, three choirs, three chamber instrumental groups, four orchestras, four kettle drums, organ and tape, Ladik uses diversified speech and singing techniques such as ululations, screeches, vocalizations on an inhale, and, briefly, vocal fry (noted especially during the ending of the third movement titled *Zbivanje / Happening* and in the fourth, *Zvučne paradigme / Sound Paradigms*).

Aside from Ladik's work and specific interventions in the Serbian and Yugoslav music scene, the last few decades of the 20th century in this context passed by without significant work in the domain of extended vocal techniques. At the turn of the millennia, voice became a more prominent tool/sound material in electroacoustic works of Serbian composers such as Srđan Hofman (1944–2021), Vladan Radovanović, and Jasna Veličković (b. 1974);[12] still there was not enough emphasis on the exploration of the voice. This trend continued in compositions of numerous contemporary authors: Tatjana Milošević Mijanović (b. 1970), Svetlana Savić (b. 1971), Ana Gnjatović (b. 1984), and Svetlana Maraš (b. 1985). One of the most notable examples is a piece where the composer Svetlana Maraš works with excerpts of language or, more precisely, speech and "paralinguistic elements" titled *Jezik* (2016).[13] Here, fragmented recordings of various human voices are subjected to electronic manipulation in a post-digital soundscape. Although not dealing directly with vocal fry as a sound category, Maraš sheds light on numerous sonic and bodily elements and situations that human speech includes.

Going forward to the end of the second and the beginning of the third decade of the 21st century, a generational change has taken place, which is also visible in the material, influences, and techniques evident in the works of new authors. This younger generation, which has grown up alongside exceptionally prolific and omnipresent popular music in the last several decades, has used these musical expressions in their own compositions. Together with polistilicism and genre eclecticism, a wide variety of vocal techniques now come into play, especially in the works of two younger Serbian composers, Tamara Knežević (b. 1993) and Jug Marković (b. 1987).

In her piece *Krvava bajka* (*A Bloody Fairytale*) for soprano, flute, clarinet, percussion, and string orchestra (2020), Tamara Knežević introduced a diverse range of vocal techniques in order to depict a horrific story told in the text. Namely, *Krvava bajka* is based on the eponymous poem by famous Serbian poet Desanka Maksimović (1898–1993) about the Nazi German massacre of several thousand people in 1941 in Kragujevac (Serbia), among which were several hundred children that German Nazi soldiers pulled out of a school.[14]

Compositional choices—especially regarding vocal elements—were made in service of this startling text and the event itself. Extended techniques and various bodily gestures were also dispersed among other performers but are the most prominent in the soprano/narrator line. Here, Knežević wanted to "tell a story" with a myriad of techniques that would "bring the gravity and significance of every Desanka's word closer to the listener" (Rossi Fest 2021). What vocal performer Mina Gligorić—who premiered the piece—underlined as the most challenging part was the constant goal of shifting between emotions to enhance dramatic expression.

The voice in this piece contains and emits the whole diapason of emotions and states of being. From powerlessness, desperation, and disbelief, to anger and, in the end, a prayerful, solemn plateau, the soprano encapsulates the whole range with its vocal and bodily techniques. In this line of sound-pictures, which starts with the representation of the careless play of children through finger-snapping, clapping, and whistling, the dramaturgy of this horrid story moves through different vocal registers or placements (falsetto, vocal fry), types of pronunciation and speech (with pitch, in a trembling voice, in a hoarse voice), nuances of breath (forceful mouth breathing, inhale vocalizations, the transition from tone to breath/noise/murmur), types of falderal singing (trembling, without singing out, with a closed mouth), extreme vocal manifestation (growl, scream, screech, distortion), singing (angrily, through the teeth), and bel canto.

In a private correspondence, Tamara Knežević described her creative relation to vocal fry:

> As a composer, I find compositional coloristics and the spectrum of color and techniques which flow into each other very important. Depending on different factors, impostation, diction, breathing… voice can produce many different colors and sounds. As I listen to music, I always imagine how the different coloring of the vocal part can make it more interesting, especially in art music, by going beyond the school setting of the voice. Vocal fry is a very interesting and versatile technique, which can be used for voice distortion, a kind of harshness with which I became familiar as a teenager in pop, rock and metal music, but we can often hear it in speech, as well. When performed correctly, with a mild glottal pressure and without muscle tension, this can be a very rewarding technique because it cannot damage the performer's vocal folds. In the piece *Krvava bajka* specifically, the focus is on the verses by our great poet Desanka Maksimović (...) Every verse is filled with shivers, sobs,

screams, inhales, trembles and whispers. When you read the poem, the hoarseness of the voice I tried to emphasize in my music comes naturally (Knežević 2022, my translation).

Thus, we can acknowledge that the use of vocal fry here is—paired with many different kinds of vocalization—treated as a tool for dramatizing the text in a contemporary music environment. Tamara Knežević employs vocal fry in the second section of her piece (measures 38–157), which is simultaneously the most dramatic and vocally most diversified part of the composition. In terms of the poem, this section is dedicated to the lyrics that, in a painfully forthright manner, speak about small children of the same age, who had the same school habits and attended the same ceremonies, were vaccinated against the same diseases and who all died on the same day.

As per instruction for performing vocal fry in the measures 97–98, 103–106 (see Fig. 12.1) and, a little further, in measure 118 (see Fig. 12.2), the vocalist should use "fry vocal" (sic) and "soil the vocals with a creak, speak like having a hoarse voice" ("запрљати вокал крчањем, изговарати налик промуклом гласу"). All of these short episodes of vocal fry are placed within the turbulent orchestral environment. The first one, set with the two-syllable word "đaka" (ђака, eng. students), is performed by sustaining the two syllables throughout the two measures. While finalizing the second syllable, the vocalist should transition from vocal fry to audible breath. Just a few measures later, the vocal fry is introduced in a series of vocal technique transitions, only in reverse order: The word "đaka" begins in the measure 103 and is performed in a whisper that becomes audible breath/noise, followed by the vocal fry on the same syllable. This fragment culminates in a pure tone because of the instruction to gradually turn audible breath into distorted voice and, finally, scream.

The third occurrence of the "fry vocal," appearing while uttering the words "pe-de-set i" (педесет и, eng. fifty and), is immediately followed by a shout on a syllable "pet" (пет, eng. five; together with the previous fragment it reads: fifty-five), which then precedes the short singing episode. Here, we can ascertain that the composer possibly did not intend for the performer to use vocal fry but instead just to create an effect using a mix between the modal voice and a creak, signified by the instruction "to soil the voice" and marking the notes heads with an x sign in the note system, although they are said to be of a symbolic nature ("написане/назначене тонске висине су симболичне").

Figure 12.1 Excerpt from *Krvava bajka*, measures 96–107.

Figure 12.2 Excerpt from *Krvava bajka*, measures 118–121.

With all these similar guidelines placed in the culmination of the second and the longest section of the piece, it seems that the performers themselves can decide the amount or the level of creak they want to use. This decision is also likely influenced by the dynamic path the vocalist should follow right before or after the vocal fry, as well as the immediate orchestral surroundings and higher tempo. At the premiere performance of *Krvava bajka* in 2021,[15] Mina Gligorić makes various choices, congruent with these circumstances. Namely, she performs pure creak phonation in the first example. In the second—a transition toward clear singing tone—she barely colors the sound in line with the dynamics, while in the last, she performs using a distorted voice, a modal voice combined with creak, i.e., the creaky voice, to better achieve the shout and accommodate the entire sound picture (in *forte* dynamics).

Vocal fry is employed with a similar strategy in Jug Marković's piece *defiant walks barefoot* (2021). This voice and electronics duo incorporating video by Sara Marković Sara and texts by the poet and writer Ana Marija Grbić premiered at CentQuatre in Paris on June 12, 2021, and was the final result of the composer's attendance at the IRCAM Cursus.[16] Performed by Marković on electronics in collaboration with the new-age vocalist, experimental performance, and sound artist Thea Soti, this piece can also be looked at as a standalone music piece, without the video component.[17]

By choosing electroacoustic media paired with experimental voice, Marković continues on the path of many who explored this setting before him, those mentioned in this chapter and many more who are not. His vocal muse and accomplice is a well-known experimenter, who shows her affinities toward extended vocal techniques, voice in a digital environment and jazz, popular music, and folk-like singing in this piece as well.[18] Creak, although very sparingly used here, is imitated through the fast clicks in the electronic part. From different discrete vocal sounds performed with a close-miking technique in the beginning, to clean singing of the understandable text that dynamically grew toward the end of the piece, voice is manipulated and counterparted with rich and mesmerizing electronic layers.[19]

Conclusion

As many other techniques and gestures that were undesirable, unpleasant, and even repulsive, creak finally found its place in music of various backgrounds and genres in the 20th and 21st centuries. The forerunners of this shift *from extreme to everyday* were experimental vocal artists of the 20th century. They introduced vocal fry and other voicings as a kind of rebellion, an anti-tradition movement that was ingrained in the performing arts of this century. This avant-garde pressure against the dominance of narrative, spoken and intelligible text, and centuries-old performing techniques in theater, music, and poetry, advocated for the prevalence of other types of language and communication. Sound and body exploration was crucial for bringing light to many possibilities of the vocal apparatus. In the second half of the century, these particularly flourished, finding echoes in Serbia and Yugoslavia of that time.

Over the next few decades, and until today, we have witnessed creak become a tool to create desired effects in purely aesthetic, experimental, or narrative/dramatic fashion in popular and, what is especially worth noticing, contemporary art music. With the strong impact of experimental, avant-garde, and popular music genres of the previous century, composers of the younger generations, which are inclined toward working with voice, have included many different techniques next to the mainstream bel canto. The next step toward further normalization of vocal fry and other "unusual" vocal phenomena—particularly in smaller musical contexts such as the contemporary Serbian one—would be the establishment of several pedagogical centers that focus on contemporary vocal performance and extended techniques. Such ideas are not too far from reality in view of the fact that some professional classical singers already started working more in vocal pedagogy in their own private setting, opening to the possibilities of teaching and performing diverse musical styles, genres, and techniques.

In the meantime, as for vocal fry, we can assume that we are only at the outset of its more varied and ardent use in art and, especially, genre-crossover music, since these liminal artistic spaces historically tended to view "new" and "dangerous" vocal techniques without the burden of traditionalistic stigmatization.

Acknowledgments

The research for this chapter was done within the Institute of Musicology of the Serbian Academy of Sciences and Arts, funded by the Ministry of Science, Technological Development and Innovation of the Republic of Serbia (RS-200176).

Endnotes

1. In this chapter, the terms creak and vocal fry will be used almost as synonyms, with a slight advantage given to creak because it more precisely denominates a type of (pulse) phonation this publication is dedicated to. Vocal fry, on the other hand, is a term that more broadly signifies a phenomenon and is slightly more common in the literature. In my previous works, I have also used vocal fry as descriptor for this vocal technique, which is the reason I will not shy away from it here.
2. Or, in terms of vocal pedagogy of extreme metal classified by famous pedagogue Melissa Cross, fry screaming, false cords scream, and death scream.
3. Dušan Cvejić was one of the founders of the Union of the European Phoniatricians. This organization was established in Belgrade, Yugoslavia, in 1971. https://www.uep.phoniatrics.eu/inside_uep/history/index.html.
4. See: https://thequietus.com/articles/23735-joan-la-barbara-interview-lcmf-london-contemporary-music-festival.
5. See, as a few examples, a short article in the Science Daily (https://www.sciencedaily.com/releases/2016/05/160524121533.htm#:~:text=The%20lowest%20register%20used%20by,according%20to%20a%20small%20study), or an example of the list of pop music songs with vocal fry: https://www.thetoptens.com/songs-with-vocal-fry-technique/.
6. For example, in Lamb of God's *Walk with me in Hell* (02:40–02:46).
7. For more detailed explication on the nature of Ladik and Király's process, see Radovanović (2021b) and Sabo (2021).
8. Although some authors explain the term as a concept that pertains to the "emergence and making (poesis) of literary speech sounds (phono) as (...) heard in early spoken recordings" (Camlot 2019: 5), in this context, phonopoetics refers to the field of sound poetry, which

was developed during the 20th century primarily in Europe and North America as a response to centuries of "silent," read poetry. A reaction to the suppression of oral and folk poetic expressions came at the beginning of the century in dadaism, futurism, cubofuturism, surrealism, as well as Kurt Schwitter's art of Merz, which insisted on live performance of their pieces. After World War II, we witnessed a proliferation of sound poetry in experimental circles in the United States, Canada, and various European countries. The artists continue exploring the limits of their voice and body in performance, getting further away from linguistic systems and semantics (McCaffery 1978; Mekaferi 2000; Perloff and Dworkin (eds.) 2009; Rasula 2015; Radovanović 2018).

9. Composition *Absurd Story* can be heard on the following link: https://tracelabel.bandcamp.com/track/absurd-story-abszurd-mese.
10. *Absurd Story* was released in 1979 on an album Ernő Király (Jugoton/Udruženje Kompozitora SAP Vojvodine, Yugoslavia) and later on Spectrum (2001, Autobus/trAce Label, France). See: https://www.discogs.com/release/3131051-Ern%C5%91-Kir%C3%A1ly-Ern%C5%91-Kir%C3%A1ly; https://www.discogs.com/release/3686274-Ern%C5%91-Kir%C3%A1ly-Spectrum.
11. This performance can be heard on the following link: https://youtu.be/uOilZ4OKluQ.
12. Such as Hofman's *Uzorci* (Samples) for flute, clarinet, and electronics (1991), Veličković's *Vriskrik.exe*, concerto for live electronics and orchestra (1997/99).
13. "Језик," pronounced Jezik and meaning "tongue," but also "language" in Serbian, is a composition made of 11 short pieces, each being centered around the theme of human speech. These are short studies in language articulation, utilized by musical means." Excerpt from the composer's comment on her website: https://www.svetlanamaras.com/jezik.html.
14. More info on: https://www.spomenpark.rs/en/about-us/the-history.
15. The translation of the poem by Sara O'Keeffe is available on the following link: www.laban.rs/lib/Desanka_Maksimovich/A_Bloody_Fairytale.html.
16. Complete performance at the Rosi Fest 2021 can be seen on the following link: https://youtu.be/eMbkdhpp21E.
17. More on Soti's stances and experiments with voice in a digital environment can be seen in Radovanović (2021c).

18. Full performance from the IRCAM event can be seen on the following link: https://www.youtube.com/watch?v=5AfPIbFtlcQ&ab_channel=Ircam.
19. For more on the diversity of Jug Marković's music, see Radovanović (2023b).

References

Abdelli-Beruh, N. B. 2024. "From deviance to the birth of a register." This collection.

Abdelli-Beruh, N. B., Wolk, L., and Slavin, D. 2014. "Prevalence of vocal fry in young adult male American English speakers," *Journal of Voice* **28**(2): 185–190.

Braggins, S. 2011. *Mystery Behind the Voice: A Biography of Alfred Wolfsohn*. Market Harborough: Troubador Publishing.

Camlot, J. 2019. *Phonopoetics. The Making of Early Literary Recordings*. Stanford: Stanford University Press.

Chao, M. and Bursten, J. R. S. 2021. "Girl talk: Understanding negative reactions to female vocal fry," *Hypatia* **36**(1): 42–59.

Cvejić, B. and Cvejić, D. 1994/2007. *Umetnost pevanja*. Beograd: Fakultet muzičke umetnosti.

Cvejić, B. and Cvejić, D. 2008. *Art of Singing*. Begrade: IP Signature.

Davidson, L. 2019. "The effects of pitch, gender, and prosodic context on the identification of creaky voice," *Phonetica* **76**(4): 235–262.

Edgerton, M. E. 2015. The 21st-Century Voice: Contemporary and Traditional Extra-Normal Voice. Lanham: Rowman & Littlefield.

Gibson, T. A., Summers, C., and Walls, S. 2017. "Vocal fry use in adult female speakers exposed to two languages," *Journal of Voice* **31**(4): 510.e1–510.e5.

Knežević, T. 2020. *Krvava bajka*, composer's manuscript, Novi Sad.

Knežević, T. 2022, May 21. Private conversation with the composer via Facebook Messenger.

Kunst, B. 2008. "Restaging the monstrous," in M. Bleeker (Ed.), *Anatomy Live: Performance and the Operating Theatre*. Amsterdam: Amsterdam University Press, 211–222.

Masnikosa, M. 2007. "Ekspresionizam," in M. Veselinović-Hofman (Ed.), *Istorija srpske muzike: srpska muzika i evropsko muzičko nasleđe*. Beograd: Zavod za udžbenike, 165–192.

McCaffery, S. 1978. "Sound poetry: A survey," in S. McCaffery and bpNichol (Eds.), *Sound Poetry: A Catalogue*. Toronto: Underwhich Editions, 6–18.

McGee, T. J. 2012. "Vocal performance in the Renaissance," in C. Lawson and R. Stowell (Eds.), *The Cambridge History of Musical Performance*. Cambridge-New York: Cambridge University Press: 318–335.

Medić, I. 2007. "The ideology of moderated modernism in Serbian music and musicology," *Musicology* **7**: 279–294.

Medić, I. 2021. "Playing catch-up: Serbian art music against the odds," in Á. Windhager (Ed.), *Shostakovich-Syndrome. The Burdened Memories of Central European Societies in the 20th Century*. Budapest: Hungarian Academy of Arts: Research Institute of Art Theory and Methodology: 43–52. Retrieved from https://hdl.handle.net/21.15107/rcub_dais_12021.

Mekaferi, S. 2000. "Glasovi dovedeni do ekstrema," trans. A. Gorobinski, *ProFemina* 23/24: 114–128.

Mikić, V. 2007. "Constituting neoclassicism in Serbia Or: How and why neoclassicism can be understood as modernism—a study of Ristić's second symphony," *Musicological Annual* **43**(2): 99–104.

Mikić, V. 2009. *Lica srpske muzike: Neoklasicizam*. Beograd: Fakultet Muzičke Umetnosti.

Milojković, M. 2019. "Saradnja Katalin Ladik sa kompozitorima elektroakustičkih muzičkih ostvarenja," *Zbornik Matice srpske za scenske umetnosti i muziku* **60**: 123–136.

Moore, P. and Von Leden, H. (1958). "Dynamic variations of the vibratory pattern in the normal larynx." *Folia Phoniatr. Logop.* **10**(4): 205–238.

Newham, P. 1997. The Prophet of Song: The Life and Work of Alfred Wolfsohn. London: Tigers Eye Press.

Nix, J. and Chappell, W. 2024. "Vocal fry, creaky voice, and singing voice pedagogy: Singing teacher attitudes and usage." This collection.

Novak, J. 2015. *Postopera: Reinventing the Voice-Body*. Farnham: Ashgate Publishing Limited.

Perloff, M. and Dworkin, C. (Eds.) 2009. *The Sound of Poetry/The Poetry of Sound*. Chicago: The University of Chicago Press.

Radovanović, B. 2018. *Eksperimentalni Glas: Savremena Teorija i Praksa*. Beograd: Orion Art.

Radovanović, B. 2021a. "Glas i tehnika/tehnologija u savremenoj muzici," PhD diss., University of Arts: Belgrade.

Radovanović, B. 2021b. "Vocal expression in works of Ernő Király: Voice as a symptom and a symbol of avant-garde artistic encounters," in

M. Milojković, N. Sovtić, and J. Baštić (Eds.), *Ernő Király – Life in Music*. Novi Sad: Academy of Arts, 87–102. Retrieved from https://hdl.handle.net/21.15107/rcub_dais_11923.

Radovanović, B. 2021c. "Challenging the human expression in a digital environment: Interview with Thea Soti," *INSAM Journal of Contemporary Music, Art and Technology* **6**: 9–16.

Radovanović, B. 2023a. Od monstruoznog do posthumanog: glas u savremenoj muzici. Beograd: Muzikološki institut SANU.

Radovanović, B. 2023b. "'The music is highly eclectic, and it should be approached accordingly': Voice in Jug Marković's compositions," *Muzikologija-Musicology* **34**: 111–131. https://doi.org/10.2298/MUZ2334111R.

Radovanović, B. 2024. "Intensifying the 'monstrous': Extended vocal techniques in extreme metal music," in *Technologien des Singens 3*, forthcoming, Munich: Allitera.

Ragan, K. (2020). *A Systematic Approach to Voice: The Art of Studio Application*. San Diego, CA: Plural Publishing.

Rasula, J. 2015. *Destruction Was My Beatrice: Dada and the Unmaking of the Twentieth Century*. New York: Basic Books.

Rossi Fest. 2021, January 1. "Rossi Fest 2021 - Koncert Portreti i sećanja, drugi deo" [YouTube video], https://youtu.be/eMbkdhpp21E.

Sabo, A. 2021. "Sound collaborations between Ladik and Király," in M. Milojković, N. Sovtić, and J. Baštić (Eds.), *Ernő Király – Life in Music*. Novi Sad: Academy of Arts, 19–37.

Salzman, E. and Desi, T. 2008. *The New Music Theater: Seeing the Voice, Hearing the Body*. New York: Oxford University Press.

Stark, J. 1999. *Bel Canto: A History of Vocal Pedagogy*. Toronto: University of Toronto Press.

Šuvaković, M. 2010. *Moć žene: Katalin Ladik / The Power of a Woman: Katalin Ladik*. Novi Sad: Muzej savremene umetnosti Vojvodine.

Swanson, F. (1984). "Changing voices: Don't leave out the boys." *Music Educators Journal* **70**(5).

Vennard, W. (1967). *Singing: The Mechanism and the Technic*. New York: Carl Fischer.

Wolfe, V., Richard, C., and Palmer, C. C. 1991. "Acoustic correlates of pathologic voice types," *Journal of Speech, Language, and Hearing Research* **34**(3): 509–516.

Wolk, L., Abdelli-Beruh, N. B., and Slavin, D. 2012. "Habitual use of vocal fry in young adult female speakers," *Journal of Voice* **26**(3): e111–e116.

Chapter Thirteen

Transmasculine creaks and cracks

Poe M. Allphin

Introduction

Creaks and cracks often play an integral role in forming the transmasculine voice, from the common use of creak, or vocal fry, as a transmasculine passing strategy, to the cracking voice in the early stages of testosterone hormone replacement therapy (often abbreviated as "T"). While creak may be an intentional sign of vocal control and agency, cracks are usually accidental and can be embarrassing. In this chapter, I explore transmasculine use of vocal fry and discuss how transmasculine vocality and presentation intersect with perceptions around age and notions of adulthood, since voice cracking is societally associated with male adolescence. Additionally, I draw from transgender archives to look at passing guides and writing by trans people about their own voices throughout history. While focusing on transmasculine

Creak: Theories and practices of pulse phonation
Edited by Francesco Venturi

ISBN 978-981-5129-29-8 (Hardcover), 978-1-003-61606-1 (eBook)
www.jennystanford.com

vocality, I place transmasculine use of vocal fry in dialogue with other trans uses of vocal fry. By surveying other transgender individuals about their voices, I situate this chapter within the realm of lived experience.

The human vocal registers are, from lowest to highest, vocal fry (or creak), modal, falsetto, and whistle. Creak is the lowest vocal register and occurs when the larynx's arytenoid cartilages are brought together, compressing the vocal folds and slowing airflow through the glottis. This creates the signature crackly or "creaky" sound of vocal fry. A 2016 study found that irrespective of gender, speakers were perceived as less feminine and more masculine when using creaky voice than when speaking in their modal register, the vocal register just above the creak, or vocal fry, register (Lee 2016: 22). Anecdotally, however, the vocal creak sound is also sometimes produced in the mid or higher range, which many trans women report using in their speech. One transfeminine person wrote in my survey on vocal fry and voice cracking (Allphin 2022), "At some point I figured out subconsciously that slipping into vocal fry in my midrange makes it sound like my overall range is higher than it is without actually raising pitch, and now I sometimes do it cautiously, especially as my voice gets tired."

Lal Zimman's research points to the idea that "gay-sounding" speech is not a single mode of speaking but, instead, a combination of "numerous deviations from the hegemonic norm," including creak, by showing that although trans men may be perceived as sounding gay, their vocal characteristics are often different from other "gay-sounding" men who are cisgender. Zimman explains that transmasculine people who go on testosterone usually experience a considerable drop in the pitch of their vocal range but often preserve learned stylistic traits associated with female voices, which has sometimes been thought to include creak. He found a high degree of correlation not only between trans voices and creak use but also between "gay-sounding" voices (in both trans men and cis gay men) and creak (Zimman 2013: 8).

In a 2020 study with 60 trans people, 13.8% of the trans men responded to the statement "my voice cracks" with the rating "always" and 55.2% of trans men with the rating "sometimes," while 31% of trans men indicated that they never experienced voice cracking. Of the trans women, 9.7% reported that they always experienced vocal cracking, 67.7% said this happened sometimes, and 22.6% never. Vocal cracking was categorized by the authors as an "indication of dysphonia," that is, a physiological speech disorder (Barra *et al.* 2020, 5). Vocal fry is likewise sometimes coded as dysphonia. A study of 17 trans people in the United Kingdom, 12 of whom were transmasculine and five of whom were transfeminine, found that both groups used creak similar amounts in their speech (Pearce 2019, 1).

Since much gendering is based on social cues such as voice pitch and vocal mannerisms, trans people have been publishing resources on how to change one's voice for decades. In 1985, Lou Sullivan published a pamphlet titled *Information for the Female-to-Male Crossdresser and Transsexual*. He informs his readers about the impact of testosterone on the voice, writing "The vocal cords thicken, the voice sounds hoarse and cracks like an adolescent boy's, and eventually deepens to a man's." He continues, "Should you decide to discontinue taking testosterone, the following side effects are not reversible: the voice will remain low…" (Sullivan 1985). In August 1989, the first issue of *Adam's Word,* a newsletter offshoot of The Adam Society, "a non-profit support group for female-to-male crossdressers, transsexuals and their significant others," carried a reprinted article from the June issue of *FTM* titled "The Sound of Your Voice." The article recounts a few tips on how to sound masculine from Carol Fleming's book by the same title. Speakers should talk slow, low, and not say too much. They should neither complain nor explain. The article's author, Emmon, writes that he has "tried this advice and it works, rather quickly. It takes conscious effort at first but the positive responses from other people more than repay it" (Adam Society 1989).

Nearly four decades later, the speaking voice is still an integral part of how someone is gendered by those around them, whether over the phone, virtually, or in person. This disproportionately impacts trans people, who are likely to have voices that do not match up with societal expectations for their presentation or their gender for at least some time during their transition. Being misgendered, or assigned incorrect pronouns or gendered labels, takes a psychological toll and is a contributing factor to poor mental and physical health for many trans people, as experiences with misgendering in healthcare situations, in addition to the access challenges that marginalized populations face, lead many trans people to avoid medical care (Dolan *et al.* 2020: 151).

Methods and demographics

Thirty trans and/or nonbinary individuals responded to an online survey on gender and the voice, circulated via Twitter. The survey opened by asking, "In terms of gender, how would you describe yourself?"[1] Seven respondents invoked the label of trans woman and/or transfeminine in describing their gender. Eight participants self-described as trans men and/or transmasculine. Many respondents described their genders with words such as nonbinary (17), agender (4), trans(gender) nonbinary (2), genderqueer (2), neutrois, genderfluid, gendervoid, genderflux, and genderneutral. The more colloquially familiar term "vocal fry" was used in place of "creak" in the survey.

Table 13.1 Gender descriptors used by survey respondents in optional demographic section

Descriptors	
Nonbinary trans woman	Nonbinary (2)
Nonbinary trans woman	Genderqueer/nonbinary/genderflux
Transfeminine close-enough-to-binary (I just tell cis people I'm a trans woman)	Transgender nonbinary (on T, indicated in other place on survey)
Trans woman	Nonbinary, AFAB
Transfeminine	Agender? Lesbian (AFAB)
Nonbinary woman (on feminizing hormone replacement therapy)	Gender? I hardly know her! (Nonbinary, AFAB)
Nonbinary/Transfem	Nonbinary/agender/gendervoid
Trans man/trans masc (on T)	Genderneutral (agender, genderqueer)
Femme trans man (on and off T)	Neutrois (on T)
Trans man	Nonbinary
Nonbinary transmasc	Agender

Descriptors	
Trans masc/trans man (on T)	Trans nonbinary (AFAB), low topical T dose
Transmasculine nonbinary (on T)	Trans male (on T)
Nonbinary transmac [sic]	Nonbinary—existing between and overlapping in masculine and feminine (AFAB)

The 15 respondents (half of total participants) who did not identify themselves as trans men or trans masc or trans women or transfem—although some respondents in these categories were also nonbinary—were not placed in either of these categories when these two groups are compared here and have solely been categorized as "nonbinary." Six of these respondents explicitly indicated that they were assigned female at birth (AFAB). In a box to disclose any demographics at will ("Beyond gender, are there other demographics you would like to share?"), 10 respondents identified themselves as white, and one as mestiza. Six identified as disabled (one of whom also specified chronic illness), six as autistic, and two as having ADHD. Four respondents identified themselves as Jewish and two as pagan. Two participants identified themselves as asexual, and one as a parent. Three respondents shared their ages (29, 30, 40).

Respondents were not asked to disclose their assigned sex at birth, as not to superimpose two main categories and, therefore, a binary on the survey from the start, however, some either disclosed this information or provided labels that typically align with a certain assignment. When voluntarily reported, this information can provide one interesting but non-definitive axis of inquiry. Six respondents included an explicit designation at birth, while many descriptors such as "transmasc(uline)" or "trans man" and "transfem(inine)" or "trans woman" often come with an implication thereof. Half of the participants (15) included this type of descriptor. Notably, all six people who reported an explicit birth assignment were AFAB. Respondents were also not asked about their hormonal history or current hormone usage, but many chose to disclose this information when pertinent to their vocal experiences.

Results

Most survey participants, regardless of gender, intentionally changed how they spoke in different situations (see Table 13.2). Less than a third of transfem respondents used vocal fry, while half of the total and half of the transmasc respondents used vocal fry. More nonbinary (nonaligned or nonspecified) respondents used vocal fry than any other group. Most transfem respondents did not have voice cracking, while most transmasc participants did experience cracking. The transfem respondents who experienced voice cracking were not the same ones who used creaky voice.

Table 13.2 Vocal fry and cracking results

	Trans women and transfem respondents	**Trans men and transmasc respondents**	**Nonbinary nonaligned/ nonspecified**	**All respondents**
Intentionally changed how they spoke in different situations	86%	75%	87%	83%
Experienced voice cracking	29%	75%	33%	43%
Used vocal fry	29%	50%	60%	50%

Of the survey respondents who reported using vocal creak, just over half of those said they used it intentionally to change pitch or to influence how their gender is perceived. The four transmasc participants who used vocal creak did use it for gender-related reasons, while of the two transfem participants who used vocal creak, one used it for gender reasons and one did not. A third of the nonbinary (nonaligned or nonspecified alignment) respondents who did report using creak used it intentionally to modulate voice/for gender reasons. Among the respondents placed into the nonbinary category, use of creak and speaking differently based on the situation was more common than in the transmasc or transfem groups of participants.

Testosterone and voice cracking

Eight survey respondents mentioned being on exogenous testosterone (T) for gender reasons. In many ways, vocal and other physical changes caused by testosterone hormone replacement therapy are similar to those that occur in testosterone-dominant adolescent puberty, leading trans people to frequently call it "second puberty." In both cases, voice cracking is followed by a voice drop of around an octave. All eight of these respondents experienced voice cracking, while five used vocal fry (four intentionally), and all but one intentionally changed how they speak in different situations. As expected, many survey respondents discussed voice cracking due to testosterone, particularly during the first year of hormone replacement therapy, one of whom wrote that this testosterone-induced vocal cracking was "often intentional for artistic purposes." Another respondent on testosterone wrote, "I notice my voice cracks most around friends, and earlier in the day," while a third explained that the minimal voice cracking they experience on testosterone only occurs when speaking in their upper register. A few people only experienced cracking while singing, especially in the morning, or when trying to project. Other self-reported reasons for voice cracking included aging, passing through a pronounced break in vocal register, "try[ing] to reach the range I used to have before I took T," and "when I've been trying to talk in a voice that's too high for too long."

Creak, disability, and neurodivergence

One nonbinary respondent wrote that their physical disability greatly impacted their voice, "My voice has gotten a lot lower and hoarser because of my disabilities and the treatment for them, and there's so much vocal fry that it's actually painful. I'm also a little nasal-sounding because I'm constantly congested." Two respondents reported using creak as a neurodivergent stim, one writing, "Mostly I use vocal fry as sort of a stim, since I'm autistic & I like the way it feels against my throat." Seven participants of various genders mentioned being autistic, either in the demographics section or as a side note in later responses. Five of these seven respondents used vocal fry, but since autism was not part of a question on the survey, it is unknown which other respondents may have been autistic and is, therefore, unclear whether this seemingly pronounced correlation between autism and creak might be significant.

Speaking sweetly, speaking safely

Several participants reported modifying their voices in certain situations for safety reasons. One nonbinary trans woman wrote, "I talk in a higher voice when I want people who might be adversarial to code me as feminine." Another nonbinary person indicated a vocal fluidity, commenting, "I make my voice higher or lower based on what binary gender it's safest for others to perceive me as." One trans man remarked, "I change pitch and sometimes inflection based on if the people I'm with know I'm [trans] or not/how safe I feel." A nonbinary woman wrote, "I don't do it purposefully, but I do probably have some version of the 'please do not 'sir' me' voice other trans women use when speaking to strangers." For trans speakers, comfort with company can be another factor in vocal pitch from one moment to the next. A transfeminine respondent wrote, "I raise [the pitch of my voice] in public and don't always practice raising it around my partner." One nonbinary person wrote, "With close friends or at home, I find I don't care quite as much if [my voice] comes out high-pitched, or if it sounds less mature."

Many respondents wrote about using a "customer service voice" or work/phone-specific vocal adaptations. Generally, the "customer service voice" tended to be raised in pitch, but a few participants reported lowering their voices in professional settings. One nonbinary person wrote, "I speak more slowly and in a lower pitch when I am in a formal professional space or when I am angry and/or trying to project more authority." A trans man wrote, "When I'm at work I definitely have a 'customer service voice,' which I inflect to be friendlier than usual." A nonbinary person described their phone voice as "higher and more feminine/polite," and one trans male respondent wrote, "When speaking with patrons at work I speak softer and higher." The responses also ranged in the degree of intentionality: While one agender lesbian wrote, "My voice is a lot higher when I'm talking to my grandparents, or doing retail, and some job interviews. It's not usually a conscious thing," another nonbinary person wrote, "in professional or academic situations I deliberately shift my voice somewhat lower." Many other survey respondents said they changed their voice or vocal mannerisms depending on the environment. A nonbinary trans woman commented, "I do a higher voice if I remember to or probably most often with randos if I think I'm passing right then." Another transfeminine person said, "I tend to raise my pitch and place my vowels forward, especially when talking to strangers. I actually try to do this more generally, but it's somehow easier with strangers (and definitely easier for short interactions)." One nonbinary transmasc respondent wrote, "I usually change the pitch of my voice…I think it gets higher when I'm talking to people [I don't know] and lower around people I know." A neutrois[2] participant explained, "I make [my voice] lower when I'm with family or feeling dysphoric, and camp it up a LOT when I'm with friends." One nonbinary transfem person wrote, "I've been increasingly aware of the disconnect between my voice and appearance, and find myself doing what I can to soften that while in the world amid strangers."

Zimman found the highest frequency of creak use in trans men who were not on testosterone, indicating that this too is a "passing" or safety strategy to avoid being misgendered or outed. He remarks, "One way that creak might be useful to trans speakers is that it does provide access to a different pitch range that people are typically not producing in their non-creaky speech" (Zimman 2019). My survey found similar results, as one trans man who used creak intentionally wrote that it was "more prevalent earlier in my transition for speaking, now for gender affirming voice work and singing." A femme trans man reflected, "I think I mostly use it in the morning to make my voice sound extra low, or when I'm singing, to try to lower my range." One nonbinary respondent commented succinctly, "I like how it sounds. That's mostly it," and another wrote, "I want my deep register to sound 'gravelly.'" Transmasculine people who are not on testosterone may use creak as a way of pushing their voices lower to be read as male or more masculine. However, their need for this type of strategy may decrease if they start testosterone and experience a voice drop, placing the resting modal voice in a male-coded register.

Conclusion and future directions

My survey suggests there is more work to be done on the intersection of voice, gender, and customer service. Another area for expansion is autistic use of creak, which was mentioned by some of the respondents. While it has previously been hypothesized that trans use of creak was more of an effect than an intention (Zimman 2012), my work suggests instead that trans people often use creak very deliberately to change the way their genders are perceived. Overwhelmingly, survey respondents were hyperaware of how they used their voices to express their genders, particularly when it came to creak. As compared to the results found by Pearce (2019), my survey found that creak was not equally common among transmasculine and transfeminine speakers, although the self-reported nature of these data, the different gender groupings, or the North America-centered responses may account for this disparity. While voice cracks may be outside the trans speaker's control, modulating the register of the voice using creak is a method that the transmasculine speaker can use to regain autonomy and express gender.

Endnotes

1. Phrasing inspired by the podcast *Gender Reveal*.
2. A term usually connoting a neutral or null gender.

References

Adam Society. (1989). "Adam's Word, Vol. 1 No. 1 (August, 1989)." Newsletter. *Digital Transgender Archive*, Retrieved from https://www.digitaltransgenderarchive.net/files/cc08hf72r (accessed March 20, 2022).

Allphin, P. (2022). Survey on trans voices conducted via twitter. Spring 2022.

Barra, B. G., Gusmão, U. M., and Araújo, A. N. (2020). "Self-perception of voice in transgender people." *Revista CEFAC,* **22**(4).

Dolan, I. J., Strauss, P., Winter, S., and Lin, A. (2020). "Misgendering and experiences of stigma in health care settings for transgender people." *Med. J. Aust.,* **212**(4): 150–151.

Lee, K. E. (2016). "The perception of creaky voice: Does speaker gender affect our judgments?" (2016). Theses and Dissertations: Linguistics. 17. Retrieved from https://uknowledge.uky.edu/ltt_etds/17.

Pearce, J. (2019). "Identity, socialization and environment in transgender speakers: Sociophonetic variation in creak and/s." *Proceedings of the 19th International Congress of Phonetic Sciences,* Melbourne, 3290–3294.

Sullivan, L. G. (1985). "Information for the female-to-male crossdresser and transsexual." *Digital Transgender Archive,* https://www.digitaltransgenderarchive.net/files/g158bh442 (accessed March 20, 2022).

Zimman, L. (2012). "Voices in transition: Testosterone, transmasculinity, and the gendered voice among female-to-male transgender people." PhD diss., University of Colorado at Boulder.

Zimman, L. (2013). "Hegemonic masculinity and the variability of gay-sounding speech: The perceived sexuality of transgender men." *J. Lang. Sexual.,* **2**(1): 1–39.

Zimman, L. (2019). Interviewed by Megan and Carrie. *Vocal Fries*. Podcast audio. https://radiopublic.com/the-vocal-fries-GOoXdO/s1!e724c (accessed July 15, 2019).

Chapter Fourteen

Mothers and creak

Andrew Cheng, Elise McClay, and H. Henny Yeung

Introduction

Creaky voice from a phonetic and phonological perspective

In North American English, the voice quality known as *creaky voice* is distinguished from other types of voice qualities in several ways pertaining to phonetics (speech sounds) and the vocal tract (Keating, Garellek, and Kreiman 2015). In acoustic–phonetic terms, creaky voice has a lower pitch than other voice qualities, and, to the average listener, it has a unique timbre that has been described as rough, harsh, and gravelly. In terms of articulation in the vocal tract, the vocal folds vibrate more closely together, and they sometimes also vibrate more slowly, or with less regularity, compared to *modal voice*.

Phoneticians, who study the properties of speech sounds, widely agree that creaky voice is influenced by both social factors

Creak: Theories and practices of pulse phonation
Edited by Francesco Venturi

ISBN 978-981-5129-29-8 (Hardcover), 978-1-003-61606-1 (eBook)
www.jennystanford.com

and factors pertaining to speech itself (Davidson 2021). For example, low, open vowels such as those in the words *dog* and *cat* are more likely to be creaky than high vowels such as those in the words *meet* and *cute*. In addition, creak is more likely to occur naturally at the ends of sentences, as the volume of air in a speaker's lungs runs low, causing airflow out of the lungs to become less regular, which contributes to creakiness.

Of course, creaky voice surfaces differently across languages: For example, in Jalapa Mazatec, a tonal language spoken in Oaxaca, the presence of creaky voice in words is independent of the speaker's pitch (Garellek and Keating 2011). In another example, Kwak'wala, a Wakashan language spoken in Western Canada, uses creak phonologically: That is, its presence or absence in a word will change the word's meaning (Werle 2010). Neither of these properties are true for creaky voice in North American English: Whether one says the word *dog* with or without the presence of creaky voice, both instances refer to the concept of a dog and the meaning does not change.

Creaky voice from a sociolinguistic perspective

While creaky voice is not phonologically meaningful in North American English, it is *socially* meaningful. Creak has been examined extensively in sociolinguistic literature, and researchers have found that speakers use creak—whether consciously or not—to express an array of social identities (Podesva and Callier 2015). For example, some women may speak with increased creakiness as a way to portray a young, confident, and upwardly mobile persona (Yuasa 2010). In this case, creaky voice is perceived to be authoritative-sounding and successful. In different communities, creaky voice will linguistically symbolize, or index, different identities and traits. A study of the voice qualities of Chicana gang members revealed that the members who used more creaky voice associated their use with "sounding hardcore" (Mendoza-Denton 2011). In yet another study, researchers found that regardless of the identities that creaky voice might evoke in particular contexts, speakers were likely to adopt it when conversing with someone else who also used it, as a way of demonstrating closeness or solidarity (Borrie and Delfino 2017).

It is important to note that in North American society and even in the scientific literature of recent decades, creak has been pathologized and stigmatized greatly (Davidson 2021). It has been saliently attributed to certain demographics, particularly young women, and has been linked to harmful stereotypes such as the "dumb blonde" and "valley girl," despite overwhelming evidence that creak is neither over-represented in these demographics nor "pathological." It has also been shown that creak is not harmful for the voice or the vocal tract.

Most studies on creak in North American English focus on its use in adults either when speaking to other adults, or when reading, or doing carefully controlled speech tasks. Very little is known, however, whether the use of creak in North American English changes when speakers adopt different registers of speech—i.e., specific ways of speaking when addressing a particular group of individuals, or in certain conversational contexts. In this chapter, we explore whether and how the presence of creak changes when speakers of North American English address infants. In the field of child language development, this is a widely studied speech register called Infant-Directed Speech (IDS), which can be contrasted with Adult-Directed Speech (ADS).

Infant-directed speech in North American English

When compared to ADS, North American English IDS usually has higher and more variable pitch, with both a wider pitch range and greater variability within that range (Fernald *et al.* 1989; Soderstrom 2007; Spinelli *et al.* 2017). IDS utterances are usually shorter and more syntactically simple than ADS utterances; the latter may have multiple clauses connected by function words like *and* or *which*, whereas the former are often simple commands, questions, and single-clause statements (Cruttenden 1994). When exploring the phonetic detail of IDS in North American English, some studies show that consonants and vowels in the IDS register are hyperarticulated and more greatly dispersed (D. Burnham *et al.* 2002; Kuhl *et al.* 1997): For example, low vowels are lower and high vowels are higher in IDS, compared to vowels in the ADS register. It is important to note, however, that these phonetic details of IDS consonants and vowels are not always observed

(Cristia and Seidl 2014; Miyazawa *et al.* 2017; McClay *et al.* 2022) or may vary according to the context in which IDS is spoken (e.g., reading versus spontaneous speech; E. Burnham *et al.* 2015).

The characteristics of IDS thus raise questions about how creaky voice might surface in the voice of someone who generally uses creaky voice in their ADS, but who is recorded speaking to an infant in IDS. For example, one might predict that if IDS is higher in pitch than ADS, it would have less creakiness due to the effects of higher pitch on the physiological mechanisms of producing creak. Similarly, if IDS utterances are shorter in duration, one might predict that there would be proportionately less creak, since speakers will not "run out" of air as often and naturally dip less into creaky voice. On the other hand, if IDS has greater vowel dispersion, one might predict that low vowels such as those in words like *dog* and *cat* are even lower in IDS than in ADS and could lead to greater creak for those specific vowels.

Our research team was interested not only in the phonetics of creaky voice in the IDS register, but also in its sociolinguistic implications. In other words, why do adults use the IDS register, and what purpose does it serve, from both social and language-learning perspectives? Research in child first-language acquisition has three broader theoretical approaches to this question. The first is a pedagogical approach, which claims that IDS may have developed to help infants acquire the sounds of their parent's language. This might explain why vowels and tones (in tonal languages, such as Mandarin) are exaggerated in IDS: Keeping the categories of different vowels and tones as acoustically distinct from one another as possible may make it easier for an infant to learn what those categories are (Kuhl *et al.* 1997; Xu *et al.* 2013). Along those same lines, research shows that IDS and pet-directed speech (i.e., speaking to one's pet puppy or kitten) both utilize heightened pitch, but only IDS utilizes vowel hyperarticulation, while pet-directed speech does not, possibly because pet owners do not need to teach their pets the different categories of English vowels (Burnham, Kitamura, and Vollmer-Conna 2002). A second approach focuses on the affective, or emotional, intent behind IDS: It may be a by-product, or consequence, of parents' positive emotions when interacting with infants. Not every language shows a hyperarticulation of phonetic sounds in IDS, and there may be acoustic–phonetic reasons why speaking with increased emotion changes the IDS register. Rather than being used for pedagogical intent, IDS may be a register for communicating simplified and exaggerated emotions, usually positive ones (Benders 2013; Trainor and Desjardins 2002). Finally, the third approach is a hybrid one, which combines elements of the first two theories. According to these approaches, IDS may serve different functions at different developmental timepoints, or in different cultural and/or communicative contexts (Kalashnikova, Carignan, and Burnham 2017). For example, IDS may be primarily an affective register in some languages or at some ages, with its pedagogical relevance becoming clearer at other ages, or in some specific contexts, e.g., efforts to communicate or teach words (Kitamura and Burnham 2003).

Here we examine cognitive, social, and linguistic reasons for IDS through the lens of creak. Our current study examined explicit attitudes about the sociolinguistic use of creak in both ADS and IDS to understand how parents in the present day reflect on their use of creak—given the climate of its linkage to a variety of social stances and attitudes in North American English—and the explicit attitudes behind how they use IDS. In doing so, we hoped to draw a potential link between a parent's understanding of the social implications of IDS and the social implications of creaky voice, particularly in mothers.

Hypothesis

Our study asked two questions: first, whether IDS is more creaky or less creaky than ADS in North American English; and second, what phonetic or sociolinguistic factors explain this difference. Our hypothesis for the first question was that IDS would be less creaky than ADS for several reasons. First, IDS could be less creaky simply due to the linguistic and physiological attributes discussed earlier. IDS uses higher pitch than ADS, on average, and it is thus less likely to have creak, which is correlated with lower pitch in North American English. On the other hand, IDS is also known to have a greater pitch range than ADS, not just higher pitch. Thus, it is possible that some portions of IDS utterances may, in fact, be creakier.

Second, because creak has no phonological meaning in English, it remains unclear whether greater exposure to this voice quality is necessary for successful first-language acquisition—if true, theories that emphasize the pedagogical role of IDS would not make clear predictions. On the other hand, even though creak does not have to be taught to children, the same argument about variability and range in pitch also applies to some aspects of IDS, such as vowels. Increased creak might come secondarily in IDS if the low vowels (e.g., *dog* and *cat*) are lower in IDS compared to ADS.

Finally, if IDS is purely affective and creak is affective to a specific persona in a specific social context—such as the upwardly mobile young woman or the valley girl—then we might suppose that a kind of "parent persona" is being employed by those who use IDS and that this persona is characterized by less creakiness. However, no research to date has asked the question of what kind of persona a parent uses when speaking to an infant, or even whether parents are aware that such a social construction exists. To that end, our current study answers our research questions with a twofold approach: First, we recorded parents speaking to their infant children and to adults to analyze the amount of creakiness in their IDS and ADS. Second, we directly interviewed the parents and asked them to comment on the social aspects of their own speech and others' speech. This way, we were able to connect the linguistic and phonetic data on the use of creaky voice in IDS by parents to their own commentary on creaky voice.

Methods

The first set of data was collected between 2018 and 2019 from 25 Canadian mothers who were living in the Vancouver metropolitan area of British Columbia. The mothers ranged in age from 22 to 40. They were recorded speaking in English first to their child and then to an adult peer who had agreed to accompany them. These data were part of another project that compared North American IDS to that of mothers living in Vanuatu (McClay *et al.* 2022).

The children ranged in age from 6 months to 1 year, 10 months. There were 18 girls and seven boys. Every mother spoke English with her child. Five of the mothers were bilingual, speaking English and another language, and three of these mothers reported using a non-English language with her child in addition to English. During the recording, all the mothers were asked to use only English with her child and with her peer. We asked each mother to play with her child using three specific toys that had unfamiliar names (Fig. 14.1). The names were *tisisi, kususu,* and *pasasa,* and these target words were designed so that we could record the mothers using a range of different English vowels in their speech, including high and low vowels. During free play, we made a note of each time the mother used one of the target words, and we recorded her until she had used each word at least eight times. We then annotated the recordings and extracted phonetic information about pitch and creaky voice from each speaker, comparing their creakiness in the IDS register with their creakiness in the ADS register, specifically in the vowels of the toy names.

Figure 14.1 These toys, named *tisisi, kususu,* and *pasasa,* were used by the mothers when talking with their infants in a structured play session. The names were designed to elicit different English vowels in speech in both registers.

The second set of data was collected in 2022 from a subset of the mothers who participated in the first data set collection. We asked the mothers to do a follow-up interview about 4 years after the initial data collection. At this time, none of the mothers would be using IDS with her child anymore. We wanted to ask for their perspectives on the use of IDS and, specifically, the use of creaky voice in IDS. Each mother who participated in the follow-up interview was asked a range of open-ended questions about IDS and creaky voice. They were then asked to listen to speech samples from the first data set and comment on the voice quality of the mothers who were speaking. Finally, we recorded them reading a few short passages, so that we could compare their voice in the present day to their voice when we first recorded them.

Results

The results from the first set of data supported our hypothesis: Mothers used significantly less creaky voice in the IDS register than in the ADS register when measured using at least one measure of creak in most, but not all mothers. In Fig. 14.2, nearly all of the mothers had a higher pitch in IDS, indicated by the higher bars in the IDS column. This difference between registers was also found to be statistically significant. Figure 14.3 illustrates a similar by-subject breakdown of the creak measurements, but instead of pitch, it shows spectral tilt (H1*–A2*), which is a type of acoustic measurement in which lower values indicate more creaky voice (Keating *et al.* 2015). That means that for Subject 6, as an example, her speech in ADS was creakier than her speech in IDS. We hypothesized that ADS would be creakier, so results like these demonstrate this pattern.

Analysis of the spectral tilt measure showed that at the group level, IDS was less creaky than ADS. However, as we can see in Fig. 14.3, there was also some variation among speakers. For example, Subjects 16, 17, and 19 are among the many subjects whose H1*–A2* was higher (less creaky) in ADS than in IDS, indicated by the red coloration. But just as there are many different types of creaky voice, there are many different acoustic

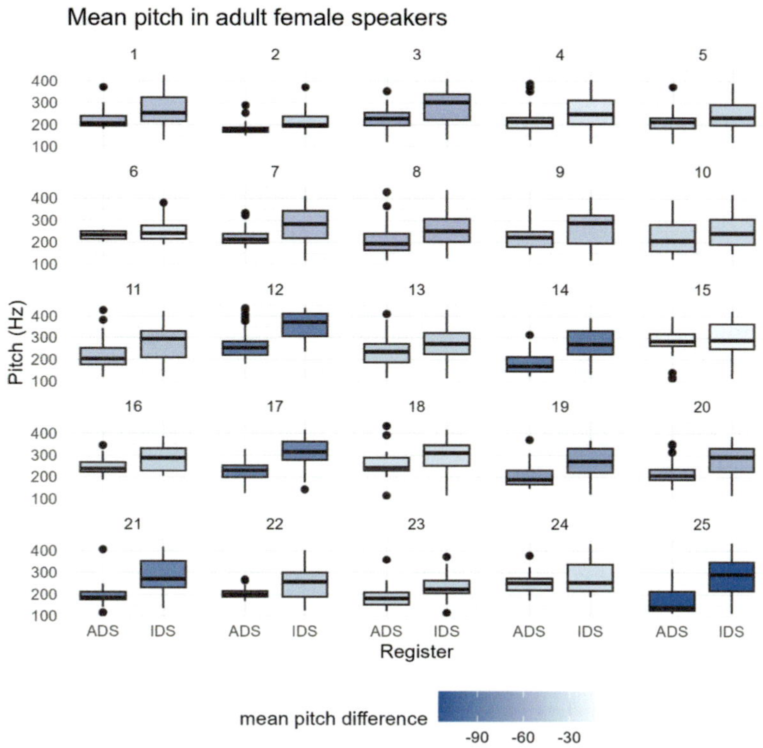

Figure 14.2 Mothers tended to use higher pitch in IDS compared to ADS.

measurements that can be used to quantify it (Keating *et al.* 2015), and many of these measurements turned out not to be significantly affected by register at all. In addition, within each of these measurements, there were additional linguistic factors that conditioned creakiness, such as the length of the vowel that was measured, whether it was a high or low vowel, and whether the target word occurred at the beginning, middle, or end of a sentence.

We turn now to two case studies of the mothers who gave a follow-up interview to take a closer look at their manifestations of creaky voice in both registers, as well as at their personal opinions on creak. Subject 6 and Subject 14, whom we will call Danya and Nadine (as pseudonyms), were similar in their language and demographic backgrounds, and both mothers used higher pitch in IDS compared to ADS. They differed, however, in which register was creakier according to spectral tilt.

Danya (Subject 6) is a white Canadian woman who was originally recorded speaking to her son, who was 16 months old at the time, and then to her friend, another adult woman. At the time of recording, Danya was 37 years old and her peer was 40 years old. Nadine (Subject 14) is a white Canadian woman who was originally recorded speaking to her son, who was nine months old and then to her sister-in-law. At the time of the recording, Nadine was 35 years old and her sister-in-law was 49 years old. Both Danya and Nadine had completed higher education in Canada and reported being monolingual in English and speaking only in English to their children.

Danya's voice was typical among participants, being in the middle of the range of mean pitches seen in our sample, and also using an average amount of creak in her ADS. Regardless of register, she employed creaky voice often at the ends of sentences, which is natural. The acoustic characteristic that set her apart was that she had a very low individual pitch range compared to other speakers. That is to say, the difference between her highest-pitched speech and lowest-pitched speech was very small, resulting in a more monotonous-sounding voice compared to other mothers. Although her pitch range was somewhat greater in IDS than in ADS on the target vowels analyzed here, the difference was quite small. The second characteristic was that in the spectral tilt measurement, she measured about the same for IDS and ADS, indicated by the neither-blue-nor-red coloration in Fig. 14.3.

When we spoke with Danya for the follow-up interview, her first thoughts on IDS were that although she had never been explicitly told what to say or how to speak to her infant child, she generally thought that she spoke more slowly to her child than to adult peers—with few other acoustic changes. She said that she never tried to speak to her child in a "baby voice," which she characterized as "softer" than ADS but thought that parents should speak as "normally" as possible to their children. In her opinion, an over-infantilization of speech toward children was somewhat condescending. Her perceptions of IDS were accurate,

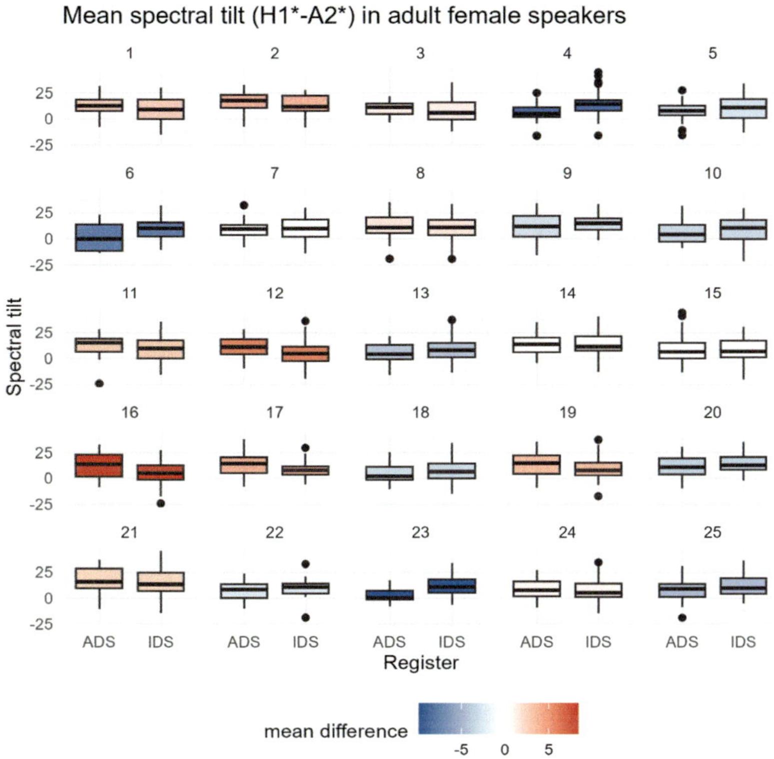

Figure 14.3 Individuals showed variation in whether ADS or IDS was creakier when quantified according to the acoustic creak measurement known as spectral tilt (H1*–A2*): some had more creak in IDS (red) and others had less creak in IDS (blue).

as far as her own speech goes: Not only was Danya one of the few speakers who did not vary their pitch much, she also spoke much more slowly in IDS compared to ADS. However, when Danya listened to speech clips containing samples of ADS and IDS, she commented that she could hear how some mothers varied more than just the speech speed, also using a higher-pitched voice, as well as lots of word repetition.

As for creaky voice, Danya was wholly unfamiliar with the term, as well as with related terms like "vocal fry." Danya was unsure whether she had ever heard people speak with creaky voice and was thus unaware of its presence in her own voice. As a result, Danya did not have much to say about her perception of others' voices when they employed creaky voice, whether when speaking to an adult or to an infant. To the extent that Danya did on occasion use creaky voice in IDS, we submit that its presence was driven largely by a purely linguistic factor: her slower speech speed in IDS. When speakers take a longer time to complete their sentences, they are more likely to "dip into" creaky voice at the ends of utterances, and Danya achieved this utterance-final creakiness in IDS despite having a canonically higher pitch.

Nadine's voice was somewhat less typical among the original speakers for having a lower-pitched voice overall, as well as having a very un-creaky voice in both ADS and IDS. As for spectral tilt, Nadine had a slightly higher value for ADS, indicating, if anything, that her IDS was creakier (though not by much). It is worth mentioning that according to other creak measurements, including one called glottal constriction, Nadine was one of the creakiest of all our participants. Yet Nadine's creaky IDS was also still quite high pitched, and she demonstrated one of the largest "jumps" in pitch from ADS to IDS out of all the participants.

When we spoke with Nadine for her follow-up interview, however, she also had never heard of the terms "creaky voice" or "vocal fry" and had no opinions about it. Although she knew that people were critical of young women's voices for some reason, she did not know what it was exactly about those voices that made them subject to such criticism and she was unaware of any creakiness in her own voice in either register. Thus, like Danya, when she listened to samples of mothers' voices containing both creaky voice and modal voice, she never commented on the amount of creak in anyone's voice. Instead, Nadine's main insight about the phonetic properties of IDS is that it was mainly due to the repetition of short words, spoken in isolation, that IDS might have a higher pitch, and even a faster speed. Nadine noticed that in many of the recordings of IDS speech, mothers would use the target words as "attention grabbers": For example, a mother would more often say the word *tisisi* quickly and excitedly, repeated several times, in order to show her infant the object in her hand or to command their attention, than use the word as a noun in a sentence, such as, "Here is the *tisisi*!" In this case, the semantic meaning of the word is almost nonexistent, and the phonetic properties of the word—i.e., higher pitch, greater speech speed, and less creak—are only used in the service of the mother's *pragmatic* goal of attention-getting. It is, of course, possible for any word to be used in this way, but for the three unfamiliar words that we asked the mothers to use, it is likely that their novelty as words—and as objects—contributed to a possible bias for being spoken in a way that enhanced attention-getting and thereby reduced creak (see Ishi, Chapter 4, for a discussion of creaky voice used for emphasis in Japanese).

Discussion and conclusion

Results from the initial dataset supported our initial hypothesis: Parents—specifically mothers—used higher pitch when speaking to their infant children compared to when speaking to adult peers, and their higher-pitched voices were also less creaky overall in IDS than in ADS. However, there were still measurable amounts of creaky voice used by the mothers in both registers: It was not the case that IDS was never creaky or that ADS was always creakier. In addition, there were multiple measures of creak, and although the mothers were creakier overall in ADS, each mother demonstrated variability along each measurement, as well as variability according to other linguistic factors such as the length of the utterance or the type of vowel spoken.

The results from the follow-up interviews made it clear that neither explicit awareness nor explicit attitudes toward creaky voice play a role in an individual mother's use of creak, or lack thereof. Our two case studies were both with mothers who varied in their use of creak in both registers but were unfamiliar with creaky voice or vocal fry and had no impressions to give on whether creak was socially meaningful in any way in their voices or in the other voices they listened to. Their unfamiliarity with creaky voice is not unusual in itself (see Abdelli-Beruh, Chapter 1, and Nix and Chappell, Chapter 2, for discussions on the historical stigmatization of vocal fry in speech language pathology (SLP) and vocal performance pedagogy, respectively), but this does mean that to the extent that a "parent persona" may be constructed by mothers when they interact with infants, creaky voice, or the lack thereof, is not a part of it. The linguist Penelope Eckert writes about the incorporation of different styles of speaking to fashion a personal linguistic identity: Such styles of speaking have an established meaning at a broader cultural level before they are used or reused for the individual (Eckert 2008). Danya and Nadine, our two case study mothers, were not aware of the broader cultural meanings that are indexed by creaky voice. If they were creating a mother's linguistic style by increasing or decreasing their use of creak in IDS, it had to have been far below the level of consciousness.

Of course, sociolinguists have long understood that most socially driven variation in language arises long before anyone notices that it exists. The linguist William Labov and his colleagues theorized that it is only after patterns of variation regarding a specific linguistic variable—e.g., who uses more creak and who uses less—are firmly established that people can begin to comment on it (Labov, Gumperz, and Hymes 1972). After that point, the association between the variant and the kinds of people who use it or do not use it takes on a life of its own as a speech stereotype (Levon 2014). As we have discussed, there are several stereotypes about the people who use creaky voice currently circulating in North American society; yet as cultural objects, stereotypes still have bounds: Not everyone is equally aware of or influenced by their presence. If the level of general societal awareness of creaky voice is still somewhat low, the level of awareness of creaky voice specifically in the IDS register is lower still.

Finally, we return to our second research question, which asked what phonetic or sociolinguistic factors contributed to the levels of creaky voice in IDS. Phonetically speaking, the use of IDS-specific prosody may have contributed to this effect. The mothers consistently used higher pitch in IDS, which reduced the amount of creak, but some of them also spoke more slowly in IDS, which may have occasionally led to greater creakiness. There were also mothers who spoke more slowly (and thus with increased creak) specifically when using the target words, not only because their IDS was slower than their ADS, but also because the target words were used for a specific attention-getting pragmatic purpose that was distinct from the rest of their IDS. The repetition and special prosodic patterns of IDS are said to promote attentional and communicative outcomes for infants (Spinelli *et al.* 2017). Overall, it is hard to rule out a purely phonetic explanation for the finding that IDS was generally less creaky and higher pitched than ADS. Furthermore, the lack of awareness of creaky voice among a subset of the mothers in this study makes it difficult to identify a link between a mother's sociolinguistic speech style and the potential use of creaky voice in IDS. The mothers were quick to point out that IDS can be slower and more syntactically simple than IDS, but they did not bring up the unique sounds of IDS, including its higher pitch, let alone the presence or absence of creaky voice.

Our study confirms that the speech of North American English-speaking mothers in the IDS register is generally less creaky and higher pitched than their ADS register. We are yet unable to determine exactly what the mechanism is that determines the amount of creak in the IDS of the mothers, apart from analyses of its correlation with other phonetic properties such as pitch. We cannot say for certain whether reduced creaky voice is part of a "parent persona" that has less to do with carefulness, clarity, or pedagogical intent in speech and more to do with a different kind of parental emotional effect. However, future research in this area could delve deeper into this question by running a controlled perceptual experiment of IDS with different voice qualities, as well as by analyzing the presence or absence of creak in languages other than English—in particular, languages that have phonologically meaningful creaky voice.

References

Benders, T. (2013). "Mommy is only happy! Dutch mothers' realization of speech sounds in infant-directed speech expresses emotion, not didactic intent." *Infant Behavior and Development*, **36**(4): 847–862.

Borrie, S. A. and Delfino, C. R. (2017). "Conversational entrainment of vocal fry in young adult female American English speakers." *Journal of Voice*, **31**(4): 513–e25.

Burnham, D., Kitamura, C., and Vollmer-Conna, U. (2002). "What's new, pussycat? On talking to babies and animals." *Science*, **296**(5572): 1435.

Burnham, E. B., Wieland, E. A., Kondaurova, M. V., McAuley, J. D., Bergeson, T. R., and Dilley, L. C. (2015). "Phonetic modification of vowel space in storybook speech to infants up to 2 years of age." *Journal of Speech, Language, and Hearing Research*, **58**(2): 241–253.

Cristia, A. and Seidl, A. (2014). "The hyperarticulation hypothesis of infant-directed speech." *Journal of Child Language*, **41**(4): 913–934.

Cruttenden, A. (1994). "Phonetic and prosodic aspects of baby talk." In *Input and interaction in language acquisition*, 135–152.

Davidson, L. (2021). "The versatility of creaky phonation: Segmental, prosodic, and sociolinguistic uses in the world's languages." *Wiley Interdisciplinary Reviews: Cognitive Science*, **12**(3): e1547.

Eckert, P. (2008). "Variation and the indexical field 1." *Journal of Sociolinguistics*, **12**(4): 453–476.

Fernald, A., Taeschner, T., Dunn, J., Papousek, M., de Boysson-Bardies, B., and Fukui, I. (1989). "A cross-language study of prosodic modifications in mothers' and fathers' speech to preverbal infants." *Journal of Child Language*, **16**(3): 477–501.

Garellek, M. and Keating, P. (2011). "The acoustic consequences of phonation and tone interactions in Jalapa Mazatec." *Journal of the International Phonetic Association*, **41**(2): 185–205.

Kalashnikova, M., Carignan, C., and Burnham, D. (2017). "The origins of babytalk: Smiling, teaching or social convergence?" *Royal Society Open Science*, **4**(8): 170306.

Keating, P. A., Garellek, M., and Kreiman, J. (2015). "Acoustic properties of different kinds of creaky voice." In *ICPhS*, 2015(1): 2–7.

Kitamura, C. and Burnham, D. (2003). "Pitch and communicative intent in mother's speech: Adjustments for age and sex in the first year." *Infancy*, **4**(1): 85–110.

Kuhl, P. K., Andruski, J. E., Chistovich, I. A., Chistovich, L. A., Kozhevnikova, E. V., Ryskina, V. L., Stolyarova, E. I., Sundberg, U., and Lacerda, F. (1997). "Cross-language analysis of phonetic units in language addressed to infants." *Science*, **277**(5326): 684–686.

Labov, W., Gumperz, J. J., and Hymes, D. (1972). "On the mechanism of language change." In *Directions in Sociolinguistics*, 312–338.

Levon, E. (2014). "Categories, stereotypes, and the linguistic perception of sexuality." *Language in Society*, **43**(5): 539–566.

McClay, E. K., Cebioglu, S., Broesch, T., and Yeung, H. H. (2022). "Rethinking the phonetics of baby-talk: Differences across Canada and Vanuatu in the articulation of mothers' speech to infants." *Developmental Science*, **25**(2): e13180.

Mendoza-Denton, N. (2011). "The semiotic hitchhiker's guide to creaky voice: Circulation and gendered hardcore in a Chicana/o gang persona." *Journal of Linguistic Anthropology*, **21**(2): 261–280.

Miyazawa, K., Shinya, T., Martin, A., Kikuchi, H., and Mazuka, R. (2017). "Vowels in infant-directed speech: More breathy and more variable, but not clearer." *Cognition*, **166**: 84–93.

Podesva, R. J. and Callier, P. (2015). "Voice quality and identity." *Annual Review of Applied Linguistics*, **35**: 173–194.

Soderstrom, M. (2007). "Beyond babytalk: Re-evaluating the nature and content of speech input to preverbal infants." *Developmental Review*, **27**(4): 501–532.

Spinelli, M., Fasolo, M., and Mesman, J. (2017). "Does prosody make the difference? A meta-analysis on relations between prosodic aspects of infant-directed speech and infant outcomes." *Developmental Review*, **44**: 1–18.

Trainor, L. J. and Desjardins, R. N. (2002). "Pitch characteristics of infant-directed speech affect infants' ability to discriminate vowels." *Psychonomic Bulletin & Review*, **9**(2): 335–340.

Werle, A. (2010). "The phonology of Wakashan languages." Manuscript, University of Victoria, Victoria, British Columbia, Canada.

Xu, N., Burnham, D., Kitamura, C., and Vollmer-Conna, U. (2013). "Vowel hyperarticulation in parrot-, dog- and infant-directed speech." *Anthrozoös*, **26:** 373–380.

Yuasa, I. P. (2010). "Creaky voice: A new feminine voice quality for young urban-oriented upwardly mobile American women?" *American Speech*, **85**(3): 315–337.

Chapter Fifteen

A case study in creak and creaky voice
The pedagogical impediments and uses

Lisa Quoresimo

I first began using creak while teaching contemporary singing techniques. Like many of my voice teacher colleagues, my only explicit training in creak had been in anti-creak bel canto techniques. However, anyone with access to music outside of the Western classical tradition, I received plenty of implicit training in other singing techniques—our soundscapes are filled with the growls and creaky sounds of blues, rock, and pop. I am currently teaching theater and voice at Southern Utah University (SUU), in a musical theater conservatory program within a state university in the United States. After many years of singing professionally, of teaching voice (with students on Broadway and in the Vienna Boys Choir), and a PhD in performance studies exploring voice and gender, one of my specialties has become the teaching of

Creak: Theories and practices of pulse phonation
Edited by Francesco Venturi

ISBN 978-981-5129-29-8 (Hardcover), 978-1-003-61606-1 (eBook)
www.jennystanford.com

healthy contemporary singing techniques. I have found ways to teach the use of creak to achieve many different effects. Most students who come to me have some training in classical vocal techniques, which emphasize an open-pharynx and dropped-larynx position for vocal production. Many students who want to sing in a contemporary sound attempt to do so by maintaining this openness while adding a highly nasal placement, by yelling on (or somewhere near) pitch, or by some combination of these. I began teaching creaky voice, often combined with a narrowed pharynx, in order to help these students produce contemporary pop, rock, and Broadway sounds in a healthy and less nasal manner. When I encountered a student at the SUU who was singing with marked nasality and was unable to control the openness of their pharynx and false vocal folds, I was forced to rethink some of my pedagogy. I had to develop exercises both to teach creak and to eliminate it. In this chapter, I will discuss the case study of this student, how I have effectively taught these techniques to other students, and what, in my experience, singers need to do in order to employ these techniques healthily.

Since the production of creaky voice requires the engagement of the supra-glottal region, it is directly opposed in this aspect to techniques of Western classical singing. The supra-glottal region can be engaged to achieve a variety of vocal techniques, whereas different schools of Western classical singing all advise openness in the pharynx, though to different degrees (Miller 1996: 10–19). Non-Western singing techniques employ various degrees of pharyngeal closure. Pop, rock, and other contemporary singing styles in Western music often employ varying levels of creak or otherwise narrowed pharynxes. In the Broadway production of *Beetlejuice*, for example, Alex Brightman regularly employed his false vocal folds to achieve a Louis Armstrong-like growl. These various uses of the pharynx to extend singing techniques beyond operatic methods are just beginning to be taught in music conservatories, as there has been a long-standing belief that their usage would lead to vocal damage.

The belief that creak is somehow wrong or damaging comes from several different sources. Elsewhere in this book, Nassima Abdelli-Beruh discusses the deeply racialized history of bias against vocal fry and other "wrong" ways of speaking. Similarly, there exists an institutionalized notion that there is one "right" way to sing. As James Potter tells us in his study of the ideology of singing styles, "the institutionalization of the profession and its pedagogy...ensured that singing could be defined and measured by certain 'standards'. It became very clear what was 'singing' and what was not, and an entire profession grew up to maintain this definition" (Potter 2006, 83). This definition of what is proper singing does not allow for the introduction of extended techniques, as Bojana Radovanovic points out in another chapter in this volume. These techniques are kept outside of the realm of institutionalized training by suggesting that, in their wrongness, they must be damaging to the voice. This belies, of course, the many singing traditions outside of bel canto and the many singers (the global majority) who employ them over a lifetime of healthy singing. Part of my work within the institution is to develop a pedagogy for the healthy use of these techniques.

A narrow view of how gender affects the voice also seems to play a role in the anxiety about the rightness and healthiness of creak. Robert Podesva observes that "the low pitch characterizing creaky voice is interpreted as resembling masculinity, due to gross tendencies for men to have lower pitched voices than women" (Podesva 2011: 427). There is a general assumption that there is a gendered pitch difference of about an octave, but in fact cultural differences are larger than gendered ones. In some cultures, the average pitch difference between male and female speakers is less than half a step, or from a white key to its neighboring black key on a piano (Quoresimo 2021). However, this strong cultural association of low pitches with masculinity may be part of what led to the moral panic that arose around a new awareness of the usage of vocal fry in young women, since male and female speakers occupy the same extremely low range of F0 when they employ the vocal fry register. Anxiety over this resulted in vocal fry in young women being referred to as a "disorder" and the effect of vocal fry in a young woman's voice as "a deterioration in voice quality" (Hornibrook *et al.* 2018: 36). However, a more recent study found "no significant relations...between creakiness or strain and vocal symptoms" (Laukkanen and Rantala 2021).

This year one of the students in my vocal studio was a senior musical theater BFA candidate (I will call them Peter). Although

I had worked with them previously as a theatrical director and as a musical director, this was my first time working with them regularly as a vocal coach. Since Peter had trained mostly under musical theater vocal coaches I had hired, their vocal pedagogy was based primarily in contemporary musical theater techniques rather than in classical vocal production. Both their singing and speaking voice had an extremely nasal placement. When I had worked with them as an MD, we had made some steps toward reducing the nasality of their singing voice. At that time, however, I was more concerned with developing good habits for vocal health than I was with finding correct placement. Peter's subsequent vocal teachers had also worked on reducing nasality, but with only minor success.

In my vocal studio, we first worked on this problem primarily by developing the control of the nasal port. We were able to get their singing voice into a less nasal placement, but the sound was thin and not resonant. In order to achieve a fuller sound, it seemed to me that Peter needed to open their pharynx more. In pursuit of this, I gave Peter the following exercise:

Yawning: If the yawn is silent, work until you can release some sound with the yawn. Check to make sure you are connecting the yawn to a deep breath. Then do a simple vocalise (1-5-1 ascending, or 5-1 descending), yawning near the pitch rather than "singing." If this proves to be difficult, it may be helpful to lay on the floor with your knees up and feet down (a position which allows the back to more fully relax) and pay attention to any tension you are gathering in your neck or shoulders.

Peter had considerable neck and shoulder tension, which was making it difficult for them to relax enough to fully open the pharynx, even while laying on their back in the above position. In addition to a wide variety of neck and shoulder stretches, I made use of yoga props to encourage the release of these muscles. On some days, we would focus on using these props just to release the tension, and on others we would vocalize while utilizing them to encourage and maintain relaxation.

Tie two tennis balls in a sock. Align the tennis balls on either side of the spine, placing them under the primary spots of tension in the neck or shoulders. This can be done either lying down, as described above, or in a half-sit with the back against the wall. At first, work on applying gentle pressure and releasing tension. Then attempt to vocalise with the tennis balls in place. This will help to develop an awareness of any connection between producing sound and the inadvertent clenching of those muscles.

While standing, take a yoga strap in both hands and extended it overhead, pulling laterally on the strap. Slowly move the strap down the back until the arms are parallel to the floor. This should be a pleasant stretch. If it begins to feel unpleasant or painful, give yourself more room on the strap. Pay particular attention to the places where there is stiffness, and move slowly back and forth in those spots a few times.

For a more restorative, less active pose, lay down on your back lengthwise on a yoga bolster with the shoulder blades just hanging off the end. A pillow or two can be placed under the head as needed. This is a good pose in which to do some gentle vocal warmups.

Scrunch up your face as small as possible, then open the mouth wide and make the face as big as possible. Repeat: Small face, big face, really small face, really big face (it helps to demonstrate as you instruct so the student doesn't feel self-conscious). Ask them to observe how cavernous the back of their mouth and throat felt as they made the really big face. Vocalize them on a triad, and instruct them to hold that space open as they vocalize.

Whenever our vocalizations involved an interval of an octave or more, Peter would tense up on the high note, widening their mouth horizontally rather than maintaining the tallness of the vowel. Many students default to this, so we work with the following exercise:

Biting the apple: Just before you sing the top note of the interval, imagine taking as big a bite of an apple as possible. If you have an actual apple on hand, hold it up to your mouth as you do so, but do not attempt to actually take a bite, and then sing.

Biting the apple proved to be so effective that one of my other students brought an apple with them to vocal juries to help remind them of this technique. Despite the previous success I had with using these exercises to help my students open their pharynx, Peter continued to struggle with this. They found some release in their tight muscles, and more opening than they had previously had, but their sound continued to present as thinly resonant and even still slightly nasally.

One of the advantages of teaching musical theater students who are also working with Shakespearean text regularly is that the spoken vocal techniques they learn often carry over into their musical theater work and vice versa. Students at SUU have the opportunity to work in our thousand-seat outdoor Globe replica, the Englestadt. When I need more volume or presence, I will often ask a student to imagine they are performing in the Englestadt on a windy night. When their Shakespearean professor is working with them in the Englestadt, he will frequently remind them to employ the techniques they have learned with me to be heard in that cavernous space. I decided to work with Peter on some of their Shakespeare soliloquies to see if their control and placement was any better in that work. As we did this, I observed that Peter's speaking voice, in addition to being placed nasally, employed a slight but consistent creak.

The use of creak and vocal fry in everyday speech is common enough in the undergraduate population at the SUU that I had not previously marked this in Peter's speaking voice as something that might be affecting their singing voice. Once I heard them continue to employ it in their Shakespearean work, however, where they are trained to speak with an open pharynx, similar to classical singing, I realized that this may be an area we should work on together.

I began by asking Peter to repeat a section of the monologue without creakiness. They were unsure about what I meant by this, so I demonstrated speaking with and without creakiness. They attempted to imitate me but were unsuccessful. They were surprised by my saying so. It turned out that, although the creak in Peter's singing voice was barely perceptible, they so regularly employed it in their speaking voice that they were unable to self-identify the sound or the feeling of creak.

Perceptually, creaky voice can be described as having roughness, growl, croak, or noise (Hornibrook *et al.* 2018; Fuks *et al.* 1998). Consider the grit in a rock-and-roll scream versus the open tone of an operatic soprano producing the same pitch. Listen side by side to early 20th-century singers Louis Armstrong and Bing Crosby—one with a famous growl and the other with a famous silky smoothness. The rock-and-roll scream and the voice of Louis Armstrong are examples of creaky voice at wide ranges of fundamental frequencies.

My work with Peter, then, was not about finding a healthier sound, but about expanding the range of possibilities for their voice. Having determined that creak was an ongoing issue for Peter, I got to work on a couple of exercises to help them feel the presence of creak and learn to control the sensation and sound:

> **Creak/silent exhale**: Produce an unvoiced, pure creak sound (a few fingers placed on the front of the throat will feel a vibration under a voiced sound, but none under an unvoiced one). Then exhale in total silence. Repeat this, going back and forth between creak and a silent exhale a few times. After a few repetitions, observe the action in the throat, the musculature you are employing, each time you move from creak to silent exhale.

Once Peter was able to employ that musculature consciously, I asked them to open as though making a silent exhale and to continue employing those muscles as they sang a median pitch (for them, F3) on Ah. The effect was immediate. They sang an open, rich, non-nasal note that startled and delighted them so much they laughed for some time after before they were able to recover their composure and attempt it again. They were able to repeat this on single notes, but they were unable to sustain it for long. We attempted the same thing with a line of dialogue, with the same effect: A word or two was manageable, but the openness was difficult to sustain. I taught Peter a quicker way to find the same opening.

> **Silent laughter**: First, give a big belly laugh. Then imagine you are in a situation where you absolutely cannot make noise, and repeat the laugh, but with the mouth closed and totally silent.

That silent laughter will pop the supra-glottal structure open in the same way as the creak/silent exhale exercise does. It functions as an excellent shortcut once the student has learned how to consciously open it. Once Peter had mastered these techniques, I would go through a song phrase by phrase, instructing them to silently belly laugh before each one. This would pop their throat into the desired position for the onset of each phrase.

After some weeks of practice, Peter was able to self-identify the feeling and sound of creakiness in both their singing and speaking voice. Once they felt confident in this, we began going back to the earlier exercises of yawning, big/little face, biting the apple, and general release of neck and shoulder tension. We also returned to an effective anti-nasal exercise:

Ng-Gah: Sing the word "sing," hold the "ng" sound for a moment, and then pop the mouth open with a "gah." Do this all on a single pitch. Then attempt the entire exercise at different places in the voice.

This exercise had previously only had moderate success in eliminating nasality for Peter. Now, however, with the combination of all of these exercises and new abilities, and with the support of the solid breath control they had developed with me and with their other teachers, they were soon singing their legit repertoire with a rich, full, and completely non-nasal sound. Once Peter was able to achieve a fully open pharynx, we began working creak exercises back into our routine to enhance their pop, rock, and contemporary Broadway sounds. A few of my standard exercises to teach healthy creak:

Balloons: Blow up a balloon, observing both the deep breath you take in preparation, and the action of the intercostal muscles as you blow it up. Then blow up the balloon as slowly as possible, while keeping the air moving continuously (slowly blowing bubbles through a straw is another way to achieve this). Observe both the depth of the inhale and the actions of the muscles all around the torso (Note: the tension in the throat you will experience is not something you will want to carry over into singing).

Some students find this so valuable they carry balloons with them at all times so that they can remind themselves as needed.

Creak Sirens: begin in unvoiced, pitchless vocal fry. Add voice on a low pitch and siren slowly upward and back down again. If you are using too much air, you will blow the ventricular folds open and will not be able to sustain the creak. Remember the feeling of blowing up the balloon slowly, and employ this as you do the creak siren. As you pass into CT-dominant mode, the nature of the creak will change slightly both in feeling and in sound.

Creaky vocalise: Slide in large intervals in creaky voice, eventually learning to sing the intervals without sliding. Keep employing the balloon breath.

Adjusting creak/airflow: Play with scales, applying more or less creakiness and more or less airflow. Observe what effect the different levels are having, and consider what styles of music the different combinations would work on. Get comfortable doing this with scales before you apply it to a song so you feel free to experiment with sounds and don't feel the pressure to make a "right" sound.

Vocal fry the song: Choose a song on which you wish to apply creak. Speak the lyrics in as much vocal fry as possible. Slowly begin adding pitch to this, experimenting with levels of airflow and creakiness.

Peter's first real success using creak was in achieving a good sound on pop music which sat in their passaggio. They had been working for some time, without much success on "Wall in My Head" from *Everybody Loves Jamie.* The addition of creak helped them find their mix, perhaps because we had worked so much on adjusting levels of airflow and creak through the passaggio. Now they were able to confidently make the adjustments they needed and apply just the right amount of creak necessary for that contemporary sound.

Later in the semester I assigned to Peter the song "Hey Little Songbird" from *Hadestown.* Although in the original Broadway Cast recording Patrick Page sings entirely in vocal fry, going as far down as F1, my intent was for Peter to sing the song an octave above that, in modal voice and in a more traditional bass/baritone range. However, having learned how to identify and control creakiness in the register below modal voice, they decided to try extending its use in this song. Peter worked on their own and came into their voice lesson one day very excited to tell me they had figured out how to sing the song in the original octave. They sang the entire song for me, and from that day on we regularly vocalized them down to F1, a full octave below their previous low note. They performed "Hey Little Songbird" successfully for their vocal jury that semester.

After my successes with Peter, I turned my attention to several other students whose speaking voices were habitually creaky, but whose singing voices did not perceptually employ creak. As I had done with Peter, we worked first on awareness and self-perception of creak, and then on full retraction of their ventricular folds, to great effect. For Peter and the rest of these students, their new awareness of creak brought them greater control over their instruments and the ability to access a much richer sound. Classic vocal pedagogy supports the use of an open pharynx, but this group of students was using creak so habitually that the usual pedagogical methods would not work on them until they could self-identify the feeling and sound of creak.

Since Peter worked on their own to make the leap into singing "Hey Little Songbird" in M0 in creaky voice, I do not yet have a method for teaching this to students. I encourage the use of vocal fry in vocalization to open up the lowest notes of a singer's range. Beyond that, I need to keep working on pedagogical methods for that register. Some work on this has appeared in the *Journal of Singing*, but this work assumes that vocal fry, as a relatively unused register, is useful as a pedagogical tool because no bad habits are engrained with its usage (Nix *et al.* 2005: 55). This assumption must be flipped, however, when students are habitually speaking in vocal fry. Once they learn to make sound without it, they are in an unused area of their voice where good habits will be easier to imprint.

Further work needs to be done to define our terminology and to discover the exact mechanisms that are producing creak and vocal fry. Is there a difference in the mechanisms employed to produce vocal fry and other kinds of creak? Are these perhaps the same mechanism—with the difference being the register—and creak is what allows us to access that lowest register? My work with Peter on "Hey Little Songbird" would indicate that this is the case. I find it convenient when teaching creak to discuss the false vocal folds, but how may the aryepiglottis and arytenoid cartilages also be employed? We have further work to do in order to develop a consistent pedagogy on this.

Once Peter became aware of their constant use of creak, they speculated that they may have been employing this subconsciously because of their anxiety around being stereotyped as having a "gay voice." Constant use of creak, they speculated, may have helped them dip down into the lower register of vocal fry. Robert Podesva writes that "much of the research on male pitch patterns has focused on the potential role of fundamental frequency as a marker of gay male identity," and observed the use of both falsetto and creak in social settings as part of "constructing a persona" (Podesva 2007: 478). It is unclear whether the use of creak was useful to Peter in constructing the persona as they wished. Interestingly enough, as we began to find an open, non-creaky sound in both their singing and speaking voices, they expressed simultaneous delight in the sound they were capable of making, and anxiety about not sounding like themselves anymore. As Hellyer points out in another chapter, this "micro-level navigation of identity" (23) may have been viewed by Peter as a threat to being seen and accepted for who and what they are. I encouraged them to keep playing and to think of this sound as something they were able to add to their actor's toolbox, and in no way a judgement on the way they expressed themselves in daily life.

References

Fuks, L., Hammarberg, B., and Sundberg, J. (1998). "A self-sustained vocal-ventricular phonation mode: Acoustical, aerodynamic and glottographic evidences." *KTH TMH-QPSR*, **3**(1998): 49–59.

Hornibrook, J., Ormond, T., and Maclagan, M. (2018). "Creaky voice or extreme vocal fry in young women." *New Zealand Medical Journal*, **131**(1486): 36–40.

Laukkanen, A. M. and Rantala, L. (2021). "Relations between creaky voice and vocal symptoms of fatigue." *Folia Phoniatrica et Logopaedica*, **73**(2): 146–154.

Miller, R. (1996). *On the Art of Singing*. Oxford University Press, USA.

Nix, J., Emerich, K., and Titze, I. R. (2005). "Application of vocal fry to the training of singers." *J Sing*, **62**(1): 53–59.

Podesva, R. J. (2007). "Phonation type as a stylistic variable: The use of falsetto in constructing a persona 1." *Journal of Sociolinguistics*, **11**(4): 478–504.

Podesva, R. J. (2011). "Gender and the social meaning of non-modal phonation types." In *Annual Meeting of the Berkeley Linguistics Society*, **37**(1): 427–448.

Potter, J. (2006). *Vocal Authority: Singing style and ideology*. Cambridge University Press.

Quoresimo, L. (2021). "Queering the voice in song." *Global Performance Studies*, **4**(1).

Chapter Sixteen

Dreamvoice
Composing from creaks, breaking through the unconscious

Ilona Krawczyk

Introduction

In my practice-as-research PhD (Krawczyk 2021), I found that a creaky voice, when expressed unintentionally, can be described as "dreamvoice": an empirical phenomenon that appears through a so-called inhibited, blocked voice, or unintentional expressions such as unexpected creaks, cracks, or singing out of tune. The dreamvoice—a notion I based on Arnold Mindell's concept of "dreambody" (1990a)—may reveal somatic sources of so-called "mistakes," and a performer's perception of their own voice's quality and capabilities, arising out of their personal life experience, and also linked to their cultural background. Hence, the unintentional expressions found in dreamvoice, such as creaky voice, may appear according to thoughts and affect (Thompson 2009), despite learned technique and training. In my research, I have developed an approach to voicework that

Creak: Theories and practices of pulse phonation
Edited by Francesco Venturi

ISBN 978-981-5129-29-8 (Hardcover), 978-1-003-61606-1 (eBook)
www.jennystanford.com

directly engages with this phenomenon in the realm of training and performance based on a post-Grotowskian practice that, to a large extent, draws on the affective reactions and memories of the performers. This process-oriented approach to performance also draws on Mindell's aforementioned concept of embodiment, which recognizes every bodily phenomenon as equally valuable. This perspective inspired me to re-evaluate what counts as a meaningful virtuosic enactment in contemporary performance. To adapt Mindell's perspective concerning embodied voice and enactment of musical material, I refer to sound art and noise practice in contemporary new music to redefine what can count as a meaningful vocal expression in contemporary performance beyond the premise of *right or wrong sound*.

In what follows, I present how dreamvoice expressed through creaks and cracks escapes the duality of validation between right/wrong sound in the realm of theater performance informed by post-Grotowskian practice. I draw from my own phenomenological experience as a performer-researcher to reveal the artistic utilization of these kinds of expressions aroused in both laboratory sessions and actual performances. On the one hand, I show how I purposefully learned these utterances and use them as artistic material in the devising process. On the other hand, based on my final PaR performance—*Dreamvoice*—I present how, with the use of processwork principles (Mindell 1990a), a performer can transform any unexpected event happening on the stage in the moment-to-moment situation. I also discuss an original form of performance-sound installation where the performers can maintain a lucid relationship with the vocal material, creatively utilize unintentional expressions in the performance dramaturgy, and apply performance states that range from casual to heightened.

Process-oriented approach: Terminology and mechanisms

In order to work with the voice through an approach that lessens the need for validation and criticism, I decided to develop a set of mindful attitudes supported by a processwork method. Amy and Arnold Mindell (2016; 2022) called these attitudes metaskills:

> Metaskills are the feeling qualities, or attitudes that bring learned skills to life and make them useful. In addition to all cognitively learned skills, Processwork, through Amy Mindell, recognised the significance of the way in which we use the skills and techniques we have learnt.

These "feeling qualities" support the surmounting of self-criticism and in turn embrace a process of embodying voice; particularly as regards the perception–expression loop. Instead of hating or fighting mistakes or blocks, through metaskills, processwork encourages one to welcome them with curiosity, compassion, and acceptance, "honouring the unknown in your own nature that made the 'mistake'" (Mindell 1993: 87). Therefore, in the process-oriented approach to voicework, instead of developing skills relating to vocal technique to "fix a problem," I focused on strengthening metaskills as I observed them: on the one hand supporting the process of acquiring new techniques and, on the other hand, allowing one to engage with the "mistakes" by unfolding them into the creative process.

Letting go of the ego and embracing the new modes of critical assessment involved in metaskills became essential for developing my process-oriented approach to voicework. Arnold Mindell proposes a mindful path for transforming the ego into a "fair observer" (1990b: 19–21). In processwork, ego is that which can observe "primary" and "secondary" processes: that is, both the I and me doing what I intend to do and express (primary processes), and distractions, internal and/or external events that arise and get in the way of what I consciously want to do (secondary processes). Secondary processes can cause the appearance of so-called double signals, i.e., unintentional, unconscious expressions of body language and voice that unfold in conflict with the primary processes. Mistakes, "unintentional parts of our music" (Arye 2001: 101), a crack in the voice, are for Mindell the double signals, the living unconscious, the dreambody trying to break through to awareness. However, there is also another strategy offered by processwork; when the ego becomes a fair observer, there is no conflict between primary and secondary processes (Mindell 1990b: 19–21). If any kind of distress or uncertainty appears, instead of fighting against these feelings and trying to "stay on track" with the flow, a performer may just acknowledge them "fairly" and move further. In such a case, the fair observer becomes a metacommunicator that makes it easier to detach from identifying with one or another process and to in turn avoid the need to validate them.

In my interpretation of processwork, the fair observer thus becomes a metaskill that supports working with the premise of *no right or wrong sound*. It helps to treat all the expressions and sounds as equally valid affirming "deep democracy" (Mindell 2008) and accepting them as a part of an "ever-changing flow" (Arye 2001: 219) and in turn developing supportive attitudes toward mistakes. Such a radical conception may raise questions about the practical utilization of a process-oriented approach to voicework, particularly in the case of performing musical material such as a song where the execution of correct notes is fundamental. Indeed, as Arye suggests, reflecting on his own experience, musical technique cannot be replaced or dismissed by processwork, but metaskills can support freedom within these structures (2001: 86–88). In developing a process-oriented approach to voicework, I treat all vocal sounds as manifestations of a performer's psychophysical state. If, using the processwork terminology, every produced sound mirrors the primary and secondary processes, I am interested in finding the sources of and somatic reasons for making unintended sounds such as creaks and cracks, but not, necessarily, to eradicate them. I search for ways of integrating their "purpose, its final goal, the direction it is leading us" (2001: 27) in a creative artistic context. Thus, as much as sounds can be correct or not according to the given task, in a process-oriented approach to voicework, I treat every sound solely as a piece of information to engage with while working with the process of embodying voice in performance. I aim to acknowledge what kind of expressions feel familiar or unfamiliar to a person and how they impact the vocal production and presence of an actor.

Another important aspect of processwork that I implemented in my inquiry involves working with sensory channels and edges. The sensory channels to which Arnold Mindell refers are different kinds of awareness: proprioception, visualization, hearing, movement, relationships, and world phenomena (Mindell 1990b: 23).[1] Arye complements Mindell's classification with the concept of "auditory subchannels," precisely referring to the components of music and sound: volume, pitch, timbre, and time to "help us to be more neutral, noninterpretive, and precise when we describe what we hear" (2001: 219–220). Distinguishing these allows the practitioner to observe where their attention is directed. For example, when attention is turned to the action of singing or speaking, the hearing channel is "occupied." If, however, while using their voice, a person makes a gesture or movement without being aware of it, this means that the movement channel is "unoccupied" in that moment. According to Mindell, an unoccupied channel is the one that can reveal a dreambody message and create an altered state. In processwork, changing channels is conceptualized as an edge, with attention shifting between different modes of perception. The edge is a state, moment, or place in which unconscious messages from the dreambody can break through into consciousness (Mindell and Mindell 1992: 43–63). In improvisation or physically demanding training, moments of thinking or feeling "I cannot do this anymore" or "I am bored" can then become opportunities to "change channels" or to "amplify the symptom," bringing to consciousness what the dreambody wants to communicate. As I present in the examples from practice-as-research, within my interpretation, changing channels can also serve as a metaskill. It can support a performer in overcoming distress in vocal performance, when a person encounters a disturbance in engaging with one of the channels.[2]

Performance devising

In my process-oriented performance-sound installation, I invited performers to execute material through diverse approaches in order to create space for the potential appearance of dreamvoice expressions such as creaky voice. To do so I adapted an "open score form" (van Eijden 2019) that allows for a great level of flexibility. The open score in avant-garde music signifies a score which does not fully regulate the final shape of the performance. Initiated by American composer John Cage, who referred to it as "indeterminacy" (van Eijden 2019), it does not predetermine pitch or rhythm but instead provides graphic and verbal indications that can be creatively interpreted by a musician. Hence, in such a score, the outcome of the performance depends on a creative dialogue between a composer and a performer.

In my inquiry into performance practice, I developed a form of open score that determined how performers interplay everyday and extra-daily modes of embodiment and vocality. This score shows a level of composition usually not notated in theater. It also gives a level of formalization that scores in post-Grotowskian practice do not usually have. Specifically, it depicts energetic states and at the same time allows performers to execute tasks with a great degree of flexibility. The open score then serves as a set of guidelines, a reference point that the performers can adapt in moment-to-moment situations, while following their psychophysical processes. As a result, the process-oriented performances are never finished products in the sense of polished or final compositions. The order and timing of elements, as well as the way performers execute them, remain substantially open.

Figure 16.1 represents the performance *Dreamvoice*, illustrating how the performers engaged in different modes of presence throughout the narrative enacted in the loop. This simplified graphical representation of the score displays a progression in which an everyday embodiment, represented by "_", was the start and end point of the durational performance structured in the loop. In between, the performers sustained both an everyday and an extra-daily presence, operating within sequences in which they altered their state to a peak "∧" with transitions between marked by "⌒". The starting points for a new loop began by picking up the first impulse coming from the sensory channels. This way, the beginning of each cycle could take any form as regards physical and vocal actions. An essential premise I applied in devising *Dreamvoice* was to begin with vocal and physical improvisations, through which patterns and associations developed and were finalized in physical actions (Richards 2003) and vocal compositions of the score. These compositions consisted of guidelines on how performers ought to relate to each other, for instance, whether to follow and tune to each other's mood and vocality or to provoke one another to vocalize independently. Therefore, the instructions were more related to the enactment of vocal effect than to music.

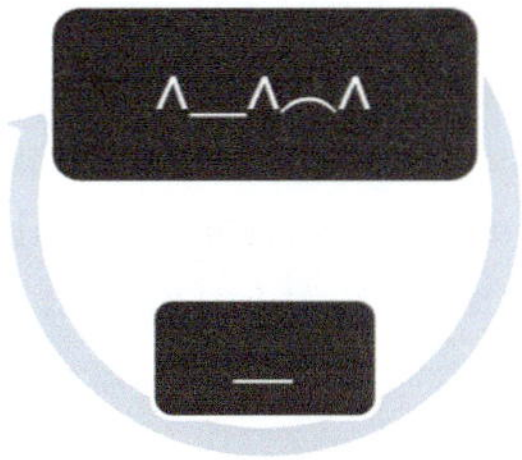

Figure 16.1 Dreamvoice.

Figure 16.2 *Dreamvoice—Mechanics.* Videographer: Ben Spatz.

I worked on the *Dreamvoice* at the University of Huddersfield in 2019 together with three practitioner-researchers: Cristina Fuentes Antoniazzi (performer and mindfulness teacher), Brice Catherin (musician and composer in new music), and David Velez (sound artist). The first task I asked performers to engage with was the recognition of where we were in our psychophysical process and with whom. Thus, the essential element to begin the performance was to establish inner and outer relationships that the voice could then express.

The structure of the performance, designed in a loop, one that required stepping away from whatever happened in the space after each cycle, served to make us observe the impact of aroused states on ourselves and to dissociate from them, particularly in cases of awakened vulnerability. Point zero was the moment of closing whatever we had opened and allowed to manifest during the run-through. It also suggested a form of performance oriented around the performer's psychophysical process, permitting various executions of the artistic material within each loop, thus privileging the exhibition of the process over a finished spectacle.

Regarding dramaturgy, the principles I applied in this process-oriented performance allowed for even greater flexibility. Despite a set order, the scenes could change and/or be skipped. Within the group of four performing artists, each of us had the freedom to initiate, interrupt, provoke, and/or support actions; thereby influencing and changing in an ad hoc manner the dramaturgy of each cycle. Notably, performers could not only change how scenes were enacted, but also the setup of the environment, e.g., manipulating set design, lights, dress code, multimedia, and the soundscape. *Dreamvoice* highlighted the processual aspect even further by displaying video material from earlier laboratory sessions (see *Dreamvoice – Mechanics* at [2:35–3:13, 6:09–7:12]). This multiplication allowed the audience to see the fluidity of performed material. The open form of the *Dreamvoice* performance-installation additionally manifested through the mixing of all vocal forms: text, songs, non-lexical and non-verbal utterances, as well as extended vocal techniques. Through this mixing, sonic images were created that challenged the perception of what is right, instrumentality and vocality, while at the same time giving space for interpretation. I spoke songs, recited text mute, performed Schubert's *Ave Maria* out of key with a creaking voice, learned growling sounds, cried and screamed and sang it all as a music piece. All to challenge what could feel wrong to me, to redefine and discover the creative potential of these "unacceptable" utterances, and to find artistic applications for them. On my own account, by allowing all these expressions to happen, I could observe fairly my reactions to them—my prejudices, my preconceptions and blockages—and go beyond them by changing my perception of the so-called "mistakes." This, in turn, supported my psychophysical process of embodying voice devoid from a conflict that often appears due to inner critique. Instead of holding on to one correct expression or pretending that a missed note was planned, I decided to consciously engage and amplify these utterances, integrating them in the dramaturgy of the performance.

By breaking down musical and textual structures, together with applying everyday embodiment in an artistic performance, I expanded the range of enactments that could be included in a contemporary, process-oriented performance. I created a setting in which performers had the freedom to transform artistic material according to their processes and in response to any impulse coming from their partners and the environment—no matter whether this was experienced as a disturbance or with ease. In such a setting, every vocal expression, even when fixed and structured according to textual or musical principles, could always be transformed through the performer's psychophysical process according to their engagement with the perception–expression loop of voice.[3] Having the performance enacted in a loop supported this approach by shifting the focus from a single virtuosic run-through to a set of cycles in which the performers could reveal the virtuosity of their transformations and their "line of life" in a different way each time. In this sense, I redefined virtuosity as the ability to perfectly execute a score to a virtuosity that lay in responding to every incoming impulse by performing and making artistic meaning from any expression appearing in moment-to-moment situations, without a time-lapse, be it intentional or manifested through a performer's unconscious.

Composing from the unconscious

Through the arrangement of a setting in the performance-sound installation described earlier, I aimed to create a space for dreamvoice expressions—such as unexpected and unintended creaks and cracks—while vocalizing. In the artistic process, I looked at so-called mistakes through the lens of their creative potential in both theatrical and personal contexts, examining the possibility of integrating such unconscious or unintentional aspects of performance into a performer's awareness. On the one hand, I investigated how to purposefully learn these utterances and use them as artistic material in the devising process. On the other hand, I examined how, with the use of processwork principles, a performer can transform any unexpected event happening on stage in moment-to-moment situations. In finding an artistic utilization for these unintended events, I drew from the aforementioned concept of "dreambody" developed in processwork, which through an amplification of symptoms shows its meaning in visions, images, and symbols linked to archetypical figures. In the realm of theater, I found this strategy to be a very useful tool, integrating the unconscious meaning of dreamvoice manifestations into the performance.

In this section, I draw from my own phenomenological experience as a performer-researcher, presenting the artistic usages of unintentional expressions aroused in both laboratory sessions and actual performance. I first discuss an example from the *Dynamic Configurations with Transversal Video (DCTV)* (Spatz 2020) laboratory session during the Judaica project founded by Ben Spatz and conducted with Agnieszka Mendel and N. Eda Erçin. This session was the origin for a scene that became one of the crucial points of the *Dreamvoice* performance-installation.[4] In this first example, I show how I engaged with the appearance of dreamvoice expressions during improvisation and how I navigated my psychophysical process from there, engaging consciously with the fair observer, amplifying signals and shifting attention between sensory channels. I then continue showing how I picked up this moment in devising *Dreamvoice* performance analyzing my associations and the symbolic meaning of the archetypical figures, as well as finding a musical structure for unintentional expressions.

As Mindell acknowledges, "dreambody (…) manifests itself as an unpredictable, uncontrollable power in everyday life" (1990a: 69). Concerning embodied voice, such an unpredictable expression of the dreamvoice can occur in a tone, rhythm, or pitch contradicting the content of the words (Mindell 1989: 63), for example speaking fast, loudly, and at a high pitch when a person asserts that one be calm.

Dreamvoice can also appear through unpredictable hoarseness, breaks, and cracks, particularly during a rehearsed and controlled vocal performance. Throughout the years of vocal training, I was taught to control and avoid my voice breaking between registers. The techniques I learned at musical school aimed to overcome this "problem" and make the transitions seamless to balance and equalize the voice in both registers (mixed voice technique). Hence, amplifying this "uncontrollable" signal—the crack in the voice appeared in *Verticality*—meant for me to act against what I knew as correct singing and do what I would normally avoid. However, approaching this expression with curiosity, from the distant position of a fair observer, encouraged me to explore the unknown—beyond the edge territory—by amplifying this "broken voice." The effect of this revolt not only allowed me to learn how to intentionally make such sound and change my viewpoint on so-called correct sound, it also allowed me to integrate the unconscious meaning of the dreamvoice into performance. As now the associations that arose from this process became symbolic images that I could use in developing the *Dreamvoice* performance-installation.

In the development process of *Dreamvoice*, a struggle over the worlds division and the paradoxical placement of the body in Christianity's figure of the messiah was approached through the figure of Mary. I see Mary as the link between the worlds, the mother of the messiah and as a woman submitted to greater forces. On the vocal/musical level, I decided then to use *Ave Maria* by Schubert, a classical composition that would at first require me to use a head register. Building on and against this register, I could then carry on with my dreamvoice expression manifested in *Verticality*: the voice cracking and breaking between registers. The decision to engage with a music piece was also a result of the need to adapt to working with a new partner and to use means of expression that were available to him. While Ben worked in *Verticality* by combining a text recited from a book with a melody from a Jewish tradition (Zalmanov 1978), in *Dreamvoice,* Brice's means of expression was through music played on a cello.[5]

Example 16.1 *DCTV lab—Verticality* (excerpt).
Videographer: Agnieszka Mendel
Video duration: 3:04
Source: https://youtu.be/5nySMNBCWBk
Date: July 31, 2017

[00:55] I hear animal-like sounds coming out from me. I am trying to suppress them, to keep myself still in human verticality. I close my eyes and try to hold my body in this vertical position as much as possible. I put my feet close together and keep vocalizing, shifting between the melody and sounds of shouting and growling. While doing this, I increase the tension in my muscles, particularly in the pelvic area. Somehow, I sense that this is the place of division and tear between the two worlds, animal and human. I place my hands on my belly and press to increase the tension even more. This tension and action affect my voice. I find an expression which is somehow in between, on the edge of the two registers and two expressions that I was using before. It is a prolonged howling. The growling and the melody are still appearing, but this third quality starts to speak through more. I begin to hear more of the words chanted by Ben. When I reach the point of exhaustion caused by my actions, I open my eyes, looking up silently, still keeping my hands on my belly.

[2:25]. I listen to Ben chanting the text of the book. I am exhausted, feeling surrendered to some power that made me go through all of this. I realize the pain of being kept in this vertical world that does not allow me to howl: the pain of division, suppression, and the cost of being vertical. I feel anger towards religion. Something that was supposed to offer salvation became an oppression—this is the image of hands suppressing my belly. As the messiah was supposed to be the one who saves us from our earthy, animalistic, sinful nature, I slowly look down at Ben and say with reproach: "mess, messiah."

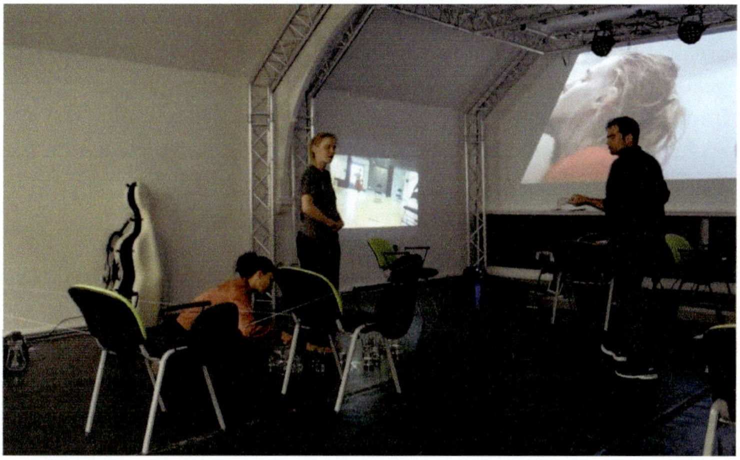

Example 16.2 *Dreamvoice—Ave Maria* (excerpt).
Video duration: 2:30
Source: https://vimeo.com/705178557/7aa9e758d3
Date: February 17, 2019

In Example 16.2, I present how the material that emerged during the Verticality improvisation transformed into a composition of the Ave Maria sequence. Given the open score of the performance, the Ave Maria became the only structured sequence musically enacted. Although the point of reference for this structure was the expression of my voice creaking between the registers, through the dreambody and dreamvoice work, each time, different aspects of my associations about the archetypal image would speak through, extending the pallet of expressions used in this sequence.

As shown in Example 16.2, due to the open formula of a process-oriented performance, I had the freedom to stage the *Ave Maria* scene according to the impulses arising in moment-to-moment situations, despite the set musical structure. This was also the case during the public presentation of *Dreamvoice* when I allowed an unexpected fall to direct the execution of the *Ave Maria* sequence, as displayed in Example 16.3. In this case, the creaking voice was not the guiding element. The movement took over as the tension I would usually create to get into the vocal pattern was too intense for my energy level. However, as the body already remembered the physical and emotional state, I got into roaring by other means at the end of the sequence.

> I begin by singing with an operatic voice in my head register to break it by pressing against my belly. I want to break the canon of beauty, the dogma of the disembodied holy Mary, the statue, the frame, the cage of verticality. I get into a playful exploration of the registers. I experiment with an operatic and white voice, letting the voice break in-between
>
> [00:30]. Listening to the sounds coming out of my mouth, the words of the song gain new meaning. Particularly the word *mortis* becomes potent to me. That is why I repeat it again and again as a mantra. My colleagues challenge me in this process with the sound of a bucket and water thrown into my face
>
> [1:03]. It makes me go over the edge and leads me into a rage. I mobilize all my body and voice into roaring so the flesh can be heard.

In Example 16.3, an unintentional expression appeared during the public performance, manifesting first in physical action. However, by following this signal in the movement channel, my vocality also changed as regards the dramaturgy I originally developed in the devising process. These physical and vocal actions evoke the image of Medea giving birth, which surprisingly, to my mind, came to be associated with Maria: *mater dei*.

Example 16.3 *Dreamvoice—Ave Maria, Medea,...* (excerpt).
Videographer: Ben Spatz
Video duration: 5:15
Source: https://youtu.be/co5pvbgZ2zU
Date: March 7, 2019

> I take the sheet with the musical score Ave Maria. It is a sign to begin this sequence. I do not have a plan or vision of how to start though. I feel tired, so I just allow myself to be carried out. Suddenly my foot slides on the water. I flip and fall on the floor
>
> [00:53]. I pick up this signal and decide to follow it to see where it takes me. I hear how my voice shakes and breaks while I amplify the signal, intentionally flipping over and over again. Nevertheless, I decide to surrender to it. I know I am on the edge of safety as there is broken glass all around me, but somehow, I still feel in control. I am approaching the edge but not passing over it. Finally, I hit the floor with my head
>
> [2:00]. It hurts, but it gives me even more drive to carry on. This is the point of a shift. I continue on the floor, giving myself completely to the water. The feeling of helplessness and being submitted to some external, inextricable forces grows in me. Associations come. I recall sensations from Medea song—desperate, tragic mother on edge between life and death, full of existential pain, hopeless in front of the forces greater than her. I get into a position of giving birth, to push this struggle and pain out.

As Mindell asserts, unconscious material "manifests mainly through visual and auditory channels in images and voices" (1990a: 69). The dreambody "hovers between body sensation and mythical visualization" (1990a: 8). Although, while devising the performance, I never connected these two mythical figures together in my conscious analysis, by following the disturbance, new associations came, provoking a completely different execution of the *Ave Maria* and guiding me toward a highly aroused state of presence. I associate the altered state achieved during the *Ave Maria, Medea* with the experience of cathartic moments in which the voice becomes a vessel, as described poetically by Ludwik Flaszen:

> You don't know if it is you who is bringing about this voice or the voice that is bringing about you, and where the source of the sound is – in your mouth, [in] your chest, in your lungs, in your stomach, in your groin, or maybe in the room's walls, ceiling, or maybe in distant celestial bodies, in the crystal spheres of the universe. (2010: 149)

In this moment, despite engaging with the unconscious, I no longer felt that it was me "doing," that I was drawing from my personal experience.[6] I did not perceive that my voice spoke of me. By working with the archetypical figures, I dissociated from the vulnerable aspects of my enactment. Although this was my individual experience, I perceived it as a collective knowledge expressed through my voice and body.

What became clear concerning the actual performance-installation and the artistic outcome of the devising process was how unconscious aspects of vocality were integrated into the performance, "seeing them in relation to their social context" (Wehr 1989: 124). As the American debate on vocal fry shows, creaky voices, or any vocal sounds that go beyond socially recognized and acceptable norms—especially performed by women—are submitted to heavy critique and factor into assessments of a persons worth at work and life; it is advised not to use these kinds of expression at all. Meanwhile, from my perspective, working with creaky voice in the performance was a liberating experience. The significant result of my engagement with dreamvoice was the creation of a strong peak moment in the score.

The dreamvoice, the creaky voice emerging from an improvisation, created a line of life that I later on applied to the musical structure of the precomposed material. Meanwhile, in the post-Grotowskian theater, the practice of using such expression within musical structures is uncommon.

The open score form, enacted in a repeating loop, generated a process-oriented performance where varying executions of the same artistic material were permitted. In such a performance, actors could stay in a lucid relation with the musical, textual, and dramaturgical structures. Considering diverse kinds of expressions as equally valuable; engaging with microtonality; the inclusion of unintentional expressions in the creative process and performance; tuning oriented not to musical correctness, but to communication; interaction between the partners and affective meaning of each sound; and, finally, allowing manifestations of dreambody and dreamvoice to take over the dramaturgy. With all these elements together, I created an environment for performance practice in which the overriding principle was the artistic application of every vocal expression and integration of their unconscious aspects that broke through in the creative process. For future research, it would be interesting to see how this practice affects the audience's perception and whether such a performance has the potential to build a new social context for vocal expressions such as creaky voice.

Endnotes

1. This classification has been slightly modified over the years by, for example, differentiating visualization from channels. However, the essential concept of sensory channels listed here has not changed.
2. This part has been published in Krawczyk and Spatz (2021), section one, that I authored.
3. The concept of *perception–expression loop of voice* comes from the extensive examination I conducted within my practice-as-research and is inspired by Jean Piaget's (1967) idea of reflexive action, defined as "a dynamic concept of circularity that brings together perception and action in a continuous process of sense-making and interaction with the environment" (Lesaffre *et al.* 2017: 63). See Krawczyk (2021) for a broad analysis of this phenomenon.
4. For more about Judaica project, see https://urbanresearchtheater.com/judaica.
5. For Ben Spatz's commentary on *Verticality*, see https://urbanresearchtheater.com/2017/07/31/verticality.
6. For more about the analysis of the personal experience versus "archetypal image," see the study on Jungian theory by Edinger (1977). For a feminist perspective on this subject, see Ulanov (1979) and Wehr (1989).

References

Arye, L. (2001). *Unintentional music*. Hampton Roads Publishing Company.

Edinger, E. F. (1977). *EGO and Archetype: Individuation and the Religious Function of the Psyche*. Penguin Books.

Flaszen, L. (2010). *Grotowski and Company*. Icarus.

Krawczyk, I. (2021). *Embodying Voice in Training and Performance: A Process-Oriented Approach.* [Data Collection]. Colchester, Essex: UK Data Service at https://reshare.ukdataservice.ac.uk.

Krawczyk, I. and Spatz, B. (2021). Dreaming voice: A dialogue. In C. Kapadocha (Ed.), *Somatic Voices in Performance Research and Beyond* (pp. 140–155). London: Routledge.

Lesaffre, M., Maes, P. J., and Leman, M. (Eds.). (2017). *The Routledge Companion to Embodied Music Interaction*. Routledge Taylor and Francis Group.

Mindell, A. (1989). *Working with the Dreamingbody*. London: Arkana.

Mindell, A. (1990a). *Dreambody: The Body's Role in Revealing the Self.* London: Arkana.

Mindell, A. (1990b). *Working on Yourself Alone*. London: Arkana.

Mindell, A. (1993). *The Shaman's Body: A New Shamanism for Transforming Health, Relationships, and Community*. San Francisco: HarperOne.

Mindell, A. (2008). Bringing deep democracy to life: An awareness paradigm for deepening political dialogue, personal relationships, and community interactions. *Psychotherapy and Politics International*, **6**: 212–225.

Mindell, A. (2016). *Metaskills: The Spiritual Art of Therapy*. Scotts Valley, California: CreateSpace Independent Publishing Platform.

Mindell, A. and Mindell, A. (1992). *Riding the Horse Backwards: Process Work in Theory and Practice.* Westminster, London: Penguin Books.

Mindell, A. and Mindell, A. (2022). Process-Oriented Psychology. Retrieved September 28, 2022, from http://www.aamindell.net/process-work

Piaget, J. (1967). *Biologie et connaissance: Essai sur les relations entre les regulations organiques et les processus cognitifs* [*Biology and Knowledge: Essay on the Relationship between Organic Regulations and Cognitive Processes*]. Paris: Gallimard.

Richards, T. (2003). *At Work with Grotowski on Physical Actions*. Taylor and Francis.

Spatz, B. (2020). *Making a Laboratory: Dynamic Configurations with Transversal Video*. Santa Barbara, California: Punctum Books.

Thompson, J. (2009). *Performance Affects: Applied Theatre and the End of Effect*. London and New York: Palgrave Macmillan.

Ulanov, A. (1979). *The Feminine in Jungian Psychology and in Christian Theology*. Evanston, Illinois: Northwestern University Press.

van Eijden, C. (2019). *Creating Open Form Scores for Improvising Musicians* [Master of Music Thesis, Royal Conservatoire the Hague].

Wehr, D. S. (1989). *Jung and Feminism: Liberating Archetypes*. Beacon Press.

Zalmanov, Š. (1978). "Yearning tune." Recording. In *Hassidic Tunes of Dancing and Rejoicing*. Smithsonian Folkways Recordings.

Chapter Seventeen

What can the pulse register do?

Francesco Venturi, Katherine Meizel, Margaret Pikes, Francesca M. Dovetto, Nicol Bana, Leah Kardos, Gino Sitson, Dario Buccino, Roberto Panzanelli, Pedro Oliveira, and Edoardo Mozzanega

Francesco Venturi

The bioacoustics of human pulse phonation proves the intertwined ties between voice, identity, and power. In contemporary societies, voices are not mere communication tools; they vocalize one's social, cultural, and even political position. While unconscious pulse phonation patterns are an integral part of all human voices, the conscious use of pulse phonation regularly serves as an embodiment of specific identity topologies, its presence in contemporary discourse often underscoring issues of exclusion and inclusion.

In the age of mediation, where representations often supersede the represented, stylized uses of "vocal fry speech patterns" continue to impact social perceptions and gradually influence how the pulse register is perceived. An evolving perception

Creak: Theories and practices of pulse phonation
Edited by Francesco Venturi

ISBN 978-981-5129-29-8 (Hardcover), 978-1-003-61606-1 (eBook)
www.jennystanford.com

changes not only how people feel about pulse phonation, but also what the pulse register can do. This results in a multiplication of possibilities, while on the other hand, increasing polarization.

Significantly, polarized feelings about pulse phonation showcase the intricate ways the voice functions as both a biological and cultural apparatus. The physiological properties of the pulse register, which remain mostly universal, are transformed when listened to with a situated ear, uttered with a situated voice. This results in myriad narratives and practices, often openly conflicting, influenced by the dualities at play, the individuals involved, and their ecosystems. Within this spectrum of vocal expressions, the pulse register represents a set of enigmatic phenomena that both embody and challenge humanity's boundaries, revealing tensions between its inherent occurrence and cultural reinterpretation.

By both highlighting and obscuring gendered and coded subjectivities, creak showcases its versatility, complicating its role in the portrayal of identity (Venturi 2021). While some dismiss it as an "affectation," this stands in contrast to its intrinsic nature, especially during early life and in response to the most varied emotions. Contradictions extend to its associations, encompassing both specific pragmatic attitudes, such as age or gender, and neurodivergent states. As individuals mature, the evolution of creak from infancy to its conditioned use in adulthood reflects a nuanced interplay of nature and nurture. Applications in music further show its versatility, from withholding emotion to amplifying the expressiveness of phrases. Unique is its capacity to convey extreme emotional polarities, from the heights of pleasure to the depths of pain. This flexibility in vocal production is further mirrored in its concrete contribution to popular songs (the typical example being Britney Spears). While it plays a crucial role in mainstream emotional conveyance, creak forges a distinct identity in genres such as Metal, under the name of "fry," setting itself apart from traditional vocal techniques. On the other hand, the pulse register is used to facilitate and improve any singing style, including Opera. Furthermore, at a somatic level, creak bridges the gap between theory and movement, fostering a proprioceptive thinking that stands in opposition to mere muscular action by mental will. Yet, perhaps its most profound duality lies in its intimate evocation of an "intimacy" juxtaposed against an "otherness."

As the following pages show, pulse phonation research (PpR) serves as a testament to the complexities of vocal expression and function, presenting them as multifaceted and versatile. Interpretations and uses have continued to evolve over the past hundred years, and pulse phonation has been the object of a continual territorialization. Deleuze's and Guattari's (1972) thinking of social relations as *territories* and their shifting grounds as processes of territorialization perfectly captures the evolving discourse surrounding the human pulse register. They theorized deterritorialization as the process wherein practices are removed from their original contexts, considering this an integral facet of language, and reterritorialization as their simultaneous incorporation into new narratives. He believed these processes allow for the birth of new possibilities.

The pages to follow represent in many ways a collection of such territorializations, polarizations, and possibilities. In a varied multivocal autoethnography, Chapters 17 and 18 bring together artists, educators, and scientists coming from three continents to reflect upon their experiences with/of pulse phonation. From child and adult education to vocology and medicine, and from live arts and poetry through to anthropology, their voices conclude this volume with a polyphonic exploration of what the pulse register can do, in a nuanced dialog that transcends scholarly observation, by bridging the gap between theory and lived experience. In so doing, what follows presents a living testament to the interplays with voice in our human experience.

Katherine Meizel

For me, among the things that creaky voice does best is to at once highlight and challenge the sonic borders of humanity, reducing speech and singing to their basic affective and acoustic components. It foregrounds the involuntarily disruptive sounds the voice produces in moments of extreme primal longing, of anxiety—in a bedroom moan, reaching for another's body, another's love; in a baby's cry, reaching for the love of a parent, or for the comfort of food. And those sounds are undeniably human. Heard largely at the onset and offset of phonation, creak distinguishes itself from most synthetic voices in 21st-century soundscapes; the digital Siris and Alexas and Hatsune Mikus that—though initially based on real voices—eschew the messiness of fry in their attacks and pitch transitions in favor of crystal-clear perfection. And creak is perhaps the most embodied of vocal gestures or qualities, stripping voice to its skeleton structure—its cartilages and tissues. It makes audible the fundamental pulse on which vocal phonation depends: a pulse that, otherwise, goes largely unheard and its rhythm unattended but is arguably second only to the heartbeat in terms of the emotional significance with which it is inscribed. And this extreme vocal representation of essential humanness, of emotion, of sexuality, is heard as both monetarily marketable and morally wrong when it emerges "out of the mouths of babes"—in the feminine and young, like the infamously creaky, sexy-infantilized Britney Spears of "...Baby One More Time."

But creak is, too, *almost* the sound made by forcing air through an excised larynx with no vocal tract to provide timbral characteristics. Voice, as it were, without a face. In this sense, it is almost unhuman: The vocal tract and the formants of vowels are certainly present, but other acoustic information associated with voice may be missing; John Nix found in a sample recording that segments of vocal fry included vowel formants but not much else (Nix 2016: 6). Without that absent information, though creak is frequently interpreted as an index of negatively coded identities, I wonder if it might also conversely be used to obscure other such signs. When creak continuously dominates a person's vocalization, is there a chance that it might strategically deny listeners some other identifying timbral aspects of voice that tend to be targets of criticism within oppressive systems of gender, race, class, ability, and age? If there is little or no modal voice to provide contrast, and relaxed, low-pitched, consistent vocal fry is the primary vocal quality listeners can perceive, then perhaps they are not perceiving other timbral and practical elements that evoke negative responses: loudness, or a high pitch, or shakiness, or a raspy timbre, or additional vocal indexes of otherness.

In this series of layered contradictions, creak sounds ultra-human and less-than-human, innocent and adult, an ur-voice and a not-voice, individualizing and identity-blurring. No wonder it has required a book.

Margaret Pikes

What is communicated when people use vocal fry, pulse phonation, or creaky voice? Hollien et al. (1966) explain that it occurs on the lower end of the frequency spectrum: But does it in fact lead to the perception of hearing a deeper voice? This may be what the speaker or vocalist believes or wishes, but what we hear when another person vocalizes is much more than a mechanical reception of frequencies. This is because the voice is not only sound waves emitted by vibrating vocal folds: an isolated scientifically measurable phenomenon. Following many other contemporary writers, I explain in my book *Owning Our Voices: Vocal Discovery in the Wolfsohn-Hart Tradition* (Pikes and Campbell 2021) that the voice cannot truly be described as a thing, but rather as a complex psychophysical activity (2021: 125).

I am a teacher who has guided people for many years in the development of a deeper connection to their expressive vocal range through engaging the voice with its roots in the body. Embodying the voice in this way can be a very valuable and empowering process stemming from the fact that the voice is an instrument that connects the mind and the body.

When we speak, we convey meaning not only through language but also through the sound of the voice itself. As babies we use preverbal sounds over a wide range of frequencies. These are the expression of as-yet unconscious emotions rooted in physicality. Breath, body, and feeling are joined holistically to express sound through the vocal folds. The baby is not yet filtering her wish to communicate through words and her small body, and big emotions are involved directly.

Language normally appears as the child develops individuality, identity, and consciousness. With this comes a separation between mind, the capacity to think, rationalize and articulate verbally, and the body's somatic, largely unconscious but nevertheless obviously essential role in the life of this child. Without wanting to delve further into the world of child development, it is relevant to note the importance of play and physical activity in maintaining a good balance between cerebral activity (the mind) and the body.

In Westerns societies especially, this vital equilibrium between psyche and soma is often hard to maintain and the resulting loss of connection between mind and body contributes, in my view, to many societal problems, including not only ill mental health—depression, eating disorders, self-harming, drug and alcohol addiction, etc.—but also to somatically induced physical disorders.

Speakers often produce vocal fry as they come to the end of their outbreath. To hold a connection to a clear voice as the lungs empty requires a certain level of muscular control. The resulting audible expression of disdain for physical effort can convey a deliberate, if unconscious, disconnection with the body. Some studies have, in fact, shown that a female creaky voice is perceived as hesitant, nonaggressive, and informal but also educated, urban-oriented, and upwardly mobile (Yuasa 2010). That such cultural meaning can be attached to this vocal mannerism in women reflects for me a worrying sociocultural tendency to negatively judge any perception of embodiment in a female voice. It links to all the historical examples where women have been led to believe that to be physically active and strong is to be unfeminine. Real vocal substance is replaced by an affectation of nonchalance.

Some men also use vocal fry to give the impression of a deeper voice. The same reluctance to employ a connection to the body can also be detected in many cases. Once again, the influence of sociocultural expectations can lead some people's attempt to speak with a deeper voice to sound more "masculine" without using the necessary vocal connection to the body.

Using a creaky voice may or may not actually be harmful to your vocal heath, but its fashionable use for the attempt to communicate a deeper voice or a relaxed and unthreatening attitude amounts to a kind of faked result. It betrays an underlying mind–body disconnection, which seems to be a growing and, in my view, unhealthy feature of contemporary society.

Francesca M. Dovetto

{<creaky voice> {<dialect> Maronna mia 'u teng' 'mpizz' 'e lengua </dialect>} </creaky voice>} [DGmaLc01N, F#72.]

(Corpus of Pathological Spoken Italian of Alzheimer's Disease, Dovetto 2023)

The voice source signal, typically defined as the airflow through the glottis, varies in a periodic way, which reflects the rapid opening and closing of the vocal folds. In running speech, the voice source signal is constantly modulated, and these modulations carry different kinds of information to the listener (e.g., as shifts in voice quality, between modal and breathy voice). Some of the modulations contribute in other ways, either to the meaning of the message, or to identify the social or individual characteristics of the speaker (Gobl and Ní Chasaide 2010). Variations in the voice source may be associated with segmental or suprasegmental aspects of the linguistic code. Creaky voice (also called laryngealized; in impressionistic labeling system "sepulchral voice") is thought to involve strong adductive tension and medial compression, but little longitudinal tension and with vigorous ventricular involvement (cf. creaky voice of suppressed rage: Fónagy 1962). Pitch has been observed to be often "deep"/"very deep" (in a five-point scale).

Creaky voice is more frequent in male voices than in female voices. In addition, the creaky voice seems to be associated with particular environments (in dialogic interaction, the creaky voice may denote a position of dominance). On the other hand, in the literature, impressionistic observations have often associated particular affective states with specific voice qualities. Gobl and Ní Chasaide (2003) illustrate how altering voice quality imparts different affective colorings to neutral utterance (e.g., for speakers of English, creaky voice is associated with boredom).

Voice quality may also differentiate between healthy and pathological groups. In particular, the creaky voice correlates with the disease condition. In medical studies, the creaky voice is considered a phonation subtended by shortened and thickened vocal folds that vibrate at a low and quasi-regular fundamental frequency with a long period of damping (Davidson 2020). A recent study (Silvestro *et al.* 2021) showed that creaky voice occurs in the acute phases of cephalalophargia. Particularly in patients with cluster headache (CH), a typical voice quality with low and croaking tone has been observed. By means of a digital voice analysis, a set of voice measures has been studied—including the traditional measures of fundamental frequency, calculations of jitter and shimmer, and noise-to-harmonics ratios as well as quantities related to the spectral tilt—in 13 CH patients and in 13 matched healthy controls. A video-laryngoscopy was performed in all the subjects. The study shows that creaky phonation can be due to a reduced capability of vocal folds to become slack or flaccid secondary to vocal cords edema subtended by laryngopharyngeal reflux affecting the phonatory mechanisms in CH patients. The laryngopharyngeal reflux probably may represent a dysautonomic sign related to the increased parasympathetic tone during in-bouts period, reinforcing the hypothesis of an extracranial autonomic dysfunction as part of the CH clinical picture. In other terms, voice characteristics and subtending laryngopharyngeal reflux may represent a hitherto neglected specific feature of CH patients that need attention in the general evaluation of the whole clinical picture.

The measures of fundamental frequency, calculations of jitter and shimmer, and noise-to-harmonics ratios, mainly related to the regularity of the vibration and to the ratio between noise and harmonics, have been investigated also in patients with neurological disorders such as Parkinson's disease and Down syndrome (Rusz *et al.* 2011; Tanaka *et al.* 2011; Kent and Vorperian 2013).

Aging also seems to cause, in both sexes, alterations in the values of jitter, shimmer, and deviation from the standard of F0 (Schötz 2006). On the other side, the cognitive and physiological-motor decline faced by elderlies inevitably leads to problems linked to alteration of all linguistic components. Among the most frequent manifestations of senile speech, there is a lower speech rate and higher incidences of processes that affect vocal fluency, timbre alteration, modification of vocal quality (a particular attention needs creaky voice phonation mode), reduction in sound intensity as well as significant phenomena of segmental and syllabic hypoarticulation. More generally, a lengthening of the vocal tract with age has been observed (Linville 2001).

In other cases, the creaky voice occurs in patients with other diseases (e.g., schizophrenia). Cresti *et al.* (2015) investigated some prototypical examples of schizophrenic speech by listening to and performing instrumental analysis on excerpts from a

Corpus of Pathological/Schizophrenic Spoken Italian (CIPPS Corpus: Dovetto and Gemelli 2013). Particular attention was given to Patient D, who presents a severe case of the disorder and idiosyncratic linguistic characteristics. The profile of the D assertive units appears to be atypical in relation to the change in voice quality for the final post-tonic (creaky voice). Actually, the prosodic emphasis given to utterances, not pragmatically justified, derives from a state of excitement, which may characterize these schizophrenic subjects, with which as well can be associated the change in vocal quality with the production of a creaky voice.

Studies on voice characteristics in Alzheimer's disease (AD), a primary form of dementia with a progressive degenerative course, have shown that AD patients have a tremulous voice with less intensity and less control of airflow. In particular, the speech of AD patients is characterized by a higher number of periods, implying that the voice vibrates at fewer cycles per second, presenting a deeper voice, slower speech, and a slower rate of speed or rhythm of the glottal pulses. This gives rise to a monotone voice (Meilán *et al.* 2014, 332).

In all these cases, rather than correlating with a position of dominance in the interactional dynamic, the creaky voice correlates with physiological effort in voice production (cephalargics) and/ or with cognitive effort in speech planning (schizophrenia, AD).

Unfortunately, there are also many disagreeing data on creaky voice compared to the results listed above; moreover, the perception of a creaky voice often does not correspond to the acoustic measures that identify a creaky voice. In fact, recent studies have shown that there are specific subcategories of creak, each with its own set of characteristics: The best acoustic measure for creaky voice entirely depends on what kinds of creak the investigator wants to identify (Keating *et al.* 2015; Wright et. al. 2019). The topic of the creaky voice in its different types, therefore, gains even more interest precisely because it can be associated with a specific pragmatic attitude, as with a specific pathology or even a characteristic that tends to be dependent on gender or age. Our studies on pathologies (Dovetto *et al.* in press) go in the direction of investigating these aspects in detail.

Nicol Bana

Since 2016, I have been teaching music education classes to toddlers and preschoolers in public and private preschools, kindergartens, family counseling, music schools, and workshop spaces. I meet an average of 250 children per school year. My area of interest and specialization is in pedagogical practices derived from Edwin E. Gordon's Music Learning Theory (MLT). In its goals and means, MLT-based music pedagogy gives a central role to the voice. Dialogs, games, and relationships are established through the voice. The sound of the human voice as a carrier of ultra-semantic, musical, and interactive meanings is the "event" upon which the lesson is based. Verbal language is suspended throughout. This wordless voice is sung and modulated with neutral syllables such as "pa" or "ba," in a musical syntax in which all elements (e.g., harmonic, rhythmic, etc.) are "signifiers" with which to play and communicate. Their meaning is not fixed and codified. In a group of children, the meaning of musical signifiers is relational and emotional. Through non-directive teaching, the MLT educator offers songs, activities, and patterns (short musical phrases). This framework allows the child to experiment vocally in a group of peers with continuity and guidance. The environment itself facilitates the creation of this exploratory space and helps the child take advantage of it. Guidance from a musically competent adult model provides direction and brings the exploration into a musical and collective context, pointing out possible directions and ensuring that the conditions are in place for the child to experiment with them. The features of this approach enable me to observe the daily processes of vocal sound production as a means of communication and play in groups of children. This enables me to appreciate the cross-cultural value of vocal exploration.

In the age group of 0–2 years, I hear a variety of spontaneous vocal qualities, including a creaky voice that seems to undergo progressive selection after age two. Whether or not children have been exposed to vocal qualities at school or at home plays a large role in this process. In the preschool years (2–6 years), selection is promoted by the creation of stereotyped voices. Children discover that their voice matters and has an identity. At this age, external references increasingly come to the fore: To express emotions, fantasy characters are brought to life, animated by stories told by adults and by media of all kinds. These vocal personalities have their own, often caricatured, fantasy, and they are, in most cases, the only experience children have with their ability to produce many different vocal qualities. That is why it is important to expose them to other voices, stimulate their imaginations, spark their curiosity, and help them find a sound for what they know but cannot communicate.

With 3–6 years old children, I work primarily with wordless musical "stories" that I express through different vocal qualities and vocal sounds, without suggesting a particular "reading." They are free to respond to this musical communication by imitating, varying, or inventing new phrases of mine. The perception of the different modes of my voice is usually quite unanimous. For example, a piercing "head" voice is generally recognized as a fairy witch; a soft and caricatured "chest" voice may be perceived as a clumsy bear or gentle giant. Growling is perceived as animalistic or monstrous.

Creaking, on the other hand, evokes different reactions: Some children perceive it as animalistic—similar to growling—others as creaking wood, still others as a silly voice. Matilde (age 2), for example, has developed a "creature" that she personifies by making high-pitched creaking sounds and moving like a funny animal. She can easily switch from this creature to a neutral voice while following, modifying, or developing musical patterns with different voices. She did not start singing in class until she discovered that it was "allowed" to play with her voice in this way. The anecdote makes me think about the implications of perceiving the creak as another voice in symbolic games and role-playing. As described earlier, creaking is less obvious for children to understand than other vocal sounds, probably because it is less common in everyday life. Therefore, such "otherness" can give voice to new narratives and create an unprecedented and raw expressive environment. Yet such creative space is strongly defined by external, stereotypical, and gendered references (e.g., the Disney princess). I wonder if imagination can take over fantasy within the fuzzy boundaries of *other voices* and open up a space where gendered narratives do not define creative action and elaboration. When the characteristics of a sound are not defined, as is the case with creaky-voice-gendered identities who appear less standardized and more malleable. In my experience, children can use this approach to overcome the divisions of "blue" or "pink."

An example of this is the case of Selena (age 5). When I first met her, her singing and movement qualities were evidently influenced by stereotypical female princess characters. Her interest in music seemed to be focused solely on imitating this kind of character. The growling made her laugh but seemed too masculine (as she told me) and did not match her "standards." The creaking sound, on the other hand, surprised her; I saw how much fun she was having finding a way to play with it, even physically (she is primarily a dancer). In subsequent sessions, she discovered ways to stop limiting herself to a fictional character. Toward the end of the school year, she proposed melodies of her own invention and qualities of movement that had never been seen before, which were very personal and not at all stereotypical.

Normally, growling seems to evoke a gendered perception that is biased toward the masculine, while creaking seems to be much less, if at all, gendered. This speaks to the great potential of *other voices* to promote inclusion and equity in the educational context. One classroom circumstance that supports this view occurs frequently with (so-called) "antagonistic" behaviors in the 3–6 age group. Inclusive exploration of voice is a helpful resource for connecting with others. Keeping in mind that there is no pedagogical formula that will solve classroom management and teacher–student relationships once and for all, vocal exploration can be a powerful resource. Once children discover that their *other voices*, whether loud, bellowing, grotesque, monstrous, or bizarre, can be used to make music and play together, it facilitates the expression of their need to express themselves individually and fit in with the group. Irene (age 4), for example, spent several encounters between selective silence and the desire to leave. Suddenly, one day, completely out of context (we greeted each other sitting in a circle on the floor), she began croaking and jumping around like a frog. I responded by doing the same, and the rest of the group joined in the game. The play was then tied back into the musical context of the lesson, and a new version of the song emerged, enriched here and there by frog-like croaking.

From then on, albeit with ups and downs, I saw Irene begin to interact with the group and participate in activities and vocal dialogs until the end of the class. Another example is Giovanni (2½ years old), who was calmly participating in class for half a school year when suddenly he started saying "No!" to everything. One day he spoke to me quite angrily, in a Louis Armstrong-like growly voice. I picked up that tone and improvised with him, accompanying it with the guitar. An interesting dialog emerged, gradually evolving from the growl to Giovanni's voice, without being reminiscent of Armstrong. It was confirmed as a complicit play. In the following sessions, I was able to find a way to welcome him back into the group.

In addition, the inclusion of extreme—that is, *extremely unexplored*—voices proved to be a valuable resource in solving classroom problems. The vocal exploration of channeled extremes was an opportunity to build relational bridges with children who seemed inaccessible. Continuity is the key differentiator in building such an educational relationship. It is critical to create a safe and defined space to channel children's emerging needs and encourage their curiosity without putting them at a disadvantage. This requires the careful use of a welcoming attitude. When free vocalization is not seen as mere exploration for its own sake, but is guided and structured to make music together, an enriching power emerges from diversity. It leads to connections between children *despite* and *because* of diversity. When the class community builds a relationship that legitimizes the use of "tutte le voci" (all voices), creative teaching is received with greater enthusiasm.

The author of this expression is Erik, a 5½-year-old boy. As we talked at the end of a preschool class, he said in amazement at the new discovery, "So we can make music with all voices." I have adopted this expression in all classes since. Among these sounds, the creaky voice holds a special place. It is not associated with a particular character; it can take on different forms and connotations—stimulating language games that encourage imagination rather than fantasy. Creak, however, is only one of the many *other voices* available to us in the educational context. They are powerful allies in setting the stage to know one's own voice. This encounter is of great creative and human value.

Leah Kardos

When I creak my voice, it is only ever me, only ever for me. When I creak, my body senses my self-sounding on the edge of something else, which is also a self. A sound of withholding. Maybe teasing. Maybe protecting. My creak is letting you in. Making you lean in closer. Magnetism.

When I creak my voice, it is sometimes a sounding of abandon. Orgasmic pleasure, or the deepest pain. The sounding of my body experiencing intensities. The edging of pitch. Slowing the vibrating air inside me, glitching identity, dragging language. Expressing grief and rapture the only way it knows how.

When I creak my voice, I do it in my sleep—no effort, no consciousness. I learned my languages as a young child, but I came creaking out of the womb. My somniloquy is only ever me, only ever for me.

The biological acoustics of my throat; the fleshy architecture of my respiratory tract is only ever me. Every sound I make is us. Microphones make possible the quietest, meekest, fiercest, most monstrous utterances. Completely animal, wholly human, and more than that. We can creak and drone with frogs, and sing in harmony with crickets. Expression can be unfiltered, untamed, spectral harmonic. My resonant filter envelope of flesh.

Gino Sitson

My first experiences with traditional music opened a universe of sources far from my usual choices and musical passions. Having been mainly interested in the human voices as a scholar or as a vocalist, I have always wondered why traditional music singers have scratchy voices or guttural timbres. Among the personal factors having impacts on the voice are the mother tongue, the region of belonging, the sociocultural level, or the way of learning a language. These factors are constitutive of a form unique to each individual. The same voice can also be modified by emotional variables in connection with the context in which it is emitted or thanks to voluntary muscular work. Voice variations can thus appear between individuals or in the same individual.

After having worked for a long time on the Gwoka music of Guadeloupe, I found quite striking similarities between this sound aesthetic and songs from sub-Saharan Africa. The use of the creaky voice is commonplace in these musical traditions. Like the music from Guadeloupe, the voices of Benin and Congo are guttural and rocky, and their timbres are rough, while certain syllables are elongated—more particularly the vowels, with a holding on a stable note, with or without vibrato.

With regard to the creaking voice mechanism—present in Gwoka—it corresponds to a very slow vibration of the vocal folds, of a very low fundamental frequency, so low that one perceives a vibration between the sound and the frequency. Among the singing voices I studied, I spotted several examples of creaky voices. This laryngeal mechanism does not seem to be used as much in Creole as in English/American. Creaky voice is generally used at the end of a musical phrase—with a slight increase in intensity to enhance the emotional expressiveness of the sung phrase and seduce the Gwoka audience at the same time. It should be noted here that some singers have a particularly rocky, creaky, or hoarse voice due to regular and heavy use of alcohol. This type of alteration in the timbre of the voice is not necessarily an element detrimental to the talent of the singer. The vocal timbre influences the audience when the latter is led to make an aesthetic judgment of the singer's voice. It corresponds to an expectation inherent in this musical culture. From a psycho-acoustic point of view, furthermore, the listening conditions—recording, live, a Cappella, indoors, outdoors, etc.—can influence the appreciation of timbre.

Dario Buccino

In my compositions, I group the various forms of pulse phonation under the generic term "DRO," which is derived from the reverse spelling of "ORD"—an abbreviation for the Italian term "ordinario," traditionally used in music as an instruction to play as usual. A DRO sound is any mode of sound production that is "anomalous" or different from the modes usually associated with the spoken or sung voice; although DRO sounds—e.g., hoarseness—are also part of speaking and singing. Shifting from one mode to another without intentionally choosing them, but following the impulses of bodily expression, we can bring the voice to life and avoid limiting it to the individual manifestations of its inexhaustible identity. Therefore, the DRO universe must be composed and performed with great finesse so that it does not degenerate into an indistinct cauldron. It is necessary to crack the surface of the voice and identify its inner currents, the parameters that govern its subtle metamorphoses and mysterious intensity. These undercurrents become the main object of composing, instead of their obvious manifestations, whose classification is thus transcended, though not necessarily censored. I follow this path with the HN system®, a music composition system for structuring performance processes that I developed in the early 1990s and with which I try to bring the affective resonance of sound to life. If pulse phonation is understood as just one possible aspect of the boundless DRO sound production flow, it becomes more natural for composer and performer to free and control their unconscious outpourings.

Sound is only crucial to musical and human expression when it transcends itself and is not identified exclusively with its materiality. Indeed, I find that there is too much sonoriality in music research of the last hundred years; too much focus on the reified sonic event. Similarly, there has been too much corporeality in recent years. The physical act seems to be limited to its motor, muscular, and postural envelope. I believe that the still-valuable 20th-century tradition that sees pure sound and presence as more important than musical sound and choreography should not be taken in a merely literal way; rather, it is fruitful in supporting the emancipation of sound and body from formalization, fossilization, and objectification—it is fruitful in rooting them in their vital origins and making their transcendence possible.

Contemporary classical composers seem more interested in the voice as a generator of sound, texture, and gesture than in the voice as a direct emotional expression. Of course, the voice is always expressive, and no composer can escape this dimension when writing for the voice, but what I think is even more crucial is to use the expressive drive not as a product but as a sound form generator. For example, when we scream in pain over a hard bump, we involuntarily generate the sound of that pain. The suffering itself generates the sound form, its mode of emission, its timbre, and its unique sonic shape. In my opinion, it is historically necessary for a composer to work with forms that reveal expressive impulses, and not with forms that are detached from their energetic and emotional source.

Vocal fry, in itself, seems to me negligible. Instead, it is extraordinarily fruitful as one of the infinite ways in which the urgencies of the body create vocal forms. If the performer can recognize the identity of this phonation in its uniqueness, he or she can naturally master it better. However, this is best accomplished only when a visceral need spontaneously meets or merges into this physio-sonic zone. On the other hand, musicians can intentionally create this specific form because it is an essential resource of their sound palette. Sound and feeling research are complementary—both fronts of work are necessary. If I were to work on sound understood only as a consequence of expressive tension, I might impoverish my palette, because sometimes our inner impulses are channeled into a limited range of behaviors that are socially shaped, whether we are aware of it or not. We know that there are vocal responses whose stereotyped nature eludes us: The ingressive phonation we use when we are astonished, frightened, or excited, for example, is in some sense a stereotype. Of course, there are also less stereotypical spontaneous sounds: when we experience extreme, almost unbearable physical pain, our voice emits high-pitched sounds that do not conform to a culturally acquired code of vocal behavior—this lesson we learn from Alfred Wolfsohn, who studied the cries of the wounded in the trenches of the Great War when he was traumatized by them.

Given the human tendency to channel impulses into habits, studying the infinite possibilities of our vocal apparatus is as valuable as any physical exercise: Not only can we express ourselves more freely, but our expressive needs can be inspired by new forms of vocalization. Our inner worlds may recognize in such modalities parts of the self that have been buried since childhood or even previous generations: "We have in our body an ancient body," said Jerzy Grotowski, who developed this principle using ancient African and Afro-Caribbean vibrational songs in the last years of his exploration of art as a vehicle. Such buried parts of the self want to be brought to light out of a deep artistic and existential need.

From a strictly musical point of view, pulse phonation—in a more or less recognizable form—becomes important when it manifests itself as foreign to a rigid classification. In my opinion, the power lies in such non-classifiability. It is like the infant's voice: it is a living vocal sound, but it is not "vocal" in the sense that it is about to become speech or is about to become song; it is a non-linguistic and non-musical sound whose dimension of immediate emotional expression is particularly evident. Likewise, the absence of a defined meaning does not mean that pulse phonation says nothing. It is neither a tabula rasa nor a non-sound or even a non-voice. On the contrary, it is a meaningful sound, but open and ambiguous in the best sense of the word. It makes the human physical voice audible and palpable without anchoring it in any identifiable musical or emotional content. And it is all the more powerful when the listener even forgets that it is "vocal fry," that is, when the first reaction is not to isolate the form perceptually and conceptually from the power of the expression that permeates it.

Roberto Panzanelli

Founded in Perugia in 2000, UmbriainCanto is an arts center specializing in the voice. For several years, it has served as a focal point for anyone interested in the subject of modern singing. Since 2010 it has become an art therapy center for the study and practice of Holistic Voice®. Over the years, UmbriainCanto has collaborated with numerous voice teachers, speech pathologists, psychologists, and therapists. Research took a major step forward when the center collaborated with the Audio Phoniatry Center at the Spoleto Hospital. The two institutions conducted joint research on voice physiology by performing acoustic analysis with videolaryngostroboscopy (VSL). UmbriainCanto has conducted pioneering studies on vocal fry and its applications in voice didactics.

First, we wanted to highlight that it is not appropriate to equate fry with the *Arytenoid voice*—also known as *Strohbass* or monk's voice—and the *Voce del consigliere* [counselor's voice], also known as pressed fry. Second, we wanted to make clear that, as advocated by Mara Behlau, fry is useful, if not fundamental, for voice training and vocal re-education. We have gone beyond Behlau by linking fry to a range of supporting sounds (e.g., growl) and a range of vocal qualities (e.g., penetrance). At the time, vocal techniques involving extreme voices were rare. Since the presentation of our studies on vocal fry at the Artistic Voice Conference (2007), several new techniques have emerged, and fry has been duly considered. The further spread of harmonic singing styles and beat-boxing has only paved the way for guttural sounds, giving further space to fry. Today, it is considered valuable by most practitioners. For example, it is used for warming up, cooling down, as onset of the voice for expressiveness, or in extreme singing techniques in relation to growl and scream. Furthermore, much of the confusion has now disappeared. Yet, in the early 2000s, fry was controversial. We published the study (Panzanelli 2009) at a time when it was suspected of being dangerous or unhealthy and our intention was to give new value to this vocalization.

Acoustic analysis of voice recordings conducted at the center confirmed the existence of three types of fry. The proposed variants were as follows: Fry 1, or "Vocal fry *strictu sensu*": simple signal characterized by a sequence of aperiodic pulses containing formants but no harmonics. It is the most typical in sonorous and stable variety, assisting vocalists in learning to produce different tones by regulating the frequency of the pulses. Fry 2, or "Fry + Voce": complex signal consisting of two overlapping sources, one pulsed—corresponding to Fry 1—and one continuous and regular—corresponding to normal voice; the tone is periodic with noise bands, resulting in a weak voice with additional vocal fry. Fry 3, or "Fry no Fry": complex signal consisting of two overlapping sources (as in Fry 2); the result is an uncertain and strained sound, typical of poorly initiated fry. This first classification was useful for two reasons. First, it showed that the human voice can be aperiodic. Second, it suggested that there is an optimal fry that is beneficial for singing practice.

Arytenoid voice, Voce del consigliere, and fry are related because they arise from and lead to vocal fry, but the first two should be considered advanced and not relevant to all voices. On the one hand, fry is nearly monotone and has little scope for modulation, while the related voices are modulable. On the other hand, *Arytenoid voice* and *Voce del consigliere* are technical and unusual, while fry is natural and usual as it belongs to all physiological voices.

Regardless of vocal fry being part of the repertoire of all healthy voices, conscious fry can be difficult for some to produce. For example, it is easier for a fit voice and more difficult for a tired voice. Even with a healthy voice, however, someone may find it difficult for psychological reasons. Common instructions from voice teachers or therapists for initiating the creak phonation are as follows: reduce air, focus on low notes, relax. I particularly agree with the first hint: It is helpful to reduce air pressure, because fry inherently requires little air. The second hint can be misleading. The experience of voice initiation in the fry register is characterized by an immediate appearance of high frequencies. As for the third indication, which is certainly helpful, it is appropriate to make a clarification. Fry requires well-calibrated muscular activity, so we do not look for muscle relaxation, but mental relaxation. According to Sherman Lowe, an opera singing teacher who uses vocal fry in his teaching, the implementation of fry occurs through imitation. I would add that when teaching in pairs, as well as in groups, as it is easier to draw on unconscious prior knowledge. Fry is, therefore, realized not only by imitation, but also by activating this field of prior knowledge.

What does a good vocal fry indicate? It indicates a good balance between the work of the laryngeal muscles and the condition of the mucous membranes of the vocal folds. It indicates a constructive and healthy use of the voice. Since the voice is an indicator of the general condition of the body, fry indicates good energy status, good health, and good perception. Improving the sound of the fry is synonymous with improving the voice: in power, in quality, and in bass enhancement. In general, fry can be improved in the following factors: pulse rate, pulse regularity, speed of initiation, and stability of sustenance. These factors are linked to each other. For example, pulse regularity is the graphic representation of stability, which is the corresponding proprioceptive sensation. Misuse of laryngeal stimuli, as well as misuse of the voice in general, can result in aphonia or, in some cases, hoarse voice.

Hoarseness is also typical of certain pathologies. Fry differs from hoarse voice in that it crackles. By the crackling, one can recognize the intact, moistened state of the mucosa of the vocal folds. If the mucous membrane is damaged, fry disappears completely.

In the Italian school of singing, the vocal qualities are brilliance, softness, and fullness (Juvarra 2007). An improvement in fry leads to an improvement in the voice, first in brilliance and second in fullness, while the softness is also positively influenced. Moreover, in many modern singing techniques, penetrance is considered the fundamental vocal quality. This quality is independent of singing, and it is defined under different names depending on the method and the environment: the "gne" of the voice (Pitteri 2022); the "twang" of the voice (Turlà 2003); the voice of the witch; etc. Penetrance is particularly emphasized and, therefore, appreciated when vowels are nasalized. However, it is a component of the voice that originates in the larynx. Improvement in fry leads to improvement in the properties of penetrance (Panzanelli, Mazzocchi, and Brozzi 2011). I consider this to be one of the most important findings from these studies.

Working on fry also leads to a general strengthening of the chest voice, which is more extended both downward and upward and is enriched with the qualities mentioned above. This general strengthening of the chest voice is especially evident in the lowest notes, as these are generally the weakest. I do not state that producing a good fry will improve the bass notes, unless I switch to *Strohbass*. Increasing penetrance by making a good fry will produce, over time, fuller basses.

Vocal fry also relates to high-energy singing with a natural voice (popular singing) and the operatic voice (opera singing). High-energy singing in natural voice corresponds to two predominant vocal qualities: belting and distortion. The two qualities are interrelated. The fry is the food and also the product of these voices. Food, because you can afford such voices only if you have a good fry. Product, because if you use these voices properly—i.e., do not abuse them—your fry will improve. The same reasoning applies to opera singing. High-energy voices can only be produced by singers who have a good fry. The key to understanding these relationships lies in penetrance.

For many years, I have presented to the public this teaching proposal, which I called Vocal Fry Method (Lombardo 2011). The central didactic hint is to adopt an extended proprioceptive philosophy, as opposed to a localized muscular tension controlled by mental will. In this panorama, the vocal fry's technique is very important because it re-educates one to the perception of high frequencies and to an extended bodily vibration.

Pedro Oliveira

DESMONTE is a performance for prerecorded voice and live electronics exploring vocal timbre at the limits of the juridical and symbolic body. The piece is part of a larger, long-term study on so-called "automated dialect recognition," a proprietary software in use by the German Office for Migration and Refugees (*Bundesamt für Migration und Flüchtlinge*, thereafter BAMF) since 2017 on cases of undocumented asylum seekers. For the past 5 years, I have been interested in critically inquiring on the ways in which this software works, by analyzing the gaps of knowledge expressed through what the German authorities say—and what they occlude—in press releases, white papers, and statements released through freedom of information requests. My process mostly enacts a facetious, lo-fi "reverse engineering" of each step undertaken by the software, with the intention of returning its operativities from the frequency to the time domain, that is, that of listening.

One of my main hypotheses is that this "dialect recognition" apparatus is not interested in measuring language or dialect, but instead what I have been calling the *timbral architecture of the voice*. It is by converting snippets of spoken language to a vectorized space, and performing a series of discrete filter operations and feature extraction from said vectors, that an approximation between the physical characteristics of the speakers' vocal cords and those found in a larger corpus of speakers might reveal the alleged "origins" of an asylum applicant. What is not said is that such measurements occur within the same methodological and technical arrangements of its own design and is, therefore, always already contingent to find that which is designed to measure to begin with. Furthermore, the fact that this entire operation is predicated in the interest of detecting "fraud" as quickly as possible reveals other discursive details of its own design: to enforce and optimize the asylum process on the BAMF's side, often against the interests of the asylum applicant and with one goal: deportation.

It is through exploring this fabricated connection that the software—and by extension the German government—creates between voice and citizenship, in the shape of timbre and origin, respectively, that DESMONTE exposes the failures of machine listening to successfully partake, if we follow Sylvia Wynter's (2003) formulation, in the "overrepresentation of Man." At the same time, it understands the spaces delimited by such failure to be generative of different pathways in which it might be possible for the voice to refuse to announce the body and thus to rethink (or un-think) the figuration of this notion of the "human" as a referent to be undone. To engage in such a proposition, I construct the piece entirely with and through the voice of Brazilian death metal singer Fernanda Lira, whose extreme fry scream technique extends and complicates the limits between timbre and voice, noise and signal, and identity and identification. In DESMONTE, a modular synthesizer uses granular synthesis techniques to "zoom in" on her fry scream to derive— rather than extract—different types of signals that take control of decisions done through the audio path, which in turn condition a palette of sounds—oscillators tuned to her voice's partials, real-time resynthesis, and spectral rearrangements of her voice through filtering—to emerge.

My interest in exploring the timbral capabilities present in micro-snippets of articulated fry screams is conceptually and technically geared toward an im/possibility of the body's disappearance from the voice, while at the same time rehearsing different ways in which the body can emerge and make itself present in and through pulse phonation. I see this presentness of the body, however, when articulated through Fernanda's fry scream, as being disconnected from extractive practices that allow for any kind of measurability. Rather, exactly by exceeding the body and making it overtly present—both as throat pulsation and as spectral technique—that this disconnection from body and the instrument of identification is exacerbated.

Death and Black metal's long history of guttural and fry scream technique and skill notwithstanding, through DESMONTE I am also interested in emphasizing the ways in which the figure of the speaker/singer is represented in such genres; I am speaking here, namely, of the figure that can be called that of an "extreme otherness," predicated on Death and Black metal's imagetic (and sonic) relationship to monstrosity, which immediately positions its fry scream away from more traditionally "academic" extended vocal techniques usually found in electroacoustic music and early sound art. However, my choice of working specifically with Fernanda's voice also emphasizes the politics that conditions this figure within a mostly masculine—and oftentimes quite misogynistic—musical genre to extend it to more encompassing conversations toward a feminist and anticolonial "monster."

This mode of reassembling the "monster"—here not visually but sonically and haptically—through the extremely skillful physicality expressed by her voice refuses to announce anything but the body "in the raw" (Ferreira da Silva 2018), which conditions the possibility of that physicality to begin with, always against identification, measurability, and taxonomization. With that, Fernanda Lira's extreme fry screaming, combined with the spectral narrative constructed in DESMONTE, holds the possibility of exceeding the body by insisting on—and in some ways enacting—the separation between body and origin, and origin and citizenship that systems such as the BAMF's dialect recognition software seek so boldly to sediment as being one and the same; as biologically immutable and always already available for measure and evaluation.

Edoardo Mozzanega

I stumbled upon the creak and words fell away.

I chanced upon the practice of creak last summer. We were in the midst of Neutopica, an experimental artistic residency I curate alongside a collective of colleagues. During this time, together with Angela Burico, a voice actress, we engaged in vocal practices. Angela led a warm-up training that began with the "creak." Surrounded by the fringes of a forest, enveloped in an almost mystical ambiance where we slept outdoors and had recently spotted wolves, the ambiance was surreal. When I heard everyone produce the creak, it dawned on me that this was not just a warm-up—it was something in its own right. In that setting, Angela introduced the creak as a kind of "degree minus one," not even zero, almost a prelude she suggests as a warm-up, a stepping stone into vocality. That initial exposure shifted my perception of the creak from a marginal or foundational act to recognizing it as a central, expressive element in itself—teeming with potential. This resonated with research I have been pursuing for years, dedicated to the figure of the tiger.

Shortly after sharing my newfound obsession with creak with a friend, she introduced me to Francesco Venturi. To my astonishment, I discovered that for him, this was not merely a marginal or preliminary practice. Instead, it represented a significant field of both practical and theoretical inquiry. This fortuitous meeting affirmed what had previously been only an intuition or an instinctive fascination for me. Through our dialog, I hope to bring fresh insights into this domain.

My work is multidisciplinary; while rooted in theater and bodywork, it often branches out into various forms of sharing. For this particular trilogy, the first chapter is titled "Dream of a Tiger," which revolves around storytelling and oral narration practices. The project was inspired by a recurring dream: I found myself in a house with placid and lazy tigers, amid an odd ambiance blending tension and relaxation. Stemming from this dream, I compiled an archive of stories concerning tigers, animism, and encounters with animal figures. This repertoire serves as a source for both personal introspection and group work, offering a springboard for improvising oral and dreamlike narratives.

When I encountered the creak while developing these practices, which are deeply intertwined with stories, a new avenue of research unfurled before me. It was a game-changer—where words, so central to the first chapter, vanished. The creak carved out a sonic, physical, and imaginative space, a cavernous realm where perhaps a preverbal intimacy with the archetypal figure of the tiger, the big cat, is possible. As the research journey unfolds, it is as if language yields to the epiphany of voice as a signifying act, plunging into profound depths.

I entered the cave and found the tigers.

Angela Burico, aiming to assist those struggling to embrace the practice, invoked the element of pleasure. She advised us to connect with the sensation of tasting something exquisite. On one hand, there was this connection to a sense of pleasure. On the other, an eerie effect arose from the sounds people emitted and the associations I made with that sound. The ambiance was certainly intensified given our proximity to the wild, but it felt as if we were touching base with a familiar yet foreign entity within us—both unsettling and enjoyable. I was captivated by this contrast between seeking pleasure while simultaneously eliciting unease. This disparity made me eager to delve deeper into the subject.

So, I requested Angela to conduct a brief workshop dedicated solely to this practice. She was quite surprised but agreed. It provided an initial opportunity to observe the practice closely. The first realization was that the creak invoked feelings of vast hollow spaces within me. This resonated instantly with the notion of a cave, an imagery we had previously invoked during our research phases, tapping into the unconscious and dreams. But how did this deep-seated space relate to tigers? One answer emerged when I delved into paleoanthropological texts. I discovered that caves, in the evolutionary history of *Homo sapiens*, represented the closest proximity to large felines. Studies suggest the plausible era when apex predators and hominids coexisted in deep, constricted environments. This was a time of profound closeness between prey and predator in the cave's darkness, possibly imprinted deep within our genetic memory. Another fascinating aspect of this proximity, as described by anthropologists and archaeologists, is that felines possess nocturnal vision, unlike hominids. In the darkness, a feline can observe you, unseen—an inherently eerie trait associated with the mysteries lurking in the shadows. This realization led to the reason why the new strand of our research, initiated a few months ago in collaboration with Francesco Venturi, is titled "Hide." In English, "Hide" signifies both the act of concealing oneself and a creature's skin or fur, but also a vantage point, a location from which one observes unseen. Within this term lies a universe: the unseen yet present, the pleasure of tactile contact, the entity that observes without being observed, all juxtaposed between unease and relaxation. Thematically, this defines our research direction. However, when we enter the rehearsal space, we must translate this into action, observing its impact and discerning where it might lead.

I did something while idling.

Following my initial experience with Angela, I began individually and later collaborated with colleagues Chiara Prodi and Francesco Ferrari to explore the creak—studying what triggers it both physically and imaginatively. A key turning point was a workshop led by Francesco Venturi. His insights, shared during the session, helped us open up to new possibilities and subtle revelations about vocal awareness. From there, we consistently reflected on the feelings elicited by the practice.

Initially, the creak gave me the sensation of hollow spaces, as if I felt empty inside. This deep, resonant sound, achieved through a state of relaxation, instilled a sense of inner vastness. It is as if I am carrying a massive secret, reminiscent of a cave that may seem small but could be enormous when lit. As a performer, the creak brings forth such feelings. Its secretive nature derives from my pleasure in it; it is like a foundational touchpoint. What attracts me to the creak is its intimate nature. Additionally, its ventriloquist quality means it can be produced with a closed mouth, making its origin ambiguous. This sound presents a dual nature: It could be perceived as a potential threat (a roar or growl) or as a point of comforting contact (like a purr or caress).

Another ambiguous aspect is that producing the creak requires the vocal cords to be relaxed. Tension must be avoided. For instance, I recently tried it before an audience and felt nervous; the ensuing tension prevented me from achieving the same relaxed creak. This is an intriguing point both conceptually and practically: where relaxation leads to a discernible result. The concept of productive relaxation resonates with our broader research on the tiger. The tiger in our trilogy thrives in such ambiguity. A tiger, while a magnificent predator, spends most of its time in restful inactivity. Contrarily, it is seen as a symbol of dynamism and vitality. For example, the term "Asian Tigers" is used for fast-developing countries in East Asia, projecting an image of relentless performance. Yet, capturing a day in the life of a tiger might be viewed as either monotonously calm or a peaceful wonder.

I don't know where to go, but I have a purpose: to spend time with the mystery.

Now, the challenge is to determine the direction for this initial discovery. Currently, I am drawn to the idea of direct, frontal performance. My initial thoughts revolve around certain intuitions, such as advancing in space with varying creak modulations, all while maintaining eye contact with the audience. I am curious about how we can adjust the tension under the observer's scrutiny. Furthermore, I am keen to understand how the creak, characterized by closely spaced pulsations, can be transformed into a rhythmic and musical element. Several questions remain. For instance, should we master the creak technically to achieve better control, or focus on the relationship with the audience, where even the effort in producing the creak becomes part of the performance? Regardless, the immediate goal is to immerse ourselves in this "phenomenon," approaching it as a mystery that retains its enigmatic nature. This means studying it in depth while preserving its innocence.

References

Cresti, C., Dovetto, F. M., and Rocha, B. (2015). "Schizophrenia and prosody. First investigations." In C. Manfredi (Ed.), *IX International Workshop Model and Analysis of Vocal Emissions for Biomedical Applications* (MAVEBA) (pp. 139–142). Firenze, Italy: Firenze University Press.

Davidson, L. (2020). "The versatility of creaky phonation: Segmental, prosodic, and sociolinguistic uses in the world's languages." *Wiley Interdisciplinary Reviews: Cognitive Science*, e1547.

Deleuze, G. and Guattari, F. (1972). *L'Anti-Œdipe. Capitalisme et Schizophrénie*. Paris: Les Éditions de Minuit.

Dovetto, F. M. and Gemelli, M. (2013). Il parlar matto. Schizofrenia tra fenomenologia e linguistica. Il corpus CIPPS (2nd ed.). Rome, Italy: Aracne.

Dovetto, F. M., Guida, A., Pagliaro, A. Ch., and Guarasci, R. (In press). "Silences and disfluencies in a corpus of patients with Alzheimer's disease (CIPP-ma)." In R. L. Rose and R. Eklund (Eds.), *Proceedings of DISS 2021*. Saint Denis, France: Université Paris VIII Vincennes.

Ferreira da Silva, D. (2018). In the Raw. Retrieved from e-flux.

Fónagy, 1962. I. Fónagy. Mimetik auf glottaler Ebene. *Phonetica*, **8**(1962), 309–320.

Gobl, C. and Ní Chasaide, A. (2010). "Voice source variation and its communicative function," in Hardcastle, W., Laver, J., Gibbon, F., (eds) *The Handbook of Phonetic Sciences*, 2 edn. Oxford: Blackwell.

Hollien, H., Moore, P., Wendahl, R. W., and Michel, J. F. (1966). "On the nature of vocal fry." *Journal of Speech Hearing Research*, **9**, 245–247.

Juvarra, A. (2007). I segre/del bel canto, Curci.

Keating, P., Garellek, M., and Kreiman, J. (2015). "Acoustic properties of different kinds of creaky voice." In *Proceedings of the 18th International Congress of Phonetic Sciences*. Glasgow, Scotland.

Kent, R. D. and Vorperian, H. K. (2013). "Speech impairment in Down syndrome: A review." *Journal of Speech, Language, and Hearing Research*, **56**(1), 178–210.

Linville, S. E. (2001). *Vocal Aging*. San Diego, CA: Singular.

Lombardo, D. (2011). *La Voce Parlata e Cantata*. Pavia, Italy: Cobert.

Meilán, J. J. G., Martínez-Sánchez, F., Carro, J., López, D. E., Millian-morell, L., and Arana, J. M. (2014). "Speech in Alzheimer's disease: Can temporal and acoustic parameters discriminate dementia?" *Dementia and Geriatric Cognitive Disorders*, **37**, 327–334.

Ní Chasaide, A. and Gobl, C. (2003). "Voice quality variation and the perception of affect: Continuous or categorical?" In *Proceedings of the 15th International Congress of Phonetic Sciences*, pp. 2409–2412.

Nix, J. P. (2016). "Why fry? An exploration of the lowest vocal register in amplified and unamplified singing." *Proceedings of the Meetings on Acoustics*, 26, 035001.

Panzanelli, R., Mazzocchi, R., and Brizi, S. (2009). Analisi e classificazione del Vocal Fry, utilizzo nella logopedia e nella didattica in La voce del cantante vol. 5, Omega, Torino.

Panzanelli, R., Mazzocchi, R., and Brozzi G. (2009) Il canto leggero: generi s2li e modi in "La voce. Fisiologia patologia clinica e terapia" Libro di Oscar Schindler ed. Piccin.

Pikes. M. and Campbell, P. (2021). *Owning Our Voices: Vocal Discovery in the Wolfsohn-Hart Tradition*. Routledge Voice Studies Series.

Rusz, J., Cmejla, R., Ruzickova, H., and Ruzicka, E. (2011). "Quantitative acoustic measurements for characterization of speech and voice disorders in early untreated Parkinson's disease." *The Journal of the Acoustical Society of America*, **129**(1), 350–367.

Schötz, S. (2006). *Perception, Analysis and Synthesis of Speaker Age* (Doctoral dissertation, Lund University). Lund.

Silvestro, M., Dovetto, F. M., Corvino, V., Apisa, P., Malesci, R., Tessitore, A., Milizia, P., Tedeschi, G., Marciano, E., and Russo, A. (2021). "Enlarging the spectrum of cluster headache: Extracranial autonomic involvement revealed by voice analysis." *Headache: The Journal of Head and Face Pain*, **61**(9), 1452–1459.

Tanaka, Y., Nishio, M., and Niimi, S. (2011). "Vocal acoustic characteristics of patients with Parkinson's disease." *Folia Phoniatrica et Logopaedica*, **63**(5), 223–230.

Turlà, E. (2003). Il Metodo Voice CraE, in La voce del cantante, vol. 2, Omega Edizioni, Torino.

Venturi, F. (2021). "Creaky voice gender." *Journal of Interdisciplinary Voice Studies*, **6**(2), 201–218. https://doi.org/10.1386/jivs_00047_1.

Vieira de Oliveira, P., J., S. (2022). DESMONTE. Retrieved November 1, 2022, from https://oliveira.work/desmonte.

Wright, R., Mansfield, C., and Panfili, L. (2019). "Voice quality types and uses in North American English." *Anglophonia*, **27**.

Wynter, S. (2003). "Unsettling the coloniality of being/power/truth/freedom: Towards the human, after man, its overrepresentation—An argument." *CR: The New Centennial Review*, **3**, 257–337.

Yuasa, I. P. (2010). "Creaky voice: A new feminine voice quality for young urban-oriented upwardly mobile American women?" *American Speech*, **85**(3), 315–37. doi:10.1215/00031283-2010-018.

Chapter Eighteen

The other side of the human voice

Francesco Venturi and Joan La Barbara

In this chapter, Joan La Barbara and Francesco Venturi are in conversation about pulse phonation and other sonic materialities, to reveal their place in her music and, more broadly, in the world we are part of. La Barbara is an American composer, performer, and sound artist known for her contributions to the field of extended vocal techniques and her interdisciplinary approach to music-making. Her distinctive style of using the voice as a musical instrument in imaginative ways has earned her recognition in the avant-garde music scenes. La Barbara's work encompasses a variety of forms, including conceptual pieces, musical works in theatrical settings, and "sound paintings," which are compositions that aurally render preexisting visual artworks and La Barbara's sensations when experiencing them (for an exploration of sound paintings, see Dohoney 2022). She has collaborated with numerous composers and musicians, and her solo recordings reflect a unique vocal style and innovative

Creak: Theories and practices of pulse phonation
Edited by Francesco Venturi

ISBN 978-981-5129-29-8 (Hardcover), 978-1-003-61606-1 (eBook)
www.jennystanford.com

compositional approach, often based on vocal multilayering. La Barbara's work integrates elements of theater and visual arts, earning her numerous awards and accolades.

In her recordings, La Barbara seeks to create "situations that cause us to reconsider attitudes about how we view the world" (La Barbara 2015). The seminal album that explored these situations is *Voice is the Original Instrument* (1976), which can be listened to as a testament to the capabilities of the human voice. In *Tapesongs* (1978), *Reluctant Gypsy* (1980), and *As Lightning Comes, in Flashes* (1983), she further refined her ability to create rich textures and complex soundscapes with the voice, which she often manipulated through delays, and other processing techniques. *Reluctant Gypsy* includes a mixture of solo vocal pieces, and *The Art of Joan a Barbara* (1985) and *The Early Immersive Music of Joan La Barbara* (released 2017) include works involving other musicians and instruments. Other notable releases include *Joan La Barbara Singing Through John Cage* [works for voice alone and with instruments] by John Cage (1990), which contains La Barbara's interpretations of Cage's vocal compositions, including his seminal works for voice and electronics; *Joan La Barbara/Sound Paintings* (1991), a showcase of her innovative use of whispers, cries, inhaled sounds, ululations, trills, and glottal clicks; *ShamanSong* (1998), which includes the sound painting ROTHKO, inspired by the Rothko Chapel and consists of multiple layers of vocal multiphonics and reinforced harmonics and bowed pianos. *In Solitude This Fear is Lived* (2011), a work inspired by Agnes Martin, incorporates elements of electronics, orchestration, field recordings, and locating the instruments around and among the audience, reflecting La Barbara's ongoing commitment to pushing the boundaries of vocal expression and the concert experience.

Francesco Venturi: Thank you so much for accepting my invitation.

Joan La Barbara: It's fascinating, and I'm delighted somebody is tackling this project. I don't know where you want to start, but just thinking about my own work, how it began, and the idea of experimenting with voice and discovering specific rather unusual sounds that came out, I can recall discovering certain sounds, and the exhaled fry is one of them. Then there's the variation of that, which is the inhaled glottal click, which is very different. The way of producing it and the result are different. In a way, using the inhaled glottal click is more musical as I can control it differently, whereas with the exhaled fry, you more or less get what you call a creaky or crackling sound. You don't get the kind of almost rhythmic differentiation possible with the inhaled click. The first time I heard it used in a popular song was in *Ain't No Sunshine When She's Gone,* composed and sung by Bill Withers. He goes into that, "I know, I know, I know, I know..." And then, eventually, he gets to a multiphonic fry. I was so excited when I heard that because I thought it had entered popular culture. This is quite a number of years ago. [Note: it was released as a single in 1971.] Those were my initial thoughts. Then looking over the document that you sent me, there is the question of vocal health and use of the fry. I've done workshops in extended technique all over the world, and in one of the workshops, when I was introducing the multiphonic fry, one of the members of the workshop was a person who worked with singers who had damaged their vocal cords. And she said we use this as an internal massage for the vocal cords. Because you often get classically trained singers who are terrified of doing extended techniques, and to hear a therapist, a vocal therapist, talk about this as a positive thing, a very positive thing, was really fascinating to me, and I often refer to that.

FV: How did you first discover creak? Did you play with it as a kid?

JLB: Probably. I think all kids do. And so, when I start making sounds, and if children are in the room, they often imitate or do similar things. When my son was about three and a half years old, and he was at my concert, I started making my sounds, and he started doing them, too. All kids play with the voice and make wonderful sounds. The problem is that we "civilize" them away when we start studying singing. We say, "well, this is a good sound," and "that is a bad sound," or "not a musical sound." So, having trained as a classical singer and then running away from classical music because I felt it was too restricting; I wanted to be able to discover what else the voice could do. So, I imitated the sounds of instruments imitating them and various other sounds. Then I did an early piece called "Hear what I feel," which was a sensory deprivation experiment I did in public because obviously, when you're in a public situation, you feel a different responsibility than when you're just in your home studio working. It was a sort of exercise or experiment, taping my eyes shut, not touching anything, and sitting in a room, isolated from the performance space for an hour. I was then led out to the performance space, still with my eyes taped shut and touched things that an assistant had put into six clear glass dishes and reacted vocally. I was curious and hoped to get some new sounds coming out. But also, I was hoping to communicate on this pre-verbal, nonverbal level with the audience to get them to sort of experience what I was experiencing and my vocalizing of it. But the creaky sound… you know, when we wake up in the morning, if you're still tired, and go [makes creak]. We all do that. It's a perfectly natural sound that we all make. So, I'm curious about what you say about the opera singers doing it. Is it a relaxation technique, or do they not even recognize that they're doing it? They're just sort of doing what I did, growling or groaning.

FV: There needs to be a recognized technique or methodology. Singing teachers use creak for several functions. John Nix and Whitney Chappell have done a wonderful job mapping the attitudes of American voice teachers toward vocal fry [see Chapter 2]. But, of course, fry singing is present in commercial music and not in opera. Traditionally speaking, creak is the enemy of bel canto [both laugh]. But it is used, for example, as a therapy for polyps. When you have a singing student who is overly breathy, you can use pulse phonation on the onset to launch a cleaner modal phonation. And yes, many people use it as a relaxation technique, also in Italy, where creak is not a cultural phenomenon, and people do not even know its name. Once, I was in the changing room, and a colleague started creaking, so I said, "ah, vocal fry!". And his reaction was: "What did you say? I don't know what it is… a former teacher told me to do so. I do it when I'm feeling stressed."

JLB: [Laughter]

FV: Did you know that we creak even before learning to laugh?

JLB: Which is also very early.

FV: Humans creak as soon as they are born. Then, as you said, a selection process takes place. Certain sounds are picked up by the language, vowels, consonants, and so on, while others are not. During this development, creak falls behind. But where? Only where it has no paralinguistic function. In Italy, France, England, the United States, New Zealand. Elsewhere, in Japan or China, creak has a function. In Mandarin, for example, it expresses irritation.

JLB: Okay!

FV: In languages like yours or mine, creak didn't make it into the race for language. So, it comes back later on as a voice game, in vocal improvisation, or unconsciously.

JLB: The sounds I make are often part of an improvisation or sometimes as often as a reaction. I had an ensemble a number of years ago, and we performed with poets and writers. We improvised a soundscape as they read their texts. Oftentimes, a sound would come out during one of these. One of them was during a reading, I believe it was by Armand Schwerner of the Tibetan Book of the Dead, Milarepa's writings, and the sound that came out of me was a multiphonic sound. Luckily, we were recording at that time, and because I had no idea what the sound was, I was able to go back, listen, and think about it. And that happens a lot. It was often in early experimental sessions that I would record and then go back and listen and try to figure out how I had made a sound and then develop a technique for making that sound. So, I can't exactly say. But I also know that sometimes when I'm talking, or in-between thoughts, I'll just [makes sustained creak], and that's vocal fry, of course. It is a way of continuing to make a vocal sound, so if you're in a conversation, somebody doesn't interrupt because they know you're still thinking. You are saying, "I'm going to continue." So, in that way, it's a very utilitarian usage of sound to keep the thought process going.

FV: Tristam Adams calls this function "human buffering sound" (2018). I creak when I'm "buffering" to sustain the sound.

JLB: Sustaining the sound without words and without meaning. But your brain is actually actively forming the next thought. It's just that your vocals continue to phonate while you're trying to come up with the next thought.

FV: Beautiful image. Pulse phonation as the sound of your brain forming a thought. You were also talking about imitating musical instruments. What fascinates me is the power of creak to replicate sounds in onomatopoeia. It allows a human to make unhuman sounds. Did you ever use creak to imitate things?

JLB: Well, sure. The vacuum cleaner, for example [both laugh]. A wonderful thing to sing along with! But also birds. I'm not imitating birds, but I'm certainly inspired by birdsongs. Last December, I did a recording outside in my backyard for the Neue Berliner Kunstverein. They had invited me to do a live concert, but then the COVID numbers rose, so we had to do it online. So, I decided to do an outdoor concert, using zoom to record. The bookends of the concert were a piece that I called "in available light," which was improvisation reflecting on some thoughts I had written in my journal. I did two versions of it, and so they bookend this live recorded performance. In the middle of it, I did a piece which is a recording. It's years of recordings of a particular work with my ensemble that I just continued to layer and remix. You'll also hear many different kinds of vocal sounds in that piece. In the piece, called *"Windows,"* there are some actual birds. There are some crows, which I had recorded. Because I was working on inspiration from the journals and writings of the American assemblage artist Joseph Cornell, and there was a certain phrase in his journals that talked about the crows raucous on the night air ["Crows helter-skeltering in rapid flight, raucous on the raw March air" (Cornell 2007)]. As I was recording outside one time, I happened to pick up some crows that were calling, so I was able to work with that recording. I didn't imitate them in that piece, but certainly, if not actually trying to imitate the birds' sounds, I often imitated the rhythm of the sounds. I live now in New York State, and I often hear owls. Sometimes, you'll hear an owl in one area, and then you'll hear another one answering, and sometimes they're microtonally different. So, I wondered if maybe they're identifying location by that microtonal separation between an absolute repetitive response. The shape of the sound is the same, but the pitch is just slightly different.

FV: Crows and other birds use pulse phonation.

JLB: The emu, for example, is this beautiful bird that is larger than an ostrich, and it has dark feathers as opposed to ostrich feathers. The emu sound is a kind of vocal fry: [makes emu call].

FV: Many other animals creak, in fact. Primates, gazelles, even crocodiles. I like to think about pulse phonation as a cross-species sound. This is something that needs to be discussed more. Diana monkeys, for example, have a "predator alarm call," which sounds like a human creak. When a leopard approaches, they go: [makes leopard alarm call].

JLB: I did a piece a long time ago called *Urban Tropics* in Miami [1988]. I was recording various sounds and wandering around the "monkey jungle," which I'm sure no longer exists because of all the hurricanes since then. But I found myself in this place, and I saw this crocodile. It was silent; it was just sitting there. And then all of a sudden, and I was recording, it opened its mouth, and this terrifying sound came out, which was, you know, it was the vocal fry, but it was a real warning. You get out of here right now. But luckily, I managed to record it. And what was the other thing I thought? Oh, this Sumatran ape, this gorgeous sound, and that had nothing to do with vocal fry. But it did have to do with something that I'm also interested in, which is this inhaled–exhaled sound.

FV: Talking about beasts, I wanted to ask you about your piece *Q-/-uatre Petites Bêtes* [from the album *Reluctant Gypsy* (1980)]. One of the four creatures that you evoke is a creaking beast. Can you tell me something more about this creature? Who is it?

JLB: The origin of the piece was the "four postcards piece" by Marcel Duchamp. It was the way he separated the words. So, it's this sort of stream-of-consciousness writing; then, at a certain point, he gets to the word "Quatre," and instead of separating it into syllables, he puts the "Q," a dash, and "-uatre" on the next line. So, it was not only the image of these four little beasts but the separation of the words that gave me this idea of these very different creatures, and I would think more alien creatures than some identifiable beasts. So imaginary creatures, and that's what I was doing; I was imagining the sounds of these four different creatures that suddenly show up in this open space at the same time, and converse, and you know, get into this, not battle necessarily, but an argument at least. And then, of course, they disappear; they "brrr" [makes noise] and go away. But I just wanted four very different sounds that I could stay with for the period of this piece that I was creating. That creaky sound, I think, was more [makes inhaled creak], which is inhaled, as opposed to [makes creak], which is exhaled. The exhale sound is more calming in a way, whereas the [makes inhaled creak]: there's something a little distressing about that sound. I know that when I teach it to people, oftentimes, they'll make a very high-pitched sound. Because they're [makes pressed creak], they're clutching, and I'll say, "no, no, no, you're not drawing the sound along your tongue." If you're inhaling a pure tone, you want to draw the sound along the roof of your mouth [makes inhaled pitched sound]. If you want an inhaled click [makes inhaled creak], you want to draw the sound slowly along your dropped tongue and into the back of your throat. So, it's a very different placement of where you achieve the sound.

FV: You have such control over the inhaled creak. It is much easier for someone to control the pulses and the patterns in exhale. You can create very precise inhaled clicks.

JLB: There's also another piece called *As Lightning Comes, in Flashes* [1981]. One of the tracks uses a lot of clicking sounds. Each of those tracks are also very different characters.

FV: You were interested in an alien quality? Something other, something different?

JLB: Yes, something different. One wasn't a bear, and one a peacock or something. I play with both those sounds, actually. I wanted them to be less identifiable with a particular creature. I wanted that when people listen to it, they could allow their minds to imagine different kinds of creatures. I'm sure if I asked 10 people to draw the creatures, they would come up with 10 very different pictures.

FV: It would be a fascinating exercise to draw those pictures. You agree that the creaky voice has a kind of otherness to it, something "different" about it.

JLB: Yeah.

FV: Do you think this has something to do with the phenomenon of vocal fry in the English-speaking world? The otherness of this voice quality?

JLB: There's certainly fear of that. There are various kinds of phobias, as you could call them. Think of, for example, speaking in tongues. There are religions where people speak non-identifiable languages. In the English-speaking world, I think we do have a fear of people who sound different and people who look different, and so, the idea of language being something we want to control. We want to be able to express ourselves. I use a lot of nonverbal language in my work because I feel the voice is so expressive beyond words. Words are almost like a limitation on the many different kinds of vocal sounds we can make. The whole idea of expressing in non-language vocalizing is where I predominantly live, where here I find myself most comfortable. That kind of exploration is really fascinating.

FV: It is. Do you think this phobia of the creaky voice comes from the fact that it reminds us of this other side of the human voice? A reminder that we are trapped in the torture house of language, in the prison of language?

JLB: Definitely.

FV: Opinions, especially in America, Britain, and New Zealand, are polarized into love or hate. Have you ever come across this polarization?

JLB: I came across it recently. I attended the vocal juries at Mannes School of Music at the end of last semester, and I just decided I was going to attend all of them. In one of the breaks between the performances, one of the teachers talked about this creaky voice and how irritating she felt it was. In classroom situation, we were teaching with masks all the time. So, we were hearing people who were wearing masks and were non-singers, so they tended not to project their voices. Then when you got closer to them, they became even quieter. There is this whole phenomenon that I have heard people talking about, particularly with young females, where they're speaking in fry almost constantly. They're just talking in this barely audible kind of sound, and it has little pitch differentiation. It's just barely getting out of their mouths. And the whole question is: why is this happening? Is it something happening in society, with what's happening to women and oppression? Is this causing them not to want to project their ideas or voices? I don't know. It's an interesting question. But I did experience this particular teacher being very irritated by so many of the students speaking in fry.

FV: The extreme and sustained vocal fry can annoy people, especially the older ones. There's a big difference between creaky voice quality, occasional creak phonation, and sustained creak phonation. The issue of vocal health and voice hygiene emerges when creak is used excessively. We all occasionally creak, but sustaining creak for the whole sentence is different. One interpretation is that people use it this way when they do not want to show emotionality. They look for an intimate vocalization to express low emotional energy. This can be true, emotional energy is actually down, or it can be fake, and you're hiding this energy because it may not be okay to show it in public. Another interpretation is that they use creak to lower the pitch of their voice to sound more authoritative. People get annoyed by these sounds because they think young women are not entitled to that authority.

JLB: That's the opposite.

FV: It is very polarized. And, as you mentioned in your email, vocal fry takes on more significance in relation to the female role in society. Would you like to elaborate on this?

JLB: I think it's a way of self-limiting. As singers, we mostly project our voices and put our sounds out there, whatever the sounds may be. But when you speak, the less pitch you put into your voice, the less demonstrative your spoken word is. At least, this is my interpretation. If you're speaking in this barely audible sound, you're keeping it close to yourself. I'm thinking about the sound, and I'm not purposefully projecting my sound. It gives the impression that you are not projecting your opinion. I have not researched this, but I feel there is something about this non-demonstrative, non-projecting of ideas.

FV: To be non-offensive.

JLB: In America, we have a terrible problem of misogyny. Really dreadful. It's in our movies. It's in the models. When you open a magazine and see models, it's very much focused on the body, and the vulnerability of the body, it seems to me. Interestingly, we have more women politicians, but not at the very top. Except for Nancy Pelosi, who was the third in the line of succession to the President: Vice President, and then Speaker of The House (of Representatives)—who happened to be a woman. She was the first female Speaker of The House, and we certainly have not managed to get there with President. You see it in the election hearing, or at least you saw with the attitudes toward Hillary [Clinton]. If she spoke her mind, she was being a "nasty woman." We can go a long time in political discussions, but we definitely have a problem with attitudes about women who speak their minds in this country.

FV: You mean these people use creak to keep their voices close to themselves? Creak as a strategy to sound affirmative but non-aggressive. That is a known interpretation [e.g., Ligon *et al.* 2019].

JLB: Be nonabrasive, non-threatening, in a way. At least, if we're talking about the exhaled, whereas the inhaled, there are certain aspects of it that, depending on how you're producing it, can be frightening.

FV: Yes, they can! [Both laugh]. I used it in quite dark pieces of music. Very powerful applications indeed. With the creak register, we can voice strange creatures. It was also used in horror movies. Did you watch the Japanese horror movie The Grudge?

JLB: I'll have to find it.

FV: The villain is identified by a creaking sound. You hear [imitates "grudge" sound] when it's coming to you, and then it kills you. Creak is the sound of the villain's grudge. Horror movies apart, would you say inhaled creak is more about music than speech?

JLB: Definitely. People do not tend to speak on the inhale. The only way I've discussed the inhale sound in speech seems to be an affirmation of something someone else has just said. When somebody says something, answering [makes inhaled noise] means agreement.

FV: Also, when you're scared: [makes inhaled noise].

JLB: Well, that's another kind. That gasp, for sure. But the affirmation sound is very easy. It's just: "yes, I agree" [makes inhaled noise]. I don't know if it is "pay attention." Sometimes when people are talking, see if you hear somebody say that. I became aware of it when I was speaking with a French woman a number of years ago. She used to run an organization called ArtServices in Paris. We were having lunch and talked about many things. And she said, "oh, the sound that you make, this inhaled sound. I think that's very difficult." I said, "no, we all do it. If I step on your foot, you'll go… [makes short, startled gasp sound] what? That's an inhaled sound. If you slow it down and you think it in pitch, it becomes an inhaled sound." As we were having this conversation, I said something, and she went [makes inhaled noise]. I said, "there. That sound!" I'm constantly aware of sound, vocal as well as non-vocal sounds, but it just seems to be like, I'm aware; I live in the country now, so I'm aware of the sound of the wind blowing through the leaves. And sometimes, it's just blowing through the tops of the leaves, as opposed to the tops of the trees and the whole tree. Then it's a very different kind of sound.

FV: Speaking of the sociopolitical part of the subject, I wanted to ask you something. As you know, vocal fry received much attention recently. Scientists remained a little quieter because perhaps they were afraid of stating, "it's about status, it's about power, it's about this, it's about that." However, media coverage was massive, leading to an ideology in popular culture, even a stigma, that sees creak as harmful to your voice; harmful to society. Do you think that vocal fry is dangerous to society, at least in the English-speaking world?

JLB: That's a really interesting question. I'm not quite sure. How would you feel? How would you express that in what would be the danger?

FV: For example, this non-aggressiveness. Using creak to keep things close to yourself, being less demonstrative. Embracing this misogynistic idea that a woman should be nonabrasive. This is a potential danger.

JLB: We are now in a situation, politically, in the United States where the whole issue of abortion and women's right to be in control of their reproductive system. I don't know if we can carry it over in the way that women express themselves so as to be non-threatening and to accept their secondary positioning in society. I don't know that I would go so far as to say that, but I think it's something that certainly could be studied to see where and when did this phenomenon start. At what point did we begin to see more and more young women speaking in this non-expressive way?

FV: Well, we all creak. Yet at the beginning of the last century, creak was considered a vocal disorder. Only in the 1960s was it raised to the status of ordinary. Afterward, it was associated with upper-class British men. Then, it was associated with transgenderism. Further later, with uneducated American young women. Popular icons like Britney Spears and Kim Kardashian made it famous, and some think that vocal fry is a bandwagon effect connected to these famous creaky voices. It became an affectation. However, for example, Elizabeth II creaks quite often, and she's anything but an uneducated young American woman. In Italy, everyone creaks a lot, regardless of gender or class. We find so much creak in young women's voices because we're looking for it into young women's voices. If we started to look elsewhere, we would find something else. There is a feminist issue here, but it does not end there. What do you think?

JLB: You know, over the course of our time here, I've heard you use it in many different ways. It's just part of the pattern of your speech. And you use it, particularly in the way we said before about continuing a thought without words. I think that the awareness of it is really fascinating, to really focus in on this particular vocal sound, and to really study the different ramifications and the different uses and where we hear it when we identify it, and whether we identified or didn't identify it, whether we just accept it as that person's voice. I think there is a vocal aberration because I can remember meeting a very old woman when I was living in Santa Fe, and she constantly spoke on the fry. I think it had something to do with a stroke and having lost the ability to actually make the vocal fluids lose. And so her entire speaking voice was on vocal fry, and it was sometimes a little difficult. You had to really listen and pay attention to what she was saying. When you spoke about considering vocal fry a kind of a bad sound or an aberration, it may have come from people thinking about those who had had strokes and something happening with the vocal cords, some freezing of the vocal folds, not being able to make them meet in a certain way.

FV: Yes, certain pathologies end up with you not being able to produce modal voice. So, you can't stop creaking. This brings us to the field of pathology. But as you said, what is essential is this opposition between conscious and unconscious creak. The question is: what do you want to achieve when you use it as a deliberate strategy? The answers are very different. English-speaking people are one thing; the rest of the world is another. In South Europe, for example, people use it unconsciously. Sometimes there are long creaking sounds. Or perhaps you have individuals who identify as male but were assigned female at birth that use it to sound more masculine. Because there's an association between low pitch and masculinity. And some think this is why young women use creak: to sound lower.

JLB: That's an interesting one. Different interpretation.

FV: They want to sound lower because lower is more manly in a masculine society. With the book, we are trying to knock down this association. What do you think? What do you see of this interpretation of using creak to speak lower, as a "fake" low voice?

JLB: That could certainly be a technique. So as opposed to just speaking on a lower pitch, using the creak to almost get there but not actually place your speaking voice at a lower pitch—which you could do, but some people may not think that they can. So yeah, I can certainly see that, that could be a strategy.

FV: It has been said that vocal fry is undermining the careers of young women. Because when they go to job interviews, they are considered less educated because they creak.

JLB: I wonder why.

FV: Certainly, there's a difference between speaking all the time in the creak register and occasionally creaking. You could have someone with naturally lower air pressure who tends to creak more, but that has nothing to do with Britney Spears and the cultural phenomenon. Yet the interviewer thinks that she is doing that for that reason.

JLB: Then we have this whole psychological thing. The interviewer has a checklist of things they are looking for, leadership qualities or lack thereof, so it depends on what the person is looking for in the job. Does the job calls for someone who will be able to demonstrate their ideas and get them out there? And if that's the case, what are you looking for in this person's vocal delivery? Can they present their ideas forcefully and effectively?

FV: Do you think people should be aware of their creaking?

JLB: I think it's almost impossible for them not to be aware since it's a sound that one must choose to make. In my experience, we don't tend to speak that way. So, I think it's a choice.

FV: It is always conscious, you mean.

JLB: In some way. And I guess it would be interesting to go around and record conversations of people and then analyze those conversations, as how they're using their voices and at what point in the conversation the creak is used or not used to make a point? Is it used to project an opinion or to not project an opinion?

FV: To project, or not to project, that is the question... [both laugh]. Speaking about creak as a technique, I wanted to ask you: what have you learned from creak at the level of music practice? Would you like to conclude by sharing some suggestions for vocalists?

JLB: Just dealing with improvisation. In my opinion, the creak sound is more of a percussion sound as opposed to a pitched sound. If I were improvising and wanted to participate in some way, either by introducing a melodic or pitched sound, I certainly would not use creak. I would use actual singing of pitch. Whereas a creak can be used in a percussive way, in short bursts, or in a sort of steady stream of clicking sounds. Moreover, it is easier for me to do the inhaled clicking and get almost individual clicks.

FV: I agree with you on this percussive nature. Pulse phonation is the place where voice and percussion meet. Such a powerful tool.

JLB: I can actually treat it almost as if it were a drum stroke. So those are just various ways that I think of it when I think of adding sound into an improvisation, and if I were to think of it as a composer, how would I introduce that sound within a texture of a composition.

FV: As a voice quality?

JLB: Also as a color. To me, the creak was gray, whereas the inhaled click would be more solid. If I were to score it, I would use black.

FV: You would say the exhaled is brighter than the inhaled creak?

JLB: It has more breath. When you inhale, you're practically not inhaling at all. You're just inhaling enough to make the sound. When you're exhaling the fry, you're definitely doing that on your air projection.

FV: I love the idea of colors. The vocal fry is gray! [Both laugh].

References

Adams, T. V. (2018). "Accelerations and speed limits: An essay on the vocal limits of semiocapitalism," *Journal of Interdisciplinary Voice Studies*, **3**(1): 21–43.

Cornell, J. (2007). *Joseph Cornell's Dreams*, Cambridge, Mass: Exact Change.

Dohoney, R. (2022). "Ekphrastic voice: On Joan La Barbara's sound paintings," *Tempo*, **76**(301): 73–85.

La Barbara, J. (2015). "Joan La Barbara | FCA Grant Recipient." Retrieved from https://www.foundationforcontemporaryarts.org/recipients/joan-la-barbara (accessed May 11, 2023).

Ligon, C., Rountrey, C., Vaidya Rank, N., Hull, M., and Khidr, A. (2019). "Perceived desirability of vocal fry among female speech communication disorders graduate students," *Journal of Voice*, **33**(5), 805.e21–805.e35.

Index